AF352906

Resilience in Soils and Land Use

Resilience in Soils and Land Use

Editor

Antonio Miguel Martínez-Graña

MDPI • Basel • Beijing • Wuhan • Barcelona • Belgrade • Manchester • Tokyo • Cluj • Tianjin

Editor
Antonio Miguel
Martínez-Graña
Geology
University of Salamanca
Salamanca
Spain

Editorial Office
MDPI
St. Alban-Anlage 66
4052 Basel, Switzerland

This is a reprint of articles from the Special Issue published online in the open access journal *Agronomy* (ISSN 2073-4395) (available at: www.mdpi.com/journal/agronomy/special_issues/ Resilience_soils).

For citation purposes, cite each article independently as indicated on the article page online and as indicated below:

LastName, A.A.; LastName, B.B.; LastName, C.C. Article Title. *Journal Name* **Year**, *Volume Number*, Page Range.

ISBN 978-3-0365-5440-2 (Hbk)
ISBN 978-3-0365-5439-6 (PDF)

Contents

About the Editor

Antonio Miguel Martínez-Graña

Doctor in Geology from the University of Salamanca. Full Professor of External Geodynamics in the Department of Geology-Sciences Faculty.

Teaching experience: in the last 25 years he has taught undergraduate courses in Geology, Geological Engineering, Agricultural Engineering and Environmental Sciences; and postgraduate in the Masters of Earth Sciences of the USAL, Master of Environmental Sciences of the USAL, Master of Water Resources of the University of Alcalá de Henares-Rey Juan Carlos and in the Master of the Department of Earth Sciences of the University of Nova Lisboa in Portugal. Evaluated in the last two Teaching Programs with excellent performance.

In Research: he is Principal Investigator (PI) of the GIR GEAPAGE. He has participated in more than 30 National and Regional research projects, in 7 of them as IP, as well as in 25 University–Company research projects, art. 83, in 20 of them as IP. He has published more than 72 JCR articles, 50 books and book chapters, 25 publications in national magazines, and participated in 45 national and international conferences. He is a member of the Editorial Committee of three international journals indexed in JCR and reviewer of more than 98 articles for international scientific journals indexed in JCR. He has participated in 8 Teaching Innovation Projects, 6 as IP. He has directed 72 Final Degree/Master Projects. At a professional level, he has produced more than 250 technical reports on Geological-Environmental Engineering.

Preface to "Resilience in Soils and Land Use"

Currently, studies on land use in territorial planning are of interest, the purpose of which was previously to analyze the aptitude of each type of land for a specific use, based on its ability to assume impacts and the potential that the land may have had. The analysis of erosive risks constitutes a parameter to take into account in said management.

The scientific community, given the enormous social interest in monitoring and controlling the environment, is developing methodologies that allow such control that is more efficient. One of the environmental factors to consider is the soil, which constitutes the support for life and is one of the basic natural elements, which is evident in the European Soil Charter, of the Council of Europe, which says, in its first point: "The soil is one of the most precious goods of Humanity. It allows the life of plants, animals and man on the surface of the Earth". This European charter also highlights the scarcity and fragility of the edaphic resource, indicating that it must be protected through a greater effort in scientific research and interdisciplinary collaboration to ensure the rational use and conservation of soil.

Antonio Miguel Martínez-Graña
Editor

Article

From Spatial Characterisation to Prediction Maps of the Naturally Occurring Radioactivity in Groundwaters Intended for Human Consumption of Duero Basin, Castilla y León (Spain)

David Borrego-Alonso [1,*], Antonio M. Martínez-Graña [2], Begoña Quintana [1] and Juan Carlos Lozano [1]

[1] Laboratorio de Radiaciones Ionizantes y Datación, Departamento de Física Fundamental, Facultad de Ciencias, Universidad de Salamanca, Espejo n°2, 37008 Salamanca, Spain
[2] Departamento de Geología, Facultad de Ciencias, Universidad de Salamanca, Plaza de la Merced s/n, 37008 Salamanca, Spain
* Correspondence: davidin@usal.es

Abstract: According to the European Council Directive 51/2013 EURATOM, the radionuclide content in human consumption waters must comply with the stated recommendations to ensure the protection of public health. The radiological characterisation assessed in Laboratorio de Radiaciones Ionizantes y Datación of Universidad de Salamanca, in more than 400 groundwater samples gathered from intakes intended for human consumption from the Castilla y León region (Spain), has provided worthwhile data for evaluating the spatial distribution of the radioactivity content in the Duero basin. For this purpose, geostatistical exploration and interpolation analysis, using the inverse distance weighting (IDW) method, was performed. The IDW prediction maps showed higher radioactivity occurrence in western and southern areas of the study region, mainly related to the mineralogical influence of the igneous and metamorphosed outcroppings in the Cenozoic sediments that formed the porous detritic aquifers of the Duero basin edges. The hydraulic characteristics promote the distribution of these radioactivity levels towards the centre of the basin as a consequence of the unconfined nature of the aquifers. Prediction maps provide a worthwhile tool that can be used for better planning and managing of groundwater monitoring programmes.

Keywords: natural radioactivity; spatial distribution; IDW; prediction maps; groundwater; drinking water

Citation: Borrego-Alonso, D.; Martínez-Graña, A.M.; Quintana, B.; Lozano, J.C. From Spatial Characterisation to Prediction Maps of the Naturally Occurring Radioactivity in Groundwaters Intended for Human Consumption of Duero Basin, Castilla y León (Spain). *Agronomy* **2022**, *12*, 2059. https://doi.org/10.3390/agronomy12092059

Academic Editor: Paul A. Ty Ferré

Received: 26 July 2022
Accepted: 26 August 2022
Published: 29 August 2022

Publisher's Note: MDPI stays neutral with regard to jurisdictional claims in published maps and institutional affiliations.

1. Introduction

It is widely known that the quality assessment of groundwater arouses great interest in environmental management for planning the use of water. Groundwater is not only a strategic resource for drinking water supply for human consumption, but also to satisfies irrigation demand, especially in arid and semi-arid regions, where intensive agriculture is one of the main economical driving forces and the seasonal variability in the precipitation regimes frequently lead to drought periods. In Spain, groundwater withdrawal is estimated at 22% of the total [1], However, there exists a wide variability range among the different administrative authorities commissioned for hydrographical planning. In Castilla y León the primary groundwater abstraction comes from the Duero Hydrographical Confederation (DHC), which extends through 82.2% of their territory, and is approximately 1220 hm^3/yr, mainly intended for agriculture and livestock farming (78%) and domestic human consumption (16%). Their use becomes especially significant in regions where the superficial supply can no longer provide enough water, due to the absence of reservoirs or during the aforementioned seasonal droughts. Hence, it is essential to accomplish exhaustive characterization assessments that ensure the quality levels of the groundwater resources intended for human consumption are in compliance with public health recommendations.

Given their critical impact on human health, the determination and spatial distribution of heavy metal concentrations have been addressed in different environments [2–5]. The naturally occurring concentration of these elements in groundwater depends on the involved processes in the transference from rock and soil minerals to aquifers. Apart from the chemical toxicity associated with the ingestion of heavy metals through the food chain and drinking water, some of them, such as uranium, radium, lead or polonium, are also radiotoxic due to the ionising radiation emitted as a consequence of their radioisotope disintegration. Consequently, regulatory frameworks developed during recent years, have increasingly focused on the radioactivity content in groundwaters intended for human consumption to protect public health from radionuclide substances.

The naturally occurring radionuclides present in groundwaters come from the three natural decay chains that form part of the crystalline structure of the bedrock minerals. Taking into account their relevance as a key resource for water availability in many worldwide regions, several works [6–10] have focused on the hydrogeochemical processes involved in the radionuclide mobilisation from mineral-bearing to the groundwater environment. Background radioactivity levels on groundwaters are often innocuous, and it is also a fact that there are regions prone to lithological features with specific hydrogeochemical settings where the radionuclide presence is enhanced [11–15], highlighting the need for quantifying and monitoring of their content. Prolonged radiation exposure due to the ingestion of drinking water containing certain radionuclide levels is likely to lead to an increase in the risk of stochastic effects on the development of severe diseases [16], related to the radiological toxicity and the chemical behaviour of uranium, radium, polonium and lead. Hence, an accurate radiological characterisation of the groundwater bodies is an important challenge to guaranteeing safe human water consumption. The current legal framework for the protection of public health regarding the radioactivity parameters was laid down in the European Council Directive 51/2013/EURATOM [17]. Spain, as a Member State of the EU, also incorporates the European Directive requirements in its legislation to ensure compliance with the quality radioactivity standards [18]. According to the safety guidelines of radioactivity in drinking water, the control of the indicative dose (ID) must be addressed by determining the activity concentrations of ^{226,228}Ra, ^{234,238}U, ^{210}Pb and ^{210}Po, as established in Annexe X of the Royal Decree (RD) 314/2016. Once the water has been radiologically characterised, the gross α- and gross β-activity determinations can be used as a screening strategy for monitoring the radioactivity levels. If gross α- and gross β-activity values are kept below their screening levels, 0.1 and 1.0 Bq/L, respectively, it may be assumed that the ID does not exceed the parametric value of 0.1 mSv, and no further costly and expertise-specific radiological determinations are necessary. Between 2017 and 2021, in the framework of the project promoted by the General Directorate of Public Health of Castilla y León, our laboratory carried out a comprehensive radioactivity characterisation of the groundwater intakes placed throughout the whole region, which comprised an analysis of the ^{238,234,235}U, ^{228,226}Ra, ^{210}Pb and ^{210}Po activity concentrations for the ID assessment. Furthermore, their respective gross α- and gross β-activities, hereinafter referred to as a_α and a_β, respectively, were also determined, due to their wide use in water radioactivity monitoring.

In the last decade, geostatistic and geospatial analysis has become a highly useful and versatile tool for understanding environmental variability and the spatial distribution of groundwater quality [19–22]. Some publications have reported that high radioactivity levels on groundwaters are severely associated with the bedrock type and the hydrogeochemical variables [22–25]. Geographic Information System (GIS) and geostatistical tools implemented in the software enable the assessment of spatial potential relationships between the groundwater radionuclide content and the environmental setting [26,27]. Despite the existence of data-driven methods used for environmental studies, such as artificial neural networks [28], we performed the inverse distance weighted (IDW) interpolation method, which allows the estimation of the unmonitored area based on the quantitative measurements carried out in neighbourhood sampled locations, to assess the spatial distribution of the radioactivity content in the aquifers of Castilla y León. The IDW interpolation method is widely used in geospatial studies related to groundwater pollution and quality assessment [29]. Despite the different interpolation methods, the most appropriate depends on the type of data parameters and their predictability is determined by multiple factors related to the distribution of the sampling points [30]. Some publications have addressed the monitoring, geospatial analysis and projecting of prediction maps of the environmental radioactivity in waters, such as ^{222}Rn in southern Belgium [31] or a_α and a_β in Turkey [32]. Our study presents a radioactivity assessment in a region where different main lithologies and permeabilities define the aquifers of Castilla y León. Given that the prediction maps of the radioactivity parameters in the study region constitute an important milestone, not only for a better understanding of the groundwater radionuclide distribution, but also because they provide a powerful tool from a management perspective for compliance with the current law. Thus, the current study was conducted to assess the spatial distribution of the main naturally occurring radionuclides using the IDW method, by considering analysed samples from hydraulically interconnected aquifers belonging to the DHC, and to provide groundwater prediction maps that can be used as a tool in the groundwater planning.

2. Materials and Methods

2.1. Study Area

Castilla y León (CyL) is situated in the NW of the Iberian Peninsula, and occupies approximately 94,225 km^2, thus, constituting the largest region of Spain and the third largest of the European Union.

Geologically, regarding the lithostratigraphic and structural features, different domains can be distinguished, as can be observed in Figure 1. The Hercynian basement is comprised mainly of igneous and metamorphic outcroppings, originating during Precambrian and Paleozoic, and constitutes the Central System (CS). Throughout the westernmost areas of the provinces of Zamora and Salamanca and the southern areas of Salamanca, Ávila, Segovia and León, acidic plutonic outcroppings and Precambrian-Cambrian metasediments, which mainly consist of slates, greywackes and sandstones, and small occurrences of quartzites, conglomerates and black-carbonaceous slates, mainly comprise the basement, which constitutes the Schist-Greywacke Complex (SGC). Igneous intrusive bodies are broadly formed by two-mica granites and other biotitic granitoids. Throughout the western Asturian-Leonese zone, and in small outcrops located in the Iberian System, located in the eastern region of CyL, Cambrian-Ordovician-Silurian sediments, with some interlaced levels of carbonates, overlying in discordance with the Precambrian metasediments, such as slates, shales, quartzites and *Ollo de Sapo* gneiss rocks predominate. The Cantabrian zone, which occupies the northernmost areas of the provinces of León and Palencia, is comprised of Precambrian and Paleozoic sediments deposited in shallow marine environments, originating in limestones and sandstones. Terrigenous and carbonated Mesozoic materials mainly constitute the basement of the Basque-Cantabrian Basin, which extends throughout the north of Palencia and Burgos provinces, and the Castillian part of the Iberian Range, represented mainly by the Cameros Basin between SE of Burgos and the north of Soria. The sedimentary sequence of the different sectors of the Cenozoic Duero Basin (DB) is influenced by the bedrock composition that surrounds the Spanish Northern Plateau. It is worth pointing out the significant impact on the composition of the siliciclastic rocks, that fill the southern and western sectors of the DB, of acidic intrusive rock debris and metamorphic sediments belonging to the CS. Finally, fine to coarse detrital grains (sands, clays, limes and conglomerates), associated with the current riverine systems, constitute the Quaternary deposits, which originated from the drainage of the DB materials and the continuous denudation of Mesozoic mountain edges [33].

Figure 1. Map of the main lithologies and sampling sites (in black dots) of the studied groundwaters of Castilla y León (Spain), belonging to the centre of DHC.

The spatial distribution of the natural radionuclides, the groundwater occurrence of which is controlled by the processes within the water–rock interactions, is, thus, highly influenced by the main lithologies represented by granitic and metamorphic outcroppings, detritic materials and carbonate rocks and their respective abilities to store, transmit and discharge groundwater [34,35]. Accordingly, 20 hydrogeological unities are defined in the centre of the DHC which are formed by groundwater masses or aquifer systems that present similar hydraulic, hydrogeological and distribution characteristics and also share the same environmental planning goals, as illustrated in Figure 2.

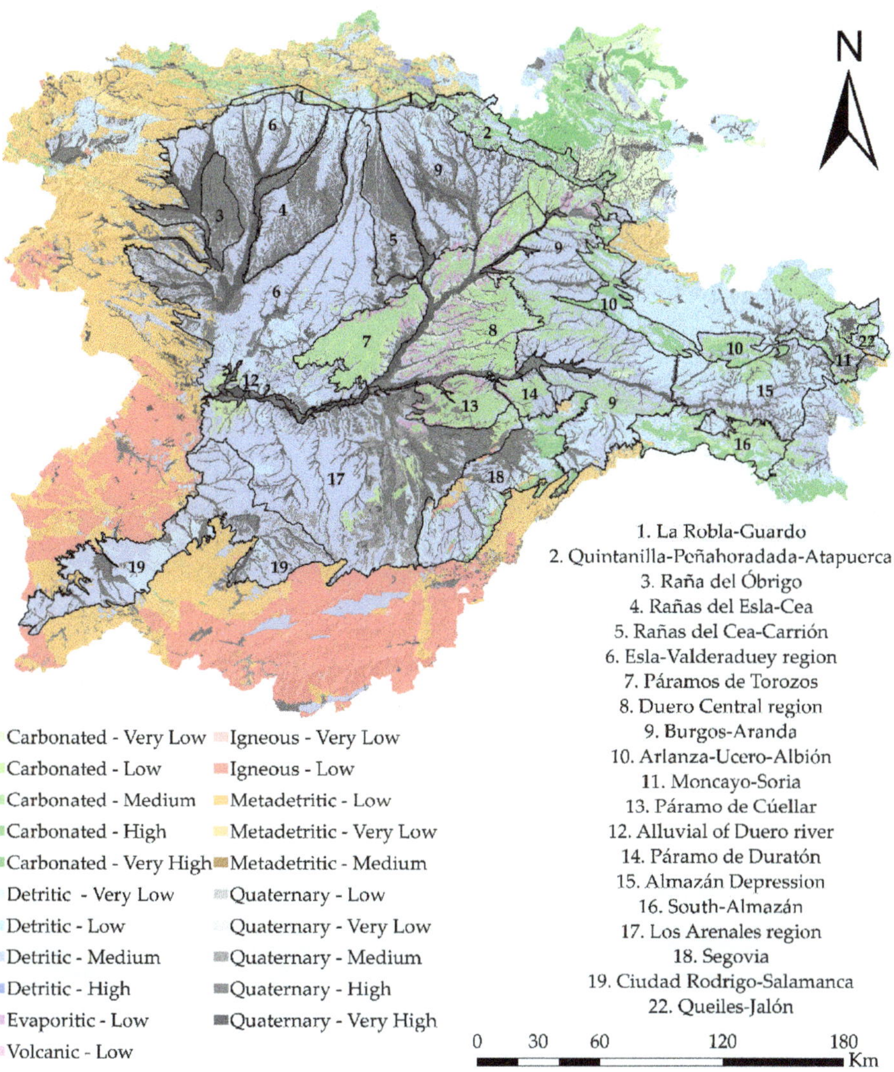

Figure 2. Permeability map of the Castilla y León region according to the main lithologies. Black lines limit the 20 hydrogeological units of the DHC studied.

Regarding permeability, there are broadly two types of productive aquifer systems associated with the rock type and permeability characteristics. On the one hand, the intergranular porous aquifers are formed in detritic materials, such as sands, gravels and conglomerates, and extend throughout the Cenozoic sediments of the DB and tectonic depressions of Ciudad Rodrigo or Amblés. These aquifers are included in the following hydrogeological unities: Obrigo, Esla-Cea and Cea-Carrión "Rañas", Esla-Valderaduey Region, Páramo de Torozos, Duero Central Region, Burgos-Aranda, Alluvials of Duero River,

Almazán Depression, Arenales, Ciudad Rodrigo-Salamanca, Valdecorneja and Amblés Valley. Theyepresent 85% of the territory and 70% of the total groundwater volume of the most important hydric resources. The Cenozoic aquifer systems are formed by a wide sands lens with a semipermeable lime-clayish matrix surrounding them. These aquifers behave as a great anisotropic interconnected aquifer complex, which presents a variable thickness of up to 2000 m depth in the basin centre. Infiltration and sideway flow are the main recharge mechanisms. Generally, the groundwater flows towards the basin centre through the current riverine drainage system. Productivity and transmitivity are highly variable, 10–40 L/s and 5–100 m^2/d, respectively. The superficial (5–15 m), low-transmissive and low-productive porous aquifers formed in the Quaternary sediments, such as rañas, alluvials and sandbanks, are, likewise, variable. On the other hand, the fissured carbonated aquifers, formed in the Mesozoic and Cenozoic limestones and dolomites, present a permeability related to fissures due to dissolution and structural processes. Mesozoic formations, which comprise the northern and western peripheral areas of the DB, constitute broadly unconfined karst aquifers recharged by freshwater runoff and interconnected groundwater lateral flow with other units, and their transmitivity and productivity range between 20–1500 m^2/d and 5–140 L/s, respectively. In the centre of the DB, limestone formations extend, giving rise to the Páramos (Cúellar, Duratón and Torozos) that constitute unconfined aquifers with a subhorizontal tabular structure, 5–50 m depth, radial flow and recharge-discharge through several springs, as well as infiltration. It can be considered that to a greater or lesser extent, all these aquifers are interconnected. It must be noted that the igneous and metasedimentary outcroppings, which extend throughout the Central System and the North-Cantabrian zone, give rise to limited and fractured regionally insignificant aquifers, hydraulically confined, and characterised by very low permeability. In the Ebro Basin, which occupies the Basque-Cantabrian zone, Cenozoic basin sediments exist with clayey interlaced levels that limit the hydraulic relation but give rise to some confined aquifers or locally significant aquifers.

2.2. Groundwater Samples

Between January 2017 and November 2021, two hundred and forty-four drinking water samples were gathered from groundwater intakes situated within the hydraulically interconnected DHC. The georeferenced location of the sampling points is shown in Figure 1. The selection of the sampling points responded to strategic goals stated by the Dirección General de Salud Pública (DGSP) of Junta de Castilla y León to comply with the guidelines for the radiological characterisation of the groundwater masses intended for human consumption.

The water samples were mainly collected by the Pharmaceutical official service of the DGSP in three 10-L PTE vessels and then transported to Laboratorio de Radiaciones Ionizantes y Datación of Universidad de Salamanca for measurement and analysis to determine the radionuclide content. Immediately after this measurement, water samples were acidified with 1 mL/L of analytical grade 65% HNO_3 for proper preservation.

2.3. Radioactivity Measurements

The activity concentration of ^{238,235,234}U, ^{228,226,224}Ra, ^{210}Pb and ^{210}Po were determined using γ-ray and α-particle spectrometry techniques. The a_α and a_β values were determined using the optimised thin source deposit and proportional counting (TSD-PC) method [36].

The γ-sources were prepared by the thermal concentration of a twenty-five-litre aliquot up to 50 mL final volume. The concentrated samples were then transferred to cylindrical 52 mm diameter and 48 mm height PTE beakers. Two low-level background HPGe detectors, manufactured by Canberra, BE5030 and GR2520 models, hereinafter referred to as BEGe and REGe, respectively, were used for the γ-ray spectrometry measurements. Both detectors were continuously ventilated with N_2. The BEGe detector was made of a p-type high-purity Ge crystal of broad-energy range with an active volume of 117 cm^3. Its relative energy peak efficiency at 1332 keV was 50% and its nominal resolution at 122 and 1332 keV were 0.75 and 2.20 keV, respectively. The detector was surrounded by a passive shielding consisting of 10 cm thick iron melted and 5 cm thick old lead, which was internally lined with 2 mm thick high-purity electrolytic Cu. The REGe detector was made of an n-type high-purity Ge crystal with an active volume of 59 cm^3. Its relative peak efficiency at 1332 keV was 22.60% and nominal resolution at 122 and 1332 keV were 0.842 and 1.79 keV, respectively. The detector was surrounded by a passive shielding made of 27.5 cm thick iron melted, internally lined with 2 mm thick high-purity electrolytic Cu. The data acquisition was done with coupled electronics consisting of an integrated module Canberra DSA1000 model, including a 16 K multichannel analyser. The γ-sources were measured for times ranging between 250,000 to 450,000 seconds. For efficiency calibration, both Monte Carlo (MC) simulation and experimental methods were used [37,38].

The acquired γ-spectra were analysed using the Gamma-Live Expert Analyser, GALEA, developed in our laboratory mainly for the analysis of spectra shaped by the natural radionuclide emissions [39]. It included the algorithm COSPAJ for the full continuum spectra fitting [40]. A genetic algorithm was also included that provided the best peak fit, even in multiplets with more than three peaks. The γ-ray emission energies and probabilities were taken from the Nuclide-Lara Library [41].

The ^{238,235,234}U activity concentrations were also determined using α-particle spectrometry. Sources were prepared through the radiochemical separation procedure which involves sequential extraction using UTEVA resin (Triskem) of purified uranium eluate and electrodeposition onto stainless steel discs at 1.8 A for 1 hour [42,43]. The method used for ^{210}Po determination was based on co-precipitated with Fe(OH)$_3$ and auto-deposition onto silver discs for 4 hours [44,45]. The activity concentration of ^{210}Po corresponded to the ^{210}Po in excess, thus, the activity concentration measured from α-particle spectrometry was corrected with the γ-ray spectrometry ^{210}Pb activity. Measurements were performed using a PIPS semiconductor detector of 450 mm^2 active area, Ortec BR-SNA-450-100 models, housed in an Ensemble Spectrometer with Alpha-Duo Modules (Ortec) for uranium sources and Canberra A450-18 AM model, coupled to low-noise preamplifiers and amplifiers, all of them housed in width NIM spectrometers Canberra 7401VR model, coupled to a multichannel analyzer Ortec 920E EtherNIM 16-Input Multichannel Buffer model, was used for polonium. Spectra and hardware were managed by the MCA Maestro Emulator and the suite Alpha-Vision both by Ortec.

According to the European framework, the water quality intended for human consumption should ensure safe radioactivity exposure for public health. Thus, the indicative dose (*ID*) was determined in all groundwater samples, following Equation (1):

$$ID = \sum_{i=1}^{n} \frac{c_i(cal)}{c_i(der)} \leq 0.1 \text{ mSv}, \tag{1}$$

c_i *(cal)* is the activity concentration determined by a radionuclide (*i*) present in the sample, c_i *(der)* is the derived concentration of a radionuclide (*i*) listed in RD 314/2016, and *n* is the number of radionuclides with concentration activities higher than the minimum detectable activity (MDA). The contribution of ^{210}Pb, ^{210}Po, ^{226}Ra, ^{228}Ra, ^{238}U and ^{234}U were considered for the *ID* assessment. The value of the derived concentration is based on the radiotoxicological properties of the radionuclides [46].

The a_α and a_β values were ascertained by following the optimised TSD-PC method developed in our laboratory for drinking water [36]. For each sample, efficiencies were directly calculated by spiking with natural uranium standard solution. Measurements were performed in a low-background gas-proportional counter (model LB770, Berthold Technologies). This equipment enables the simultaneous measurement of the α and β emissions. Underground location, the anticoincidence guard detector and the passive shielding based on thick lead blocks ensured low background.

2.4. Statistical Analysis of the Samples

The spatial and geostatistical analysis was performed using ArcGIS software (v 10.9) which included the 3D Analyst Toolkit for applying the interpolation methods and the semivariogram analysis. The inverse distance weighted (IDW) interpolation technique determined the unsampled locations from a linear weighted function based on the inverse distance between the sampled location, in which values were known. This interpolation method takes into account that the nearest measured values have more weight in the prediction than those further away. The predicted surface area was computed using the Equation (2):

$$Z\left(S_O\right) = \sum_{i=1}^{N} \lambda_i Z\left(S_i\right), \tag{2}$$

where $Z\left(S_i\right)$ is the measured value in the sample point i; λ_i is the unknown weighting for the measured value in the sample point i; $\left(S_O\right)$ is the location of the prediction; and N is the number of measured values [47].

The weight of the set of measured points in the prediction model diminishes with the distance. Despite other non-deterministic interpolation techniques, such as geostatistical techniques like Kriging, often being applied, the current work considered the IDW technique to be most proper interpolation technique to also consider the spatial distribution of the sampled locations and build prediction surface maps with some measure of certainty and predictive accuracy [48]. The absence of a higher density of the grid sampled points in such an extensive territory led to discarding of Kriging interpolation.

The IDW interpolation method was conducted within the boundaries of the hydraulically connected aquifers. It was adjusted to 12 site points with a variable search radius, and with a grid size of 50 m. The power parameter, which determines the significance of the model, was defined with a value of 2. The break values were established to represent in red colour the interpolated areas where the activity concentration of the specific radionuclide exceeded their contribution to the *ID* parametric value stated. Despite ^{210}Po also being considered for *ID* calculation, we dide not apply IDW interpolation to it because there were a lot of water samples having ^{210}Po activity concentration below MDA. Moreover, only 9 samples exceeded 10 mBq/L, which is a magnitude order lower than the activity concentration that leads to presenting an *ID* higher than the stated parametric value, Table 1. In the case of the a_α and a_β, classification shown in the map legends was performed regarding whether values exceeded the screening value or not, established at 0.1 Bq/L and 1.0 Bq/L, respectively [17,18].

Table 1. Derived concentrations corresponding to the main natural radionuclides present in the studied samples for *ID* calculation according to Annexe X of RD 314/2016.

Radionuclide	^{210}Pb	^{226}Ra	^{228}Ra	^{234}U	^{238}U
Derived concentration (Bq/L)	0.2	0.5	0.2	2.8	3.0

3. Results

Exploration data analysis was performed to gain a better understanding of the spatial distribution, as well as to identify outliers and trends. For each radionuclide and parameter studied, the histogram provided all the statistics and the frequency distribution shape. The right tail shown in Figure 3a indicated a relative lack of large activity concentration values, evidencing that they were not normally distributed. All of them followed a log-normal decreasing continuous probability distribution as can be seen in Figure 3b, where the QQ plot for a_α is shown, in which normality was then demonstrated, given that most of the data were close to the 45-degree line. In Table 2, where non-log-normal transformed statistic values are summarised, the skewness values, as well as the differences observed between the mean and median, indicated high asymmetry of the data distribution and the interpolation method improved the modelisation of the dataset A three-dimensional view of the data was given by the trend analysis tool which allowed us to identify global spatial trends in the dataset. According to the distribution pattern observed in Figure 3c, the spatial autocorrelation of high values increased longitudinally from east to west and latitudinally southward, but the latitudinal component had greater weight than the longitudinal one. The variability of the spatial autocorrelation of the sampling points was analysed through the semivariogram cloud (Figure 3d). The theoretical base was the first Tobler's law, which states that the nearest points are more similar than those that are outermost and are then more predictable and less variable. On the x-axis the distance separating each pair of points was plotted, while the y-axis refers to the relative semivariogram values, which corresponded to the squared difference between each pair of sampling points. As the distance between the pair of sampling points increased, the semivariogram value also increased. The semivariogram cloud shown in Figure 3d indicated that most of the data were spatially autocorrelated.

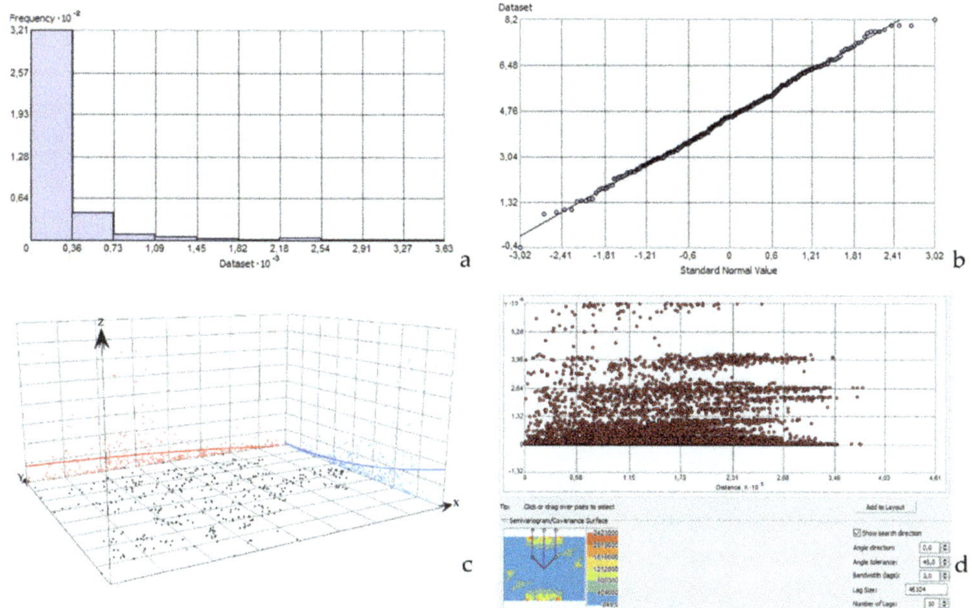

Figure 3. Geostatistical exploration corresponding to the a_α analysis: (**a**) histogram; (**b**) log-transformed QQ normal plot; (**c**) trend analysis; (**d**) semivariogram.

Table 2. Statistical analysis of activity concentration of the main natural radionuclides and radioactivity parameters in the interconnected groundwaters of Castilla y León.

	^{234}U	^{238}U	^{226}Ra	^{228}Ra	^{210}Pb	ID	a_α	a_β
Mean ($\bar{x}$)	131	58.2	24.04	9.62	16.1	20.8	215	228
Median (Me)	51.4	21.8	5.30	4.10	4.84	11.6	108.0	132
Minimum (min)	0.07	0.04	0.13	0.03	0.37	0.33	0.67	1.83
Maximum (max)	1616	705	1162	196	512	318	2407	1547
Standard deviation (S)	208	91.1	86.80	18.9	47.4	28.5	323.3	254
Kurtosis	20.0	16.8	101.08	58.4	64.7	53.1	20.184	9.45
Skewness	3.57	3.25	9.16	6.59	7.14	5.72	3.64	2.35

3.1. Radionuclide Analysis and Their Relationship with the Lithological Context and Permeability

IDW provided the prediction maps for the main natural radionuclides as well as the a_α, a_β and *ID*, which are represented in Figure 4a–f. The spatial distribution pattern predicted for uranium radionuclides, [234]U and [238]U, showed a similar defined area, as can be seen in Figure 4a,b, but with a greater occurrence of [234]U. Although in the mineral grains both radionuclides were frequently in secular equilibrium, the presence of [234]U in groundwater was often much higher than the [238]U concentration because of the different physicochemical processes involved in their mobilisation. Given that uranium solubility is enhanced under oxidising conditions at near-neutral pH, typical in shallow aquifers, the weathering of mineral-bearing host rock releases the uranium nuclides to groundwater, which is the only mechanism controlling [238]U release from crystal lattice to water phase. However, when [238]U decays by alpha emission the recoil process may eject [234]Th into the mineral lattices close to the water–rock interaction boundary, producing recoil damage tracks where is located. The rapid decay of [234]U from [234]Th (half-life 24.1 d), promotes the [234]U preferential leaching [49].

As seen in Figure 4a, [238]U activity concentrations were higher than 100 mBq/L, which corresponded to the recommended screening level for a_α, and had been predicted in the southern and central aquifers of the DB region. The concentrations of [238]U extended mainly throughout the hydrogeological units of Los Arenales and Segovia, from the southern edges towards the centre of the DB, reaching values up to and exceeding 700 mBq/L in some points (Table 3). Furthermore, in the surrounding areas of the Duero river, the presence of [238]U was predicted in significantly high concentrations, mainly in the south of the carbonate aquifers of Páramo de Torozos and Duero Central region, as well as in the detritic hydrogeological units of Alluvial of the Duero and Esla-Valderaduey.

Table 3. Minimum, maximum and median radioactivity values of the analysed samples according to their respective hydrogeological unit (HU). Results from HU nº 14 are not included because samples were not gathered from there.

HU		[238]U mBq/L	[234]U mBq/L	[228]Ra mBq/L	[226]Ra mBq/L	[210]Pb mBq/L	*ID* μSv	a_α mBq/L	a_β mBq/L
1	Min–Max	0.91–36.8	1.53–66.8	0.36–11.2	0.18–11.6	0.84–5.9	0.48–12.9	11.1–91.6	17.5–241
	Median	12.4	49.4	6.5	5.5	2.8	5.5	51.6	57.0
2	Min–Max	3.69–7.89	3.94–6.61	0.39–0.43	0.25–0.31	0.84–2.16	1.03–1.92	14.3–19.5	21.7–35.0
	Median	5.79	5.28	0.41	0.28	1.50	1.47	16.9	28.4
3	Min–Max	0.72–37.0	1.48–91.7	2.10–6.26	0.92–5.96	3.88–8.60	4.88–11.12	7.12–115	30.1–150
	Median	12.5	28.3	3.82	2.95	4.39	7.39	43.1	96.7
4	Min–Max	0.3–42.2	0.86–81.3	0.49–9.28	0.66–24.8	1.52–6.56	5.03–13.6	28.9–117	50.4–175
	Median	24.4	33.9	4.57	3.86	4.08	12.3	112	134
5	Min–Max	0.92–37.5	1.99–53.1	0.42–3.90	1.40–3.36	1.12–9.76	2.04–10.6	8.95–107	46.0–136.0
	Median	18.6	20.7	1.2	1.95	5.3	6.57	42.7	74.4
6	Min–Max	0.04–169.8	0.07–247	0.17–196	0.41–251	1.16–166	2.46–172	6.68–817	20.0–1547
	Median	21.6	44.6	5.2	11.8	4.52	12.1	127	134
7	Min–Max	15.5–400	19.9–786	2.10–27.2	2.41–73.0	2.90–15.6	6.32–72.3	71.8–1174	82–1425
	Median	66.2	79.4	9.92	45.2	6.36	28.5	157	492
8	Min–Max	5.60–321	7.94–434	0.46–33.1	1.32–32.0	0.96–10.1	3.01–29.4	17.3–481	20.9–909
	Median	29.1	49.7	3.42	7.20	2.44	6.93	93.7	160
9	Min–Max	0.13–146	0.46–251	0.31–18.8	0.17–25.2	0.88–24.1	2.05–74.5	19.8–436	29.1–792
	Median	16.7	25.3	3.14	3.96	3.98	10.16	70.8	100
10	Min–Max	1.20–31.3	1.05–124	0.40–48.1	0.54–122	0.85–51.8	0.89–76.7	24.7–672	28.0–956
	Median	13.0	27.3	6.76	6.29	6.01	6.54	41.2	80.3

Table 3. *Cont.*

HU		^{238}U mBq/L	^{234}U mBq/L	^{228}Ra mBq/L	^{226}Ra mBq/L	^{210}Pb mBq/L	ID µSv	a_α mBq/L	a_β mBq/L
11	Min–Max	4.97–50.3	22.1–372	1.61–13.1	1.48–39.6	0.78–11.6	5.17–27.1	38.5–452	52.0–1191
	Median	16.9	31.8	2.18	1.86	3.27	7.96	81.95	160
12	Min–Max	17.6–170	108–487	3.10–53.0	4.58–77.2	1.20–15.0	13.1–63.4	174–681	312–819
	Median	152	296	9.75	6.48	8.56	30.8	482	600
13	Min–Max	12.1–18.4	14.9–72.3	1.85–2.92	4.36–7.13	1.44–10.7	5.95–12.2	28.8–65.0	37.0–141
	Median	16.6	15.8	2.09	4.86	1.72	10.4	60.5	77.0
15	Min–Max	9.62–22.6	40.3–66.8	2.42–11.2	3.10–11.6	1.20–4.80	8.17–12.9	51.6–91.7	57.6–241
	Median	13.9	50.7	7.05	8.58	2.58	9.37	74.0	129
16	Min–Max	9.76–48.6	13.2–106	0.88–27.1	0.48–16.9	1.45–24.0	2.39–23.1	26.5–128	37.7–198
	Median	15.9	47.0	1.61	3.29	2.43	4.89	92.6	100
17	Min–Max	5.84–705	7.52–1616	0.03–34.4	1.76–61.2	1.08–122	6.73–94.9	70.8–2407	59.5–1325
	Median	94.3	237	6.28	15.1	6.32	32.4	300	310
18	Min–Max	8.24–485	11.0–1100	0.51–29.7	0.70–62.4	0.72–23.1	1.86–69.0	12.5–1847	49–781
	Median	18.5	45.4	3.84	2.32	3.68	8.86	74.0	112
19	Min–Max	1.76–199	2.07–891	0.37–164	0.16–738	1.76–156	8.74–318	6.07–1223	51.4–1015
	Median	43.6	136	11.7	16.2	7.12	23.2	254	189
20	Min–Max	0.21–48.2	0.48–69.4	0.26–4.45	0.13–17.5	1.92–6.08	0.33–10.6	0.67–158	1.83–179
	Median	5.56	5.97	1.66	0.72	2.32	2.34	4.10	25.8

Similarly to ^{238}U, the spatial distribution of high ^{234}U concentrations was found in the southcentral aquifers of the DB, mainly belonging to Los Arenales, Esla Valderaduey, Ciudad Rodrigo-Salamanca and throughout the course of the Duero river, as illustrated in Figure 4b. ^{234}U concentrations exceeding 100 mBq/L were mainly associated with detritic aquifers, but also some groundwaters hosted in the carbonate rocks developed in the Páramo de Torozos, Páramo de Duratón and Páramo de Cúellar, and the outcroppings in the central-north area of Duero Central region.

The outcropping of U-rich granitoid and other intrusive igneous rocks, as well as metamorphosed Hercynian rocks throughout the CS, gave rise to limited and confined fractured aquifers with ^{238}U and ^{234}U concentrations that could even exceed 1300 and 1900 mBq/L, respectively, in some points of these confined aquifers, from previous analysis performed in our laboratory. High naturally-occurring uranium concentrations in soils developed on the Hercynian basement have been reported in the west of Salamanca province [2]. The freshwater infiltration through the fractured basement, runoff in the mountain system of S^{a} de Ávila and Gredos, and streamflow infiltration from surface bodies recharge the local confined aquifers, where oxic conditions promote the uranium leaching from mineral-bearing to groundwater, leading to an increase in the uranium concentration. The uranium spatial distribution predicted throughout these aquifers could indicate that the Cenozoic detritic deposits comprising the southern part of the DB are influenced by the U-rich mineral composition characteristic of the CS. Porous and moderate to highly productive, predominantly oxic aquifers [50], Cenozoic and Quaternary aquifers of the DB, and the Mesozoic fissured and moderately productive carbonate ones, are recharged through surface infiltration of runoff freshwater, which easily promotes the leaching and dissolution of U-minerals. Furthermore, the general groundwater flux from recharge mountain areas towards the Duero River, as well as the streamflow of fluvial drainage across the Quaternary deposits, may also positively affect uranium mobilisation to groundwater. According to these hydrogeological settings, the IDW method provided a reliable prediction map of the uranium concentrations in the DB aquifers.

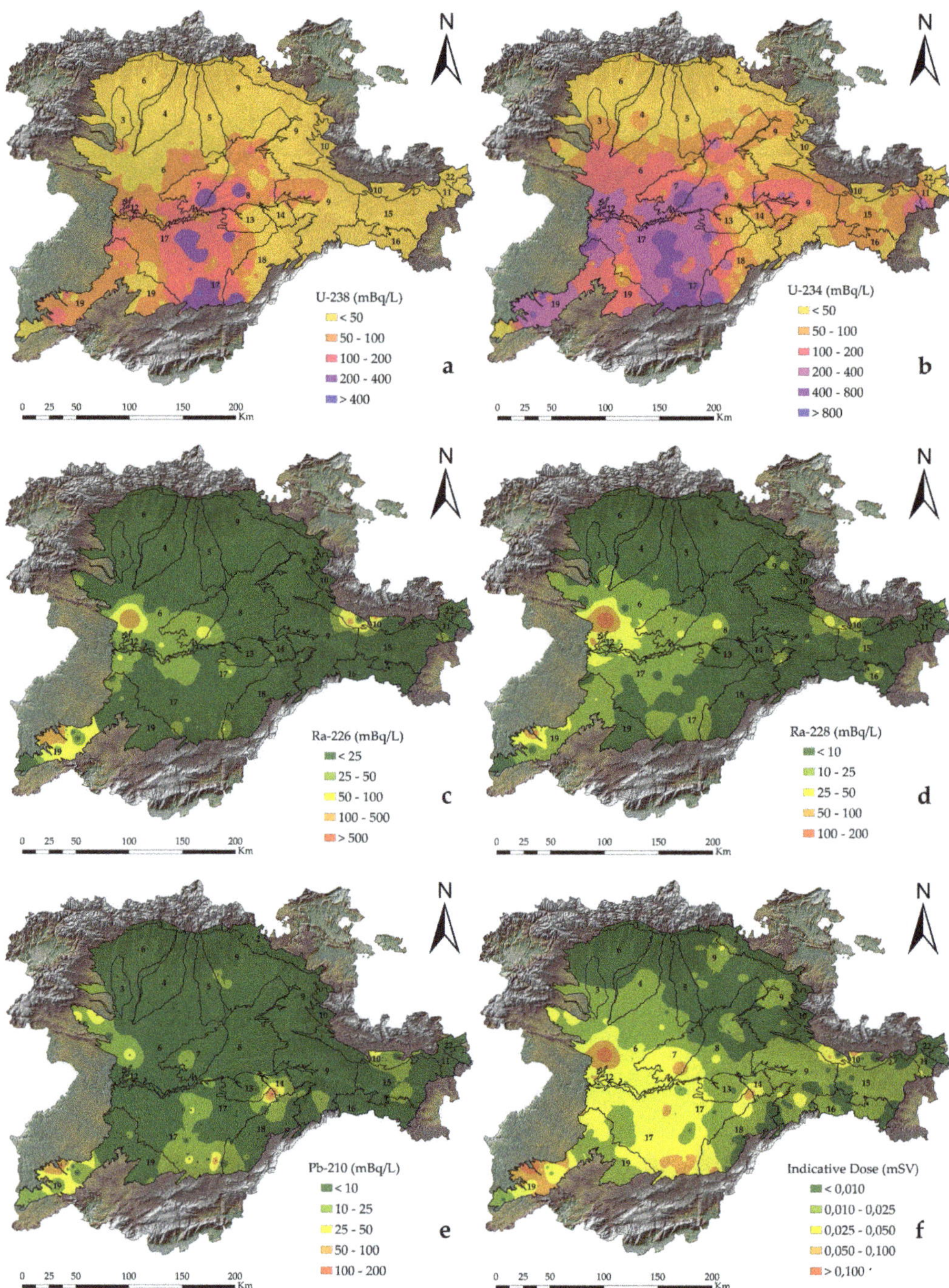

Figure 4. Spatial distribution of the main natural radionuclides in groundwater and the *ID*. IDW interpolation technique was applied: (**a**) ^{238}U activity concentration; (**b**) ^{234}U activity concentration; (**c**) ^{226}Ra activity concentration; (**d**) ^{228}Ra activity concentration; (**e**) ^{210}Pb activity concentration; (**f**) *ID*. The hydrogeological unit names corresponding to the numbers illustrated are referred to in Figure 2.

Despite ^{226}Ra (half-life 1.6×10^3 y) and ^{228}Ra (half-life 5.75 y) having the same chemical behaviour, they come from different disintegration chains, ^{238}U and ^{232}Th decay series, respectively. Thus, the head chain radionuclide concentrations in aquifer rocks and their widely varying half-lives can affect the different occurrences of ^{226}Ra and ^{228}Ra in groundwater [51,52]. As seen in Figure 4c, the interpolation map of ^{226}Ra occurrence predicted groundwater concentrations between 100 to 500 mBq/L, values which correspond to a_α screening level and the maximum reference concentration (MRC) in drinking waters, respectively, extending through three main locations:

- In the central to the west of the Zamora province, and in the western area of the Esla-Valderaduey unit, ^{226}Ra content in groundwaters from the marginal area of Tierra de Campos facies was formed by low to medium permeable detritic rocks, and varied from 100 to 250 mBq/L.
- Extending through the southern areas of the Ciudad Rodrigo-Salamanca medium and low permeable detritic aquifers there was up to, or exceeding, 500 mBq/L in Gallegos de Argañan (Salamanca).
- In carbonate aquifers of the Arlanza-Ucero-Albión unit, in the Cameros Basin, throughout the Lobos river canyon and the surrounding area of Cabrejas de Pinar, the ^{226}Ra concentration values ranged between 100 and 120 mBq/L.

According to the mechanism that may control the presence of ^{226}Ra in groundwater, the IDW method provided a consistent spatial distribution. The diverse lithological context and the occurrence of high ^{226}Ra concentration may be mainly due to both mineralogical bedrock composition and the chemical properties that govern the water–rock interaction. On the one hand, the elevated uranium content observed in felsic rock and shale groundwater masses enhances the entry of ^{226}Ra into the water by the decay of dissolved parent radionuclides, alpha recoil effect, and desorption from mineral grains. On the other hand, the analogue chemical behaviour of radium with other crustal cations, such as Ca or Ba, may lead to it being included in the carbonate complexes and then mobilised to water through dissolution, adsorption or ion exchange with the reactive surface of other colloids present in groundwater. It can be observed in Figure 4c that in the moderately productive carbonate aquifers, in particular, which extended through the Duero Central region, Páramo de Cúellar and Páramo de Duratón, the ^{226}Ra concentration predicted was <25 mBq/L.

Regarding the ^{228}Ra concentration, given that it decays from ^{232}Th, which in most aquifer conditions tends to precipitate as insoluble hydroxides, high values exceeding MRC (200 mBq/L), were not expected in the study area. As shown in Figure 4d, the prediction map showed a co-occurrence of significant ^{228}Ra concentration in the same areas where ^{226}Ra-rich was found, corresponding to the southwest of the Esla-Valderaduey region aquifers and in the western of the Ciudad Rodrigo-Salamanca unit, with maximum concentrations of 196 and 164 mBq/L, respectively, as reported in Table 3. Nevertheless, in the IDW map, there was no high ^{228}Ra content predicted in the easternmost areas of the Ciudad Rodrigo-Salamanca hydrogeological unit, thus groundwaters of this area must be affected by other local conditions (i.e., groundwater chemical composition and/or ^{232}Th concentration in the rock mineral-bearing) that led to minimizing the presence of ^{228}Ra.

According to the IDW method (Figure 4e), and the concentration values from the analysed groundwater intakes, ^{210}Pb > 200 mBq/L (MRC) was not found in the DB. Significant high concentrations, between 100 and 200 mBq/L, were predicted only in the Ciudad Rodrigo-Salamanca and Los Arenales hydrogeological units, with maximum values exceeding 150 mBq/L at some sampling points, where groundwaters are stored in the Cenozoic and Quaternary medium permeability detritic lithologies. The ^{210}Pb presence in near-neutral pH groundwaters is mainly related to the alpha recoil of ^{214}Pb and its solubility enhancement under reducing conditions [25,53]. Moreover, as ^{210}Pb is a progeny of ^{226}Ra, its presence may be associated with a high ^{226}Ra presence in groundwater. The similarities observed between the spatial distribution of ^{226}Ra and ^{210}Pb in the Ciudad Rodrigo-Salamanca unit, as illustrated in Figure 4c,e, suggest that the same hydrogeological settings are the controlling processes involved in the water–rock interaction and are also the same.

3.2. Radioactivity Parameters Used for Drinking Water Monitoring

As has been discussed in the previous section, the spatial distribution of the uranium isotopes, ^{238}U and ^{234}U, had relatively high concentrations, exceeding 100 mBq/L, in wide areas comprising several hydrogeological units of the DB. Both radioisotopes disintegrate by α emission, so, consequently, their occurrence in groundwater contributes to the a_α value. This relation between uranium concentration and a_α spatial distribution gave rise to a similar prediction map, as illustrated in Figures 4a,b and 5a. Prediction areas with values higher than 100 mBq/L occupied a wide extension of the Castilla y León territory, and, in the southern boundaries of the Los Arenales, Segovia and Ciudad Rodrigo-Salamanca hydrogeological units. even exceeded by one magnitude order, with a maximum of 2407, 1847, 1223 mBq/L, respectively (Table 3). As can be observed in Figure 5a, in the southern areas of the Los Arenales region and the Segovia HU, the high concentration seemed to follow the groundwater flow pattern towards the centre of the DB, which suggests transference of α emitters, mainly uranium radionuclides, through the unconfined DB aquifers of high concentrations, from the highly influenced mineralogical detritic aquifers to the Duero River. The shape of the interpolated a_α matched with the predicted areas of high uranium concentration, as illustrated in Figure 4a,b. In the shallow and well-oxygenated aquifers, formed in alluvial and terraces deposits of the current riverine system occupying the central areas of the DB, a_α values were found. Regarding a_α prediction in the central carbonate aquifers, especially significant was the concentration in Paramo de Torozos HU, which was higher than 1100 mBq/L, due to the dissolution of the limestones leading to an increase in the concentration of uranium isotopes in groundwater. This correlation between uranium occurrence in the Spanish carbonate aquifers has been previously demonstrated [10,15].

Regarding the a_β prediction map, the occurrence of higher values than their respective screening level, 1000 mBq/L, was limited to small regions within the DB, as seen in Figure 5b. The long-lived ^{238}U-daughters, ^{234}Th and ^{234m}Pa, that decay from β-emission, were in secular equilibrium with their parent in the analysed samples. Both β-emitting radionuclides were the main contributors to a_β, especially in the sandy aquifers of the Esla-Valderaduey region and the Páramo de Torozos HU, with maximum values of 1547 and 1425 mBq/L, respectively. It should be noted that a_β hot spots were also influenced by the contribution of the highest ^{228}Ra and ^{210}Pb concentrations, such as those occurring in groundwaters from the southwestern of the Ciudad Rodrigo-Salamanca HU.

According to the *ID* prediction map shown in Figure 4f, the radioactivity content in groundwaters from the southwestern areas of Ciudad Rodrigo-Salamanca HU, as well as the aquifers placed in the southwestern of the Esla-Valderaduey region, predicted *ID* that exceeded their corresponding parametric value. The maximum *ID* values of 0.32 and 0.17 mSv corresponded to both hydrogeological units, respectively, as reported in Table 3. This meant that the consumption of these groundwaters may pose a risk to public health. The suitability of the screening levels as a monitoring strategy is out of the scope of this research, and, due to the low selectivity of screening levels, especially the a_α not really appropriate. The interpolation maps do, however, provide evidence that performing specific radionuclide analysis to properly evaluate the radioactivity quality of the groundwaters for human consumption, is necessary.

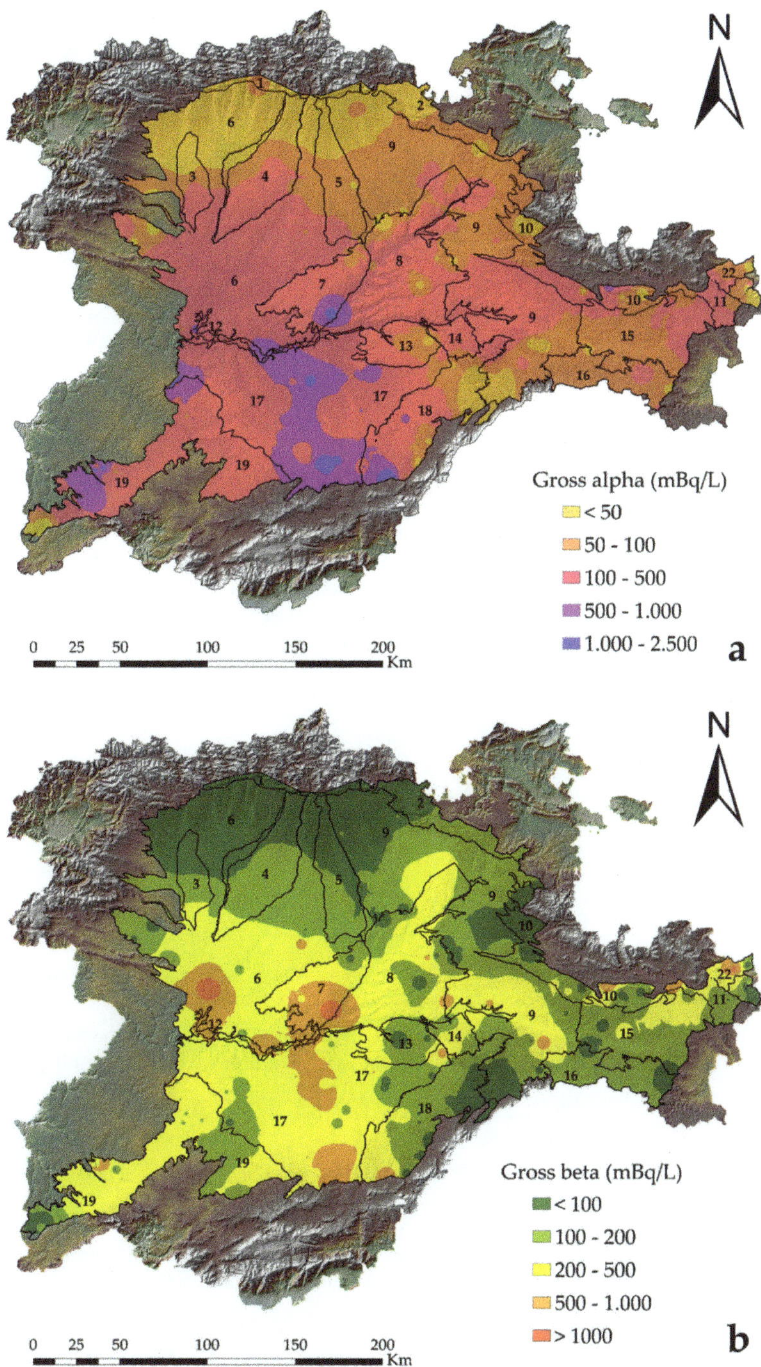

Figure 5. Spatial distribution of the gross alpha activity (**a**) and gross beta activity (**b**) screening levels for the hydraulically unconfined aquifer system of the Duero Basin. The hydrogeological unit names corresponding to the numbers illustrated are referred to in Figure 2.

4. Conclusions

As part of more extensive research concerning the groundwater radiological characterisation of CyL, this paper addressed the use of a proper interpolation method, IDW, for elaborating radioactivity prediction maps to provide an overview concerning radioactive spatial distribution in the hydraulically interconnected aquifers of the DHC. The major occurrence of ^{234}U and ^{238}U activity concentrations were found in the southern area of the DB, mainly promoted by the groundwater flux toward the Duero River, leaching the minerals of the dentritic materials, which form a moderate to highly productive aquifer system, the composition of which is influenced by the igneous and metamorphosed rocks of the CS. Despite the presence of the radionuclides from the natural U- and Th-decay series in groundwater being controlled by several interconnected hydrogeochemical and physical parameters, such as redox, pH, adsorption/desorption rates, water composition, alpha recoil, etc., and water-rock interaction, the geostatistical analysis performed showed that the radioactivity content in DHC groundwaters is related to the hydrogeological context.

It should be underlined that further hydrogeological evaluation is necessary for a better understanding of the behaviour of radionuclides in groundwaters, which was out of the scope of this study. The prediction maps generated a worthwhile qualitative overview concerning the spatial distribution of radionuclides in the studied area. Maps can contribute towards improving radioactivity quality control planning, and in identifying hazardous-prone areas, in which the consumption of groundwaters may pose a potential risk to human health. According to the current drinking water framework, and given the radionuclide spatial distribution content in the DB, the use of the prediction maps of a_α and a_β may not be suitable as a radioactivity screening strategy for cost-effective management in groundwater quality control.

Author Contributions: Conceptualization, D.B.-A.; methodology D.B.-A., B.Q., A.M.M.-G. and J.C.L.; software, D.B.-A.; validation, D.B.-A., B.Q., A.M.M.-G. and J.C.L.; formal analysis, D.B.-A.; investigation, D.B.-A. and B.Q.; resources, D.B.-A. and B.Q.; data curation, D.B.-A., B.Q. and A.M.M.-G.; writing—original draft preparation, D.B.-A.; writing—review and editing, D.B.-A.; visualization, D.B.-A.; supervision, D.B.-A., A.M.M.-G., B.Q. and J.C.L.; project administration, B.Q.; funding acquisition, B.Q. All authors have read and agreed to the published version of the manuscript.

Funding: This research received no external funding.

Institutional Review Board Statement: Not applicable.

Informed Consent Statement: Not applicable.

Data Availability Statement: The data presented in this study are available on request from the corresponding author.

Acknowledgments: The authors wish to acknowledge the staff of the Dirección General de Salud Pública of Junta de Castilla y León for the planning, sampling and discussion of the results during the course of the project. This work was promoted and funded by the Dirección General de Salud Pública (Consejería de Sanidad de la Junta de Castilla y León), in the framework of the research project "Radiological characterisation of groundwater intakes intended for human consumption".

Conflicts of Interest: The authors declare no conflict of interest.

References

1. De Stefano, L.; Martínez-Santos, P.; Villarroya, F.; Chico, D.; Martínez-Cortina, L. Easier said than done? The establishment of baseline groundwater conditions for the implementation of the water framework directive in Spain. *Water Resour. Manag.* **2013**, *27*, 2691–2707. [CrossRef]
2. Santos-Francés, F.; Gil-Pacheco, E.; Martínez-Graña, A.; Alonso-Rojo, P.; Ávila-Zarza, C.; García-Sanchez, A. Concentration of uranium in the soils of the west of Spain. *Environ. Pollut.* **2018**, *236*, 1–11. [CrossRef] [PubMed]
3. Luque-Espinar, J.A.; Pardo-Igúzquiza, E.; Grima-Olmedo, J.; Grima-Olmedo, C. Multiscale analysis of the spatial variability of heavy metals and organic matter in soils and groundwater in Spain. *J. Hydrol.* **2018**, *561*, 348–371. [CrossRef]
4. Wang, Z.; Su, Q.; Wang, S.; Gao, Z.; Liu, J. Spatial distribution and health risk assessment of dissolved heavy metals in groundwater of eastern China coastal zone. *Environ. Pollut.* **2021**, *290*, 118016. [CrossRef] [PubMed]

5. Santos-Francés, F.; Martínez-Graña, A.; Ávila-Zarza, C.; Criado, M.; Sáncez-Sánchez, Y. Soil quality and evaluation of spatial variability in a semi-arid ecosystem in a region of the southeasterrn Iberian Peninsula. *Land* **2022**, *11*, 5. [CrossRef]
6. Godoy, J.M.; Godoy, M.L. Natural radioactivity in Brazilian groundwater. *J. Environ. Radioact.* **2006**, *85*, 71–83. [CrossRef]
7. Labidi, S.; Mahjoubi, H.; Essafi, F.; Ben Salah, R. Natural radioactivity levels in mineral, therapeutic and spring waters in Tunisia. *Radiat. Phys. Chem.* **2010**, *79*, 1196–1202. [CrossRef]
8. Akweetelela, A.; Kgabi, N.; Zivuku, M.; Mashauri, D. Environmental radioactivity of groundwater and sediments in the Kuiseb and Okavango-Omatako basin in Namibia. *Phys. Chem. Earth Parts A/B/C* **2020**, *120*, 102911. [CrossRef]
9. Rojas, L.V.; Santos Junior, J.A.; Corcho-Alvarado, J.A.; Amaral, R.S.; Rölin, S.; Milan, M.O.; Herrero, Z.; Francis, K.; Cavalcanti, M.; Santos, J.M.N. Quality and management status of the drinking water supplies in a semiarid region of Northeastern Brazil. *J. Environ. Sci. Health Part A* **2020**, *55*, 1247–1256. [CrossRef]
10. Pérez-Moreno, S.M.; Guerrero, J.M.; Mosqueda, F.; Gázquez, M.J.; Bolívar, J.P. Hydrochemical behaviour of long-lived natural radionuclides in Spanish groundwaters. *Catena* **2020**, *191*, 104558. [CrossRef]
11. Shabana, E.I.; Kinsara, A.A. Radioactivity in the groundwater of a high background radiation area. *J. Environ. Radioact.* **2014**, *137*, 181–189. [CrossRef] [PubMed]
12. Sherif, M.I.; Lin, J.; Poghosyan, A.; Abouelmagd, A.; Sultan, M.I.; Sturchio, N.C. Geological and hydrogeochemical controls on radium isotopes in groundwater of the Sinai Peninsula, Egypt. *Sci. Total Environ.* **2018**, *613*, 877–885. [CrossRef]
13. Stackelberg, P.; Szabo, Z.; Jurgens, B.C. Radium mobility and the age of groundwater in public-drinking-water supplies from the Cambrian-Ordovician aquifer system, north-central USA. *Appl. Geochem.* **2018**, *89*, 34–48. [CrossRef]
14. Ulrich, S.; Gillow, J.; Roberts, S.; Byer, G.; Sueker, J.; Farris, K. Hydrogeochemical and mineralogical factors influencing uranium in background area groundwater wells: Grants, New Mexico. *J. Hydrol. Reg. Stud.* **2019**, *26*, 100636. [CrossRef]
15. Ratia, J.M.; Hernando, A.P.; Aguilar, C.; Ballarín, F.B. Role of lithology in the presence of natural radioactivity in drinking water samples from Tarragona province. *Environ. Sci. Pollut. Res.* **2021**, *28*, 39333–39344. [CrossRef] [PubMed]
16. World Health Organization. *Management of Radioactivity in Drinking Water*; World Health Organization: Geneva, Switzerland, 2018.
17. Council Directive 2013/51/EURATOM of 22 October 2013 Laying down Requirements for the Protection of the Health of the General Public with Regard to Radioactive Substances in Water Intended for Human Consumption. Available online: http://data.europa.eu/eli/dir/2013/51/oj (accessed on 11 February 2022).
18. RD 314/2016. Real Decreto 314/2016, de 29 de julio, por el que se modifican el Real Decreto 140/2003, de 7 de febrero, por el que se establecen los criterios sanitarios de la calidad del agua de consumo humano, el Real Decreto 1789/2010, de 30 de diciembre, por el que se regula la explotación y comercialización de aguas minerales, naturales y aguas de manantial envasadas para consumo humano, y el Real Decreto 1799/2010, de 30 de diciembre, por el que se regula el proceso de elaboración y comercialización de aguas preparadas envasadas para el consumo humano. *Boletín Oficial Estado* **2016**, *183*, 53106–53126.
19. Paul, R.; Brindha, K.; Gowrisankar, G.; Tan, M.L.; Singh, M.K. Identification of hydrogeochemical processes controlling groundwater quality in Tripura, Northeast India using evaluation indices, GIS, and multivariate statistical methods. *Environ. Earth Sci.* **2019**, *78*, 470. [CrossRef]
20. Panneerselvam, B.; Paramasivam, S.K.; Karuppannan, S.; Ravichandran, N.; Selvaraj, P. A GIS-based evaluation of hydrochemical characterisation of groundwater in hard rock region, South Tamil Nadu, India. *Arab. J. Geosci.* **2020**, *13*, 837. [CrossRef]
21. Keshavarzi, A.; Tuffour, H.O.; Brevik, E.C.; Ertunc, G. Spatial variability of soil mineral fractions and bulk density in Northern Ireland: Assessing the influence of topography using different interpolation methods and fractal analysis. *Catena* **2021**, *207*, 105646. [CrossRef]
22. Asikainen, M.; Kahlos, H. Anomalously high concentrations of uranium, radium and radon in water from drilled wells in the Helsinki region. *Geochim. Cosmochim. Acta* **1979**, *43*, 1681–1686. [CrossRef]
23. Banks, D.; Royset, O.; Strand, T.; Skarphagen, H. Redioelement (U, Th, Rn) concentrations in Norwegian bedrock groundwaters. *Environ. Geol.* **1995**, *25*, 165–180. [CrossRef]
24. Siegel, M.D.; Bryan, C.R. Radioactivity, Geochemistry, and Health. In *Treatise on Geochemistry*, 2nd ed.; Henrich, D.H., Turekian, K.T., Eds.; Elsevier: Amsterdam, The Netherlands, 2014; Volume 11, pp. 191–256. [CrossRef]
25. Szabo, Z.; Stackelberg, P.E.; Cravotta III, C.A. Occurrence and geochemistry of lead-210 and polonium-210 radionuclides in public-drinking-water supplies from principal aquifers of the United States. *Environ. Sci. Technol.* **2020**, *54*, 7236–7249. [CrossRef] [PubMed]
26. Hu, K.; Huang, Y.; Li, H.; Baoguo, L.; Chen, D.; White, R.E. Spatial variability of shallow groundwater level, electrical conductivity and nitrate concentration, and risk assessment of nitrate contamination in North China Plain. *Environ. Int.* **2005**, *31*, 896–903. [CrossRef] [PubMed]
27. Skeppström, K.; Olofsson, B. A prediction method for radon in groundwater using GIS and multivariate statistics. *Sci. Total Environ.* **2006**, *367*, 666–680. [CrossRef] [PubMed]
28. Valencia-Ortiz, J.A.; Martínez-Graña, A.M. A neural network model applied to landslide susceptibility analysis. *Geomat. Nat. Hazards Risk* **2018**, *9*, 1106–1128. [CrossRef]
29. Gong, G.; Mattevada, S.; O'Bryant, S.E. Comparison of the accuracy of Kriging and IDW interpolations in estimating groundwater arsenic concentrations in Texas. *Environ. Res.* **2014**, *130*, 59–69. [CrossRef] [PubMed]
30. Kumar, S.; Sangeetha, B. Assessment of groundwater quality in Madurai city by using geospatial techniques. *Groundw. Sustain. Dev.* **2020**, *10*, 100297. [CrossRef]

31. Zhu, H.C.; Charlet, J.M.; Poffijn, A. Radon risk mapping in southern Belgium: An application of geostatistical and GIS techniques. *Sci. Total Environ.* **2001**, *272*, 203–210. [CrossRef]
32. Kobya, Y.; Taşkin, H.; Yeşilkanat, C.M.; Çevik, U.; Karahan, G.; Çakir, B. Radioactivity survey and risk assessment study for drinking water in the Artvin province, Turkey. *Water Air Soil Pollut.* **2015**, *226*, 49. [CrossRef]
33. Junta de Castilla y León. *Mapa Geológico y Minero de Castilla y León, Escala 1:400.000*; SIEMCALSA: Valladolid, Spain, 1997.
34. IGME. *Mapa Geológico de la Peninsula Ibérica, Baleares y Canarias a Escala 1:1.000.000*; IGME: Madrid, Spain, 1995.
35. Del Pozo Gómez, M. *Mapa Litoestratigráfico, de Permeabilidad e Hidrogeológico βde España Continuo Digital a Escala 1:200.000. Convenio de Colaboración Entre el Ministerio de Medio Ambiente y el Instituto Geológico y Minero de España para la Realización de Trabajos Técnicos en Relación con la Aplicación de la Directiva Marco del Agua en Materia de Agua Subterránea*; IGME: Madrid, Spain, 2009.
36. Borrego-Alonso, D.; Quintana, B.; Lozano, J.C. Revisiting methods for the naturally-occurring radioactivity assessment in drinking waters. *Appl. Radiat. Isot.* **2022**. submitted.
37. García-Talavera, M.; Neder, H.; Daza, M.J.; Quintana, B. Towards a proper modelling of detector and source characteristics in Monte Carlo simulations. *Appl. Radiat. Isot.* **2000**, *52*, 777–783. [CrossRef]
38. Daza, M.J.; Quintana, B.; García-Talavera, M.; Fernández, F. Efficiency calibration of an HPGe detector in the 46.54–2000 keV energy range for the measurement of environmental samples. *Nucl. Instrum. Methods Phys. Res. A* **2001**, *470*, 520–532. [CrossRef]
39. Quintana, B.; Pedrosa, M.C.; Vázquez-Canelas, L.; Santamaría, R.; Sanjuán, M.A.; Puertas, F. A method for the complete analysis of NORM building materials by γ-ray spectrometry using HPGe detectors. *Appl. Radiat. Isot.* **2018**, *134*, 470–476. [CrossRef] [PubMed]
40. Quintana, B.; Fernández, F. Gamma-ray spectral analysis with the COSPAJ continuum fitting routine. *Appl. Radiat. Isot.* **1998**, *49*, 9–11. [CrossRef]
41. LNHB. Nucléide-Lara, Library for Gamma and Alpha Emissions. 2020. Available online: http://www.nucleide.org/Laraweb/ (accessed on 11 February 2022).
42. Hallstadius, L. A method for the electrodeposition of actinides. *Nucl. Instrum. Methods Phys. Res.* **1984**, *223*, 266–267. [CrossRef]
43. Horwitz, E.P.; Dietz, M.L.; Chiarizia, R.; Diamond, H.; Essling, A.M.; Graczyk, D. Separation and preconcentration of uranium from acidic media by extraction chromatography. *Anal. Chim. Acta* **1998**, *266*, 25–37. [CrossRef]
44. Flynn, W.W. The determination of low levels of polonium-210 in environmental materials. *Anal. Chim. Acta* **1968**, *43*, 221–227. [CrossRef]
45. Férnandez, P.L.; Gómez, J.; Ródenas, C. Evaluation of uncertainty and detection limits in ^{210}Pb and ^{210}Po measurement in water by alpha spectrometry using ^{210}Po spontaneous deposition onto silver disk. *Appl. Radiat. Isot.* **2012**, *70*, 758–764. [CrossRef]
46. World Organization of Health. *Guidelines for Drinking Water Quality: 4th ed Incorporating the First Addendum*; World Organization of Health: Geneva, Switzerland, 2017.
47. ESRI. *Creating, Editing and Managing Geodatabases for ArcGis Desktop*; Virtual Campus, ESRI Educational Service, Environmental Systems Research Institute, Inc.: Redlands, CA, USA, 2014.
48. Li, J.; Heap, A.D. A review of comparative studies of spatial interpolation methods in environmental sciences: Performances and impact factors. *Ecol. Inform.* **2011**, *6*, 228–241. [CrossRef]
49. Osmond, J.K.; Cowart, J.B. U-series nuclides as tracers in groundwater hydrology. In *Environmental Tracers in Subsurface Hydrology*; Cook, P.G., Herczeg, A.L., Eds.; Springer: Boston, MA, USA, 2000; pp. 145–173. [CrossRef]
50. Giménez-Forcada, E.; Smedley, P.L. Geological factors controlling occurrence and distribution of arsenic in groundwaters from the southern margin of the Duero Basin, Spain. *Environ. Geochem. Health* **2014**, *36*, 1029–1047. [CrossRef]
51. Sturchio, N.C.; Banner, J.L.; Binz, C.M.; Heraty, L.B.; Musgrove, M. Radium geochemistry of groundwaters in Palaeozoic carbonate aquifers, midcontinent, USA. *Appl. Geochem.* **2001**, *16*, 109–122. [CrossRef]
52. Vengosh, A.; Coyte, R.M.; Podgorski, J.; Johnson, T.M. A critical review on the occurrence and distribution of the uranium- and thorium-decay nuclides and their effects on the quality of groundwater. *Sci. Total Environ.* **2022**, *808*, 151914. [CrossRef] [PubMed]
53. Jurgens, B.C.; Parkhurst, D.L.; Belitz, K. Assessing the lead solubility potential of untreated groundwater of the United States. *Environ. Sci. Technol.* **2019**, *53*, 3095–3103. [CrossRef] [PubMed]

Article

Soil Chemical Properties and Trace Elements after Wildfire in Mediterranean Croatia: Effect of Severity, Vegetation Type and Time-Since-Fire

Iva Hrelja [1],*, Ivana Šestak [1], Domina Delač [1], Paulo Pereira [2] and Igor Bogunović [1]

[1] Department of General Agronomy, Division of Agroecology, Faculty of Agriculture, University of Zagreb, 10000 Zagreb, Croatia; isestak@agr.hr (I.Š.); ddelac@agr.hr (D.D.); ibogunovic@agr.hr (I.B.)

[2] Environmental Management Center, Mykolas Romeris University, Ateities g. 20, LT-08303 Vilnius, Lithuania; pereiraub@gmail.com

* Correspondence: ihrelja@agr.hr

Abstract: Natural landscapes in the Mediterranean ecosystem have experienced extensive changes over the last two centuries due to wildfire activity. Resulting interactions between climatic warming, vegetation species, soil natural, and meteorological condition before and after a wildfire create substantial abrupt landscape alterations. This study investigates the evolution (2 days, 3, 6, 9, and 12 months after a fire) of topsoil (0–5 cm) chemical properties in burned Cambisols (Zadar County, Croatia) with respect to different wildfire severities (HS—high severity, MS—medium severity, C—unburned) and vegetation species (*Quercus pubescens* Willd. and *Juniperus communis* L.). Soil pH, electrical conductivity (EC), calcium carbonates ($CaCO_3$), total organic carbon (TOC), total nitrogen (TN), total sulphur (TS), copper (Cu) and zinc (Zn) were significantly higher in HS than in MS and C. Total soil potassium (TK), Fe and Ni were significantly higher in C than in HS. The increase of TOC and TN was more pronounced in *Quercus p.* than *Juniperus c.*, especially in the first three months. Soil pH, EC, $CaCO_3$, TOC, TN, and TS were most affected by wildfire severity. The distinction between C, MS and HS categories was less visible 9 and 12 months post-fire, indicating the start of the recovery of the soil system. Post-fire management and temporal recovery of the soil system should consider the obvious difference in soil disturbance under HS and MS between vegetation species.

Keywords: soil organic matter; recovery; post-fire management; *Quercus pubescens* Willd.; *Juniperus communis* L.

Citation: Hrelja, I.; Šestak, I.; Delač, D.; Pereira, P.; Bogunović, I. Soil Chemical Properties and Trace Elements after Wildfire in Mediterranean Croatia: Effect of Severity, Vegetation Type and Time-Since-Fire. *Agronomy* **2022**, *12*, 1515. https://doi.org/10.3390/agronomy12071515

Academic Editor: Antonio Miguel Martínez-Graña

Received: 23 May 2022
Accepted: 22 June 2022
Published: 24 June 2022

Publisher's Note: MDPI stays neutral with regard to jurisdictional claims in published maps and institutional affiliations.

1. Introduction

Wildfires are an inevitable occurrence in ecosystems worldwide and are considered one of the main causes of environmental change [1,2]. A warmer and drier climate, along with land abandonment and afforestation, is leading to an increase in catastrophic wildfires [3–6].

Soils are an important component of the ecosystem and can be significantly altered by wildfires and their recurrence [7]. The changes in soil systems after wildfires are mainly caused by high temperatures, duration, and severity of the wildfire [8]. Several authors have indicated how the increase in wildfire severity can lead to degradative long-term cumulative effects on soil chemistry [9–11], while low and moderate severity fires increase soil organic matter (SOM), total nitrogen (TN), and total carbon (TC) [1,12]. Regardless of the fire severity, organic matter burning leads to rapid mineralisation and an increase of soil nutrients and trace elements such as iron (Fe), aluminium (Al), zinc (Zn), nickel (Ni), cadmium (Cd), and lead (Pb) [13]. Their concentrations depend mainly on the severity of the fire and the type of vegetation burned [8,14,15]. The biological toxicity and potential bioaccumulation of trace elements in the ecosystem after a wildfire are of major concern for the contamination and environmental sustainability of downstream lands and water bodies [13,16,17].

Understanding the relationships between forest management, vegetation species, fire severity, and post-fire recovery of forest ecosystems is especially important for developing fire governance and management strategies to increase socio-ecological resilience in a rapidly changing environment affected by the changing climate [18]. Fire behaviour is determined by the quantity and quality of the fuel load. Therefore, the differences in fire behaviour between grasslands, shrublands, and forests cannot be ignored [19]. Low fuel loads in grasslands and shrublands result in low fire severity and rapid spread, while forests generally have slower fire spread and moderate to high fire severity [20,21]. Different vegetation species in burned forest land have been shown to have significant effects on soil properties. Lombao et al. [11] studied the effects of wildfires on soil under *Eucalyptus* and *Quercus* vegetation and indicated that vegetation was the most important factor controlling the overall quality of burned soils and that wildfire is also an important source of variation in soil quality.

Vegetation species is an important determinant of the composition of ash incorporated into the soil over time, and because of the close relationship between fire severity, ash, and soil interaction in the post-fire period, it is critical to conduct research that considers the overall influence of these interacting factors (vegetation species × fire severity × time) on soil quality recovery following a wildfire. Few studies have examined the phenomenon of fire severity in relation to different vegetation species [22–24], while the interaction between these factors and the evolution of chemical properties and trace elements in burned soils after a fire is not well understood. To fill this knowledge gap, this study aims to: (a) determine the short-term evolution of soil chemical properties after a wildfire and their temporal dynamics and (b) compare the impact of vegetation species (*Quercus pubescens* Willd. and *Juniperus communis* L.) on soil properties in the environmental conditions of Mediterranean Croatia.

2. Materials and Methods

2.1. Study Area and Soil Sampling

The study was conducted in Zadar County, Croatia (44° 05′ N; 15° 22′ E; 72 m a.s.l.), characterized by hot-summer Mediterranean (Csa) climate according to the Köppen–Geiger classification [25]. The average annual temperature is 14.9 °C, and the average annual precipitation is 879.2 mm in the period from 1971 to 2000 [26]. Most of the vegetation in the area consists of *Quercus pubescens* Willd., *Pinus halpensis* Mill., *Pinus pinaster* Ait., *Pinus pinea* L., and *Juniperus communis* L. The soil type of the study area is chromic Cambisol [27] with calcareous parent material and sandy clay loam to clayey texture. These soils are characterized by their permeability and good drainage. The general soil properties of the study area are listed in Table 1.

Table 1. General soil properties in the investigated area.

Horizon	Texture (%)			pH	SOM	P_2O_5	K_2O
	Sand	Silt	Clay		(%)	g kg^{-1}	
0–10 cm	56.0	17.8	26.2	6.5	3.2	0.002	0.19
10–60 cm	41.0	12.2	46.8	6.5	0.8	0.002	0.22

The wildfire affected ~13.5 ha of a mixed forest of *Quercus pubescens* Willd. and *Juniperus communis* L. in August 2019. Fire severity was moderate to high, as determined by visual inspection of burned vegetation and ash characteristics using the methodology described in Pereira et al. [28] and Úbeda et al. [29]. According to the characteristics of the burned area, the experimental design recognized three categories of sampling areas: C—control, unaffected by fire; HS—high severity, which came from sites where the foliage and trunk were completely burned, and the soil was covered with white ash; and MS—medium severity, where the foliage and trunk were partially burned, and the soil was covered with black ash. Additionally, each category was subdivided by the two vegetation species, i.e., each of the three severity categories contained thirteen (13) sample areas under *Quercus pubescens* Willd. and seven (7) sample areas under *Juniperus communis* L. Flat terrain and an area with average wind strength of 2 Beaufort (1.6–3.3 m/s) were selected to minimize the intensity of immediate ash transport due to terrain slope and wind. The final experimental design is shown in Figure 1. Soil samples (with the ash layer removed) were collected at a depth of 0–5 cm using a spade and georeferenced using a Trimble GeoXH handheld device (GeoExplorer ® 6000 series, Trimble GmbH, Raunheim, Germany). Samples were collected two days (0 MAF), 3 months (3 MAF), 6 months (6 MAF), 9 months (9 MAF), and 12 months (12 MAF) after wildfires. During each sampling period, 60 soil samples were collected (20 per severity category), resulting in a total of 300 samples collected.

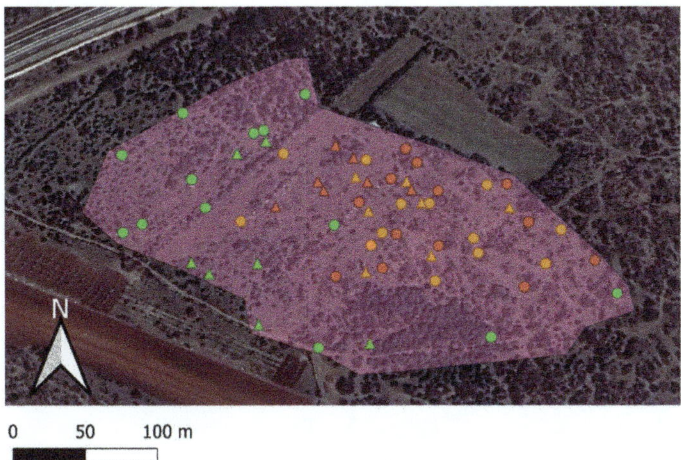

Figure 1. Study area and experimental design. Different shapes denote vegetation species (circle indicate sample under *Quercus pubescens* Willd.; triangle indicates sample under *Juniperus communis* L.), and different colours denote wildfire severity (green—Control; orange—Medium severity; red—High severity). *Source:* Google Earth.

2.2. Laboratory Analysis

Soil samples were air-dried and sieved through a 2 mm sieve. Subsequently, soil pH, electrical conductivity (EC), carbonates ($CaCO_3$), total organic carbon (TOC), TN, and total sulfur (TS) were determined by standard laboratory methods: pH and EC were determined electrometrically, using Beckman's φ 72 pH meter, in H_2O (1:2.5) and a HACH-CO150 conductometer (300–1900 μS), respectively. $CaCO_3$ content was determined via volumetric Scheibler calcimeter and TOC, TN, and TS content by dry combustion using a CHNS analyzer (Elementar Analysensysteme GmbH, Langenselbold, Germany). Total soil potassium (TK) and trace element (Fe, Zn, Ni, and Cu) contents were determined by X-ray fluorescence using a VantaTM C series handheld XRF analyzer (Olympus Scientific Solutions Americas, MA, USA).

2.3. Statistical Analysis

Before statistical analysis was carried out, the collected data were checked for normality using Q–Q plots and transformed when needed to meet this assumption in the statistical analysis. Z-scores were subsequently calculated for each variable in order to detect outliers, which were removed if the score exceeded 3 standard deviations [30]. Data were then analyzed using a factorial ANOVA to determine the percent of variation attributable to each of the factors: fire severity, time, and vegetation species. Tukey's HSD test was applied when significant differences were found ($p < 0.05$). In the case of EC, $CaCO_3$, TK, and Cu, where data normality could not be achieved, a non-parametric Kruskal–Wallis analysis was applied, and a multiple comparison of mean ranks for all groups was used as a post-hoc test.

A principal component analysis (PCA) was also conducted to examine the relationship between selected soil variables and wildfire severity during the study period. PCA analysis was performed separately for each sampling period in the R 3.6.2 environment [31] using plyr (v1.8.5) and car (v3.0-6) packages and visualized using ggbiplot (v0.55) package. Statistica 12.0 software [32] was used for factorial ANOVA analysis, and additional graphs were created using Plotly Chart Studio [33].

3. Results

3.1. Meteorological Observations

During the study period, total average precipitation was 959.8 mm, 80.6 mm above the 30-year average (879.2 mm). Precipitation was higher than the average in November and December 2019 and June 2020, while it was lower in January through May 2020 (Figure 2). The monthly average temperature was slightly lower (14.5 °C) than average (14.9 °C) during the study year. The largest difference was observed in January 2020, when the average temperature was 2 °C lower than the 30-year average. The first major precipitation events were recorded about 30 days after the fire, on September 24 and 26, with 19.1 and 44.9 mm of precipitation, respectively.

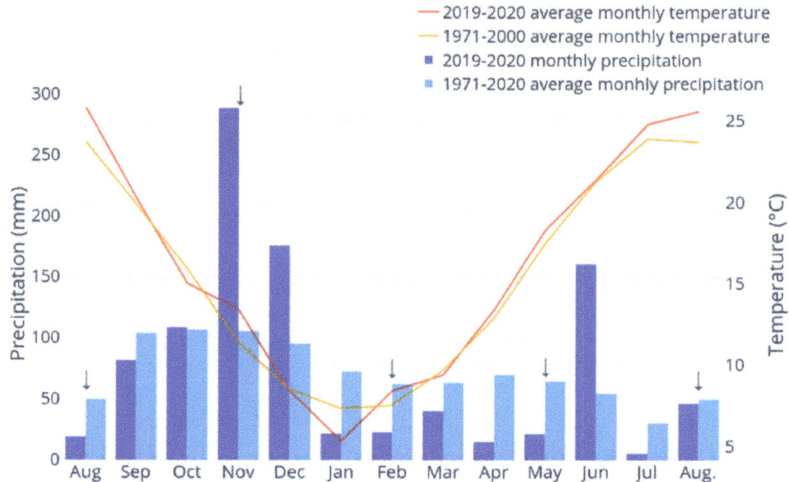

Figure 2. Monthly 30-year average of precipitation and temperature (measured in the period from 1971 to 2000) and during the investigated year (2019–2020). The arrows indicate the dates on which sampling campaigns were conducted.

3.2. Basic Chemical Properties

Factorial and Kruskal–Wallis ANOVA and post-hoc results of soil properties are presented in Tables 2–4. All properties studied were significantly affected by fire severity. pH, EC, TOC, TN, TS, and Zn were significantly different for vegetation, while all properties (with the exception of TN) were significantly different between sampling periods (Table 2). The difference between fire severity × vegetation interaction was significant for pH, Fe, and Zn, while the difference between fire severity × time interaction was significant only for pH and TS. Additionally, the difference between vegetation × time was significant for TS. Fire severity × vegetation × time was not significant for any soil property.

Table 2. Factorial ANOVA (for pH, TOC, TN and TS, Fe, Ni, and Zn) and Kruskal–Wallis (for EC, CaCO$_3$, TK, and Cu) test results and mean values for investigated soil properties according to fire severity, time, vegetation, and their interactions.

	pH	EC (µS/cm)	CaCO$_3$ (%)	TOC (%)	TN (%)	TS (%)	TK (g/kg)	Fe (g/kg)	Ni (mg/g)	Cu (mg/kg)	Zn (mg/g)
FS	***	***	***	***	***	***	***	***	**	***	***
V	**	**	n.s.	***	***	***	***	n.s.	n.s.	n.s.	*
T	***	**	***	*	n.s.	n.s.	***	***	***	**	**
FS × V	*	–	–	n.s.	n.s.	n.s.	–	*	n.s.	–	**
FS × T	n.s.	–	–	n.s.	n.s.	**	–	n.s.	n.s.	–	n.s.
V × T	n.s.	–	–	n.s.	n.s.	**	–	n.s.	n.s.	–	n.s.
FS × V × T	n.s.	–	–	n.s.	n.s.	n.s.	–	n.s.	n.s.	–	n.s.
Fire severity											
C	6.47 ± 0.02 c	106.2 ± 11.91 b	0.07 ± 0.04 b	3.84 ± 0.17 b	0.28 ± 0.01 b	0.07 ± 0.0 b	19.48 ± 0.12 a	29.78 ± 0.23 a	67.63 ± 0.79 a	28.39 ± 0.8 b	73.05 ± 0.82 b
MS	6.71 ± 0.02 b	114.1 ± 11.91 b	0.08 ± 0.04 b	4.06 ± 0.17 b	0.29 ± 0.01 b	0.07 ± 0.0 b	18.83 ± 0.12 ab	28.64 ± 0.23 b	65.59 ± 0.79 ab	29.82 ± 0.8 b	70.39 ± 0.82 b
HS	7.42 ± 0.02 a	277.2 ± 11.91 a	0.53 ± 0.04 a	5.34 ± 0.17 a	0.38 ± 0.01 a	0.08 ± 0.0 a	18.25 ± 0.12 b	28.01 ± 0.23 c	64.12 ± 0.79 b	33.46 ± 0.8 a	74.23 ± 0.82 a
Time											
0 MAF	6.90 ± 0.03 b	266.9 ± 15.38 a	0.15 ± 0.05 ab	4.95 ± 0.22 a	0.33 ± 0.01 a	0.05 ± 0.0 c	21.24 ± 0.15 a	30.38 ± 0.3 a	73.21 ± 1.02 a	33.13 ± 1.04 a	72.75 ± 1.06 a
3 MAF	6.85 ± 0.03 b	140.1 ± 15.38 ab	0.37 ± 0.05 a	4.23 ± 0.22 ab	0.32 ± 0.01 a	0.08 ± 0.0 b	16.89 ± 0.15 d	27.83 ± 0.3 c	63.28 ± 1.02 b	28.79 ± 1.04 cb	74.47 ± 1.06 a
6 MAF	7.12 ± 0.03 a	168.1 ± 15.38 ab	0.33 ± 0.05 a	4.59 ± 0.22 ab	0.33 ± 0.01 a	0.08 ± 0.0 b	19.42 ± 0.15 bc	28.78 ± 0.3 b	64.51 ± 1.02 b	31.13 ± 1.04 acb	73.71 ± 1.06 a
9 MAF	6.76 ± 0.03 c	132.8 ± 15.38 b	0.14 ± 0.05 b	4.30 ± 0.22 ab	0.31 ± 0.01 a	0.09 ± 0.0 a	20.62 ± 0.15 ac	29.33 ± 0.3 b	65.08 ± 1.02 b	30.69 ± 1.04 acb	70.52 ± 1.06 bc
12 MAF	6.70 ± 0.03 c	121.4 ± 15.38 b	0.12 ± 0.05 b	4.00 ± 0.22 b	0.30 ± 0.01 a	0.08 ± 0.0 b	16.10 ± 0.15 d	27.73 ± 0.3 c	62.83 ± 1.02 b	29.04 ± 1.04 b	71.33 ± 1.06 ac
Vegetation											
Quercus p.	6.83 ± 0.01 b	181.5 ± 8.14 a	0.24 ± 0.03 a	4.90 ± 0.11 a	0.36 ± 0.01 a	0.08 ± 0.0 a	18.80 ± 0.11 a	28.68 ± 0.21 a	65.29 ± 0.54 a	30.03 ± 0.75 a	73.29 ± 0.56 a
Juniperus c.	6.90 ± 0.02 a	150.2 ± 11.09 b	0.21 ± 0.03 a	3.93 ± 0.16 b	0.28 ± 0.01 b	0.07 ± 0.0 b	18.91 ± 0.08 a	28.93 ± 0.16 a	66.28 ± 0.74 a	31.09 ± 0.55 a	71.83 ± 0.77 b

Abbreviations: FS—Fire severity; V—Vegetation; T—Time; C—Control; MS—Medium severity; HS—High severity; MAF—Months after fire Significant differences at: $p < 0.05$ *, $p < 0.01$ **, $p < 0.001$ ***, n.s. not significant at $p < 0.05$ Different letters represent significant ($p < 0.05$) differences between fire severity, sampling time and vegetation values following ± indicate standard deviation.

Table 3. Results of factorial ANOVA (for pH, TOC, TN, and TS) and Kruskal–Wallis test (for EC and CaCO$_3$, and TK) during the study period.

	C -Q	MS -Q	HS -Q	C -J	MS -J	HS -J
0 MAF						
pH ($-\log[H^+]$)	6.57 ± 0.06 c	6.77 ± 0.06 b	7.23 ± 0.06 aB	6.61 ± 0.08 b	6.76 ± 0.08 b	7.45 ± 0.08 aA
EC (µS/cm)	204.9 ± 32.09 b	202.9 ± 32.09 b	558.1 ± 32.09 a	118.4 ± 43.74 b	122.4 ± 43.74 b	394.6 ± 43.74 a
CaCO$_3$ (%)	0.05 ± 0.01 b	0.06 ± 0.01 b	0.35 ± 0.01 a	0.08 ± 0.12 b	0.07 ± 0.12 b	0.30 ± 0.12 a
TOC (%)	4.43 ± 0.44 b	4.41 ± 0.44 b	7.77 ± 0.44 aA	3.79 ± 0.61 a	4.89 ± 0.61 a	4.41 ± 0.61 aB
TN (%)	0.30 ± 0.03 b	0.30 ± 0.03 b	0.54 ± 0.03 aA	0.26 ± 0.04 a	0.28 ± 0.04 a	0.28 ± 0.04 aB
TS (%)	0.05 ± 0.00 b	0.05 ± 0.00 b	0.06 ± 0.00 aA	0.05 ± 0.00 a	0.05 ± 0.00 a	0.05 ± 0.00 aB
TK (g/kg)	21.71 ± 0.32 a	21.04 ± 0.32 ab	20.65 ± 0.32 b	21.65 ± 0.43 a	21.34 ± 0.43 ab	21.04 ± 0.43 b
3 MAF						
pH ($-\log[H^+]$)	6.39 ± 0.06 c	6.64 ± 0.06 b	7.35 ± 0.06 a	6.43 ± 0.08 c	6.79 ± 0.08 b	7.51 ± 0.08 a
EC (µS/cm)	104.1 ± 32.09 b	99.3 ± 32.09 b	249.1 ± 32.09 a	83.1 ± 43.74 b	101.1 ± 43.74 b	203.9 ± 43.74 a
CaCO$_3$ (%)	0.08 ± 0.01 b	0.09 ± 0.01 b	0.96 ± 0.01 a	0.06 ± 0.12 b	0.09 ± 0.12 b	0.96 ± 0.12 a
TOC (%)	4.15 ± 0.44 bA	4.34 ± 0.44 bA	6.22 ± 0.44 aA	3.36 ± 0.61 aB	3.39 ± 0.61 aB	3.92 ± 0.61 aB
TN (%)	0.31 ± 0.03 bA	0.31 ± 0.03 bA	0.48 ± 0.03 aA	0.26 ± 0.04 aB	0.26 ± 0.04 aB	0.29 ± 0.04 aB
TS (%)	0.08 ± 0.0 aA	0.08 ± 0.0 a	0.08 ± 0.00 a	0.07 ± 0.00 aB	0.07 ± 0.00 a	0.07 ± 0.00 a
TK (g/kg)	17.8 ± 0.32 a	17.11 ± 0.32 a	15.59 ± 0.32 b	17.54 ± 0.43 a	17.16 ± 0.43 a	16.13 ± 0.43 b
6 MAF						
pH ($-\log[H^+]$)	6.71 ± 0.06 b	6.82 ± 0.06 b	7.70 ± 0.06 a	6.77 ± 0.08 c	7.02 ± 0.08 b	7.71 ± 0.08 a
EC (µS/cm)	105.3 ± 32.09 b	112.9 ± 32.09 b	242.9 ± 32.09 a	104.2 ± 43.74 b	116.9 ± 43.74 b	326.6 ± 43.74 a
CaCO$_3$ (%)	0.07 ± 0.01 b	0.12 ± 0.01 b	0.92 ± 0.01 a	0.07 ± 0.12 b	0.14 ± 0.12 ab	0.66 ± 0.12 a
TOC (%)	4.18 ± 0.44 b	4.04 ± 0.44 b	5.87 ± 0.44 a	3.83 ± 0.61 ab	3.70 ± 0.61 b	5.93 ± 0.61 a
TN (%)	0.31 ± 0.03 b	0.30 ± 0.03 b	0.44 ± 0.03 a	0.29 ± 0.04 ab	0.27 ± 0.04 b	0.41 ± 0.04 a
TS (%)	0.07 ± 0.00 b	0.07 ± 0.00 b	0.08 ± 0.00 aB	0.07 ± 0.00 b	0.07 ± 0.00 b	0.09 ± 0.00 aA
TK (g/kg)	20.31 ± 0.32 a	19.48 ± 0.32 ab	18.56 ± 0.32 b	20.28 ± 0.43 a	19.38 ± 0.43 ab	18.52 ± 0.43 b
9 MAF						
pH ($-\log[H^+]$)	6.41 ± 0.06 c	6.70 ± 0.06 b	7.14 ± 0.06 aB	6.32 ± 0.08 c	6.57 ± 0.08 b	7.39 ± 0.08 aA
EC (µS/cm)	98.7 ± 32.09 b	126.5 ± 32.09 b	209.0 ± 32.09 a	66.3 ± 43.74 b	87.4 ± 43.74 b	208.8 ± 43.74 a
CaCO$_3$ (%)	0.12 ± 0.01 ab	0.06 ± 0.01 b	0.26 ± 0.01 a	0.09 ± 0.12 b	0.07 ± 0.12 b	0.26 ± 0.12 a
TOC (%)	4.11 ± 0.44 bA	4.67 ± 0.44 ab	5.88 ± 0.44 a	3.07 ± 0.61 bB	3.75 ± 0.61 a	4.35 ± 0.61 a
TN (%)	0.29 ± 0.03 bA	0.33 ± 0.03 abA	0.43 ± 0.03 aA	0.23 ± 0.04 bB	0.27 ± 0.04 abB	0.30 ± 0.04 aB
TS (%)	0.09 ± 0.00 b	0.10 ± 0.00 abA	0.10 ± 0.00 a	0.09 ± 0.00 a	0.09 ± 0.00 aB	0.09 ± 0.00 a
TK (g/kg)	21.29 ± 0.32 a	20.06 ± 0.32 b	19.88 ± 0.32 b	21.33 ± 0.43 a	20.61 ± 0.43 b	20.54 ± 0.43 b
12 MAF						
pH ($-\log[H^+]$)	6.25 ± 0.06 c	6.52 ± 0.06 b	7.25 ± 0.06 a	6.29 ± 0.08 b	6.47 ± 0.08 b	7.44 ± 0.08 a
EC (µS/cm)	94.0 ± 32.09 b	100.3 ± 32.09 b	215.1 ± 32.09 a	83.4 ± 43.74 b	71.6 ± 43.74 b	163.8 ± 43.74 a
CaCO$_3$ (%)	0.02 ± 0.01 b	0.05 ± 0.01 b	0.34 ± 0.01 a	0.02 ± 0.12 b	0.02 ± 0.12 b	0.29 ± 0.12 a
TOC (%)	3.85 ± 0.44 b	4.21 ± 0.44 bA	5.36 ± 0.44 aA	3.65 ± 0.61 a	3.22 ± 0.61 aB	3.70 ± 0.61 aB
TN (%)	0.29 ± 0.03 b	0.31 ± 0.03 b	0.41 ± 0.03 aA	0.27 ± 0.04 a	0.25 ± 0.04 a	0.26 ± 0.04 aB
TS (%)	0.08 ± 0.00 a	0.08 ± 0.00 a	0.08 ± 0.00 aA	0.08 ± 0.00 a	0.07 ± 0.00 a	0.07 ± 0.00 aB
TK (g/kg)	16.62 ± 0.32 a	15.9 ± 0.32 a	15.97 ± 0.32 a	16.23 ± 0.43 a	16.23 ± 0.43 a	15.67 ± 0.43 a

Abbreviations: C—Control; MS—Medium severity; HS—High severity; Q—*Quercus pubescens* Willd.; J—*Juniperus communis* L.; MAF—Months after fire. Values represent the sampling means followed with standard deviation. Different letters indicate significant ($p < 0.05$) differences between fire severity (lowercase letters) in one vegetation group and vegetation (uppercase letters).

Table 4. Results of factorial ANOVA (for Fe, Ni and Zn) and Kruskal–Wallis test (for Cu) during the study period.

	C -Q	MS -Q	HS -Q	C -J	MS -J	HS -J
			0 MAF			
Fe (g/kg)	31.00 ± 0.61 a	30.76 ± 0.61 a	28.15 ± 0.61 b	31.56 ± 0.83 a	30.38 ± 0.83 a	30.42 ± 0.83 a
Ni (mg/kg)	75.46 ± 2.09 a	73.92 ± 2.09 a	68.00 ± 2.09 a	74.71 ± 2.85 a	72.71 ± 2.85 a	74.43 ± 2.85 a
Cu (mg/kg)	29.69 ± 2.13 b	34.85 ± 2.13 ab	34.38 ± 2.13 a	31.29 ± 2.90 b	29.00 ± 2.90 ab	39.57 ± 2.90 a
Zn (mg/kg)	72.38 ± 2.17 a	73.08 ± 2.17 a	72.46 ± 2.17 a	73.57 ± 2.96 a	71.00 ± 2.96 a	74.00 ± 2.96 a
			3 MAF			
Fe (g/kg)	28.39 ± 0.61 a	28.29 ± 0.61 a	26.28 ± 0.61 b	28.56 ± 0.83 a	27.53 ± 0.83 a	27.96 ± 0.83 a
Ni (mg/kg)	65.38 ± 2.09 a	64.15 ± 2.09 a	59.00 ± 2.09 b	63.43 ± 2.85 a	63.29 ± 2.85 a	64.43 ± 2.85 a
Cu (mg/kg)	25.31 ± 2.13 a	30.00 ± 2.13 a	28.00 ± 2.13 a	28.14 ± 2.90 a	26.71 ± 2.90 a	34.57 ± 2.90 a
Zn (mg/kg)	73.23 ± 2.17 b	76.38 ± 2.17 bA	79.23 ± 2.17 a	74.00 ± 2.96 a	69.00 ± 2.96 aB	75.00 ± 2.96 a
			6 MAF			
Fe (g/kg)	29.42 ± 0.61 a	28.71 ± 0.61 a	27.63 ± 0.61 b	30.65 ± 0.83 a	28.39 ± 0.83 a	27.85 ± 0.83 b
Ni (mg/kg)	66.77 ± 2.09 a	66.08 ± 2.09 a	61.38 ± 2.09 b	68.71 ± 2.85 a	61.71 ± 2.85 a	62.43 ± 2.85 a
Cu (mg/kg)	28.15 ± 2.13 a	31.54 ± 2.13 a	31.08 ± 2.13 a	30.00 ± 2.90 a	27.71 ± 2.90 a	38.29 ± 2.90 a
Zn (mg/kg)	71.85 ± 2.17 a	69.54 ± 2.17 a	76.62 ± 2.17 a	77.29 ± 2.96 a	68.43 ± 2.96 b	78.57 ± 2.96 a
			9 MAF			
Fe (g/kg)	30.50 ± 0.61 a	28.72 ± 0.61 b	28.54 ± 0.61 c	30.68 ± 0.83 a	28.31 ± 0.83 b	29.24 ± 0.83 a
Ni (mg/kg)	65.00 ± 2.09 a	62.92 ± 2.09 a	62.85 ± 2.09 a	70.14 ± 2.85 a	65.14 ± 2.85 a	64.43 ± 2.85 a
Cu (mg/kg)	26.15 ± 2.13 b	32.15 ± 2.13 ab	30.85 ± 2.13 a	28.86 ± 2.90 b	28.86 ± 2.90 ab	37.29 ± 2.90 a
Zn (mg/kg)	72.00 ± 2.17 a	69.69 ± 2.17 a	73.31 ± 2.17 a	72.00 ± 2.96 a	67.00 ± 2.96 a	69.14 ± 2.96 a
			12 MAF			
Fe (g/kg)	27.82 ± 0.61 a	28.57 ± 0.61 a	27.5 ± 0.61 a	29.25 ± 0.83 a	26.70 ± 0.83 b	26.52 ± 0.83 c
Ni (mg/kg)	60.00 ± 2.09 b	65.38 ± 2.09 a	63.00 ± 2.09 b	66.71 ± 2.85 a	60.57 ± 2.85 a	61.29 ± 2.85 a
Cu (mg/kg)	27.46 ± 2.13 a	32.23 ± 2.13 a	28.54 ± 2.13 a	28.86 ± 2.90 a	25.14 ± 2.90 a	32.00 ± 2.90 a
Zn (mg/kg)	70.23 ± 2.17 b	72.77 ± 2.17 b	76.53 ± 2.17 aA	74.00 ± 2.96 a	67.00 ± 2.96 a	67.43 ± 2.96 aB

Abbreviations: C—Control; MS—Medium severity; HS—High severity; Q—*Quercus pubescens* Willd.; J—*Juniperus communis* L.; MAF—Months after fire. Values represent the sampling means followed with standard deviation. Different letters indicate significant ($p < 0.05$) differences between fire severity (lowercase letters) and vegetation (uppercase letters).

Wildfires caused an increase in soil pH, EC, and soil $CaCO_3$ content. Throughout the study period, pH was significantly higher in HS and MS than in C. Soil samples collected under *Juniperus communis* L. had significantly higher pH values than those collected under *Quercus pubescens* Willd. (Table 2). Both MS -Q and HS -Q (collected under *Quercus pubescens* Willd. Vegetation) had significantly higher values than C-Q throughout the monitoring period, with the exception of 6 MAF, where a significant difference in pH was observed only in the HS -Q samples (Table 3). There was no significant difference between C-J and MS -J (taken under *Juniperus communis* L. vegetation) in 0 MAF and 12 MAF, while significantly higher pH values were observed mainly in HS -J. The overall significantly higher pH values in samples MS and HS persisted until 12 MAF (Table 3).

Soil samples collected under *Quercus p.* had significantly higher EC values than samples collected under *Juniperus c.*, but no significant differences in $CaCO_3$ were observed in samples from the two vegetation species (Table 2). Both EC and $CaCO_3$ behaved similarly and were significantly higher in HS than in MS and C samples at all sampling times, regardless of the vegetation species under which they were taken (Table 3).

Overall, TOC, TN, and TS content were significantly higher in HS than in MS and C samples collected during the study period and significantly lower in soil samples under *Juniperus c.* than under *Quercus p.* (Table 2). Immediately after a wildfire, the higher content of TOC, TN, and TS were evident in HS -Q samples compared to HS -J samples. One year after a wildfire, both MS -Q and HS -Q had significantly higher TOC content than MS -J and HS -J, respectively. Significantly higher TOC and TN content in soil samples under

Quercus p. were observed in HS than in C during all sampling periods. Under *Juniperus c.*, significantly higher content of TOC and TN values were observed in HS than in C at 9 MAF. Furthermore, 6 MAF TOC and TN under *Juniperus c.* were significantly higher in HS than in MS. Additionally, the overall increase in TOC and TN content was significant in HS -Q and MS -Q compared to C-Q throughout the study period, while HS -J and MS -J showed a slight relative increase compared to C-J; however, it was significant only at 6 and 9 MAF (Table 3). In soil samples under *Quercus p.*, a significantly higher TS content was observed in HS compared to C 0, 6, and 9 MAF. Under *Juniperus c.*, a significantly higher TS content was observed in HS compared to C 6 MAF.

Furthermore, TK content was significantly lower in HS in the first 6 months after a fire, while at 9 MAF, whereas it was significantly lower in both HS and MS than in C 9 months after fire (Table 3). No significant differences in TK content were observed in the soil samples taken under different vegetation species (Table 2).

3.3. Soil Trace Elements

During the 12-month observation period, significantly lower Fe and Ni content was observed in HS and MS than in C. HS showed significantly higher Cu and Zn content than MS and C (Table 2). Zn content was significantly higher under *Quercus p.* (Table 2). Under *Quercus p.*, Fe content was significantly lower in HS than in C and MS during 0 to 9 MAF, while under *Juniperus c.*, on 6 and 12 MAF, Fe content was significantly lower in HS than in MS and C. Moreover, the significant effect of fire severity on Ni content was found mainly in *Quercus p.* on 3 and 6 MAF (Table 4). Ni concentrations were significantly lower in HS than in MS and C (Table 2).

The significantly lower Cu content was observed at 0 and 9 MAF in C than at MS, which was collected under both *Quercus p.* and *Juniperus c.* (Table 4). In addition, significantly higher Zn content was observed under *Quercus p.* at 3 and 12 MAF in HS than in C and MS. Among *Juniperus c.*, at 6 MAF, significantly more Zn was seen in C and HS than in MS.

3.4. Multivariate Analysis

For each sample date, a matrix was constructed with three principal components, with the corresponding proportion of variance explained by them, as well as the variables with the highest loadings (Table 5). The relationship between PC1 and PC2 for each corresponding sampling time is shown in Figure 3. The variables with the highest loadings do not differ much between sampling dates. These are TOC, TN, and EC in PC1, which explain 35 to 54% of the variance in all sampling dates, and Ni, Cu, Zn, and Fe in PC2, which account for 17 to 24% of the total variance throughout the sampling period. The remainder of 11 to 15% of the variance can be attributed to high pH, $CaCO_3$, and TS loadings in PC3.

Table 5. Summary of the results of principal component analysis.

	0 MAF			3 MAF			6 MAF			9 MAF			12 MAF		
	PC1	PC2	PC3	PC1	PC2	PC3	PC1	PC2	PC3	PC1	PC2	PC3	PC1	PC2	PC3
Eigenvalue	2.38	1.38	1.19	2.25	1.58	1.30	2.43	1.38	1.12	2.18	1.61	1.14	1.96	1.59	1.29
Proportion of variance	0.51	0.17	0.13	0.46	0.23	0.15	0.54	0.17	0.11	0.43	0.24	0.12	0.35	0.23	0.15
Cumulative proportion	0.51	0.68	0.81	0.46	0.69	0.84	0.54	0.71	0.82	0.43	0.67	0.79	0.35	0.58	0.73
Eigenvectors															
pH ($-\log[H^+]$)	0.269	0.257	**−0.453**	**0.324**	0.128	**−0.441**	**0.308**	0.232	**−0.354**	**0.342**	0.022	**−0.499**	**0.295**	0.196	**−0.488**
EC (μS/cm)	**0.336**	0.133	−0.191	**0.399**	0.075	−0.212	**0.381**	0.076	−0.041	**0.405**	0.051	−0.297	**0.434**	0.114	−0.195
$CaCO_3$ (%)	**0.328**	0.132	**−0.285**	**0.364**	0.103	**−0.318**	**0.302**	0.226	**−0.391**	**0.278**	0.098	**−0.411**	0.273	0.182	**−0.356**
TOC (%)	**0.370**	0.018	**0.308**	**0.380**	0.049	**0.355**	**0.366**	−0.050	**0.329**	**0.415**	0.084	0.278	**0.470**	−0.021	0.240
TN (%)	**0.384**	0.036	**0.293**	**0.382**	0.078	**0.332**	**0.360**	−0.040	**0.359**	**0.412**	0.102	0.305	**0.463**	−0.045	0.272
TS (%)	**0.363**	0.070	**0.341**	0.180	0.229	**0.533**	**0.328**	0.033	0.264	**0.329**	0.210	0.323	**0.291**	−0.221	**0.452**
TK (g/kg)	**−0.325**	0.232	0.110	**−0.371**	0.061	0.116	**−0.340**	0.017	0.104	**−0.331**	0.220	−0.206	−0.237	−0.216	0.040
Fe (g/kg)	**−0.331**	0.349	0.117	−0.256	**0.476**	0.027	**−0.299**	**0.404**	0.228	−0.208	**0.511**	−0.009	−0.025	**−0.586**	−0.135
Ni (mg/kg)	−0.231	**0.495**	0.090	−0.245	**0.456**	−0.036	−0.264	**0.467**	0.103	−0.164	**0.477**	−0.050	−0.011	**−0.484**	**−0.316**
Cu (mg/kg)	0.087	**0.501**	**−0.371**	−0.014	**0.477**	−0.308	0.070	**0.496**	**−0.359**	0.094	**0.321**	−0.343	0.073	−0.137	**−0.372**
Zn (mg/kg)	0.125	**0.470**	**0.456**	0.130	**0.491**	0.169	0.136	**0.506**	**0.459**	0.049	**0.535**	0.239	0.261	**−0.470**	−0.055

The loadings that significantly contribute to each principal component are shown in bold.

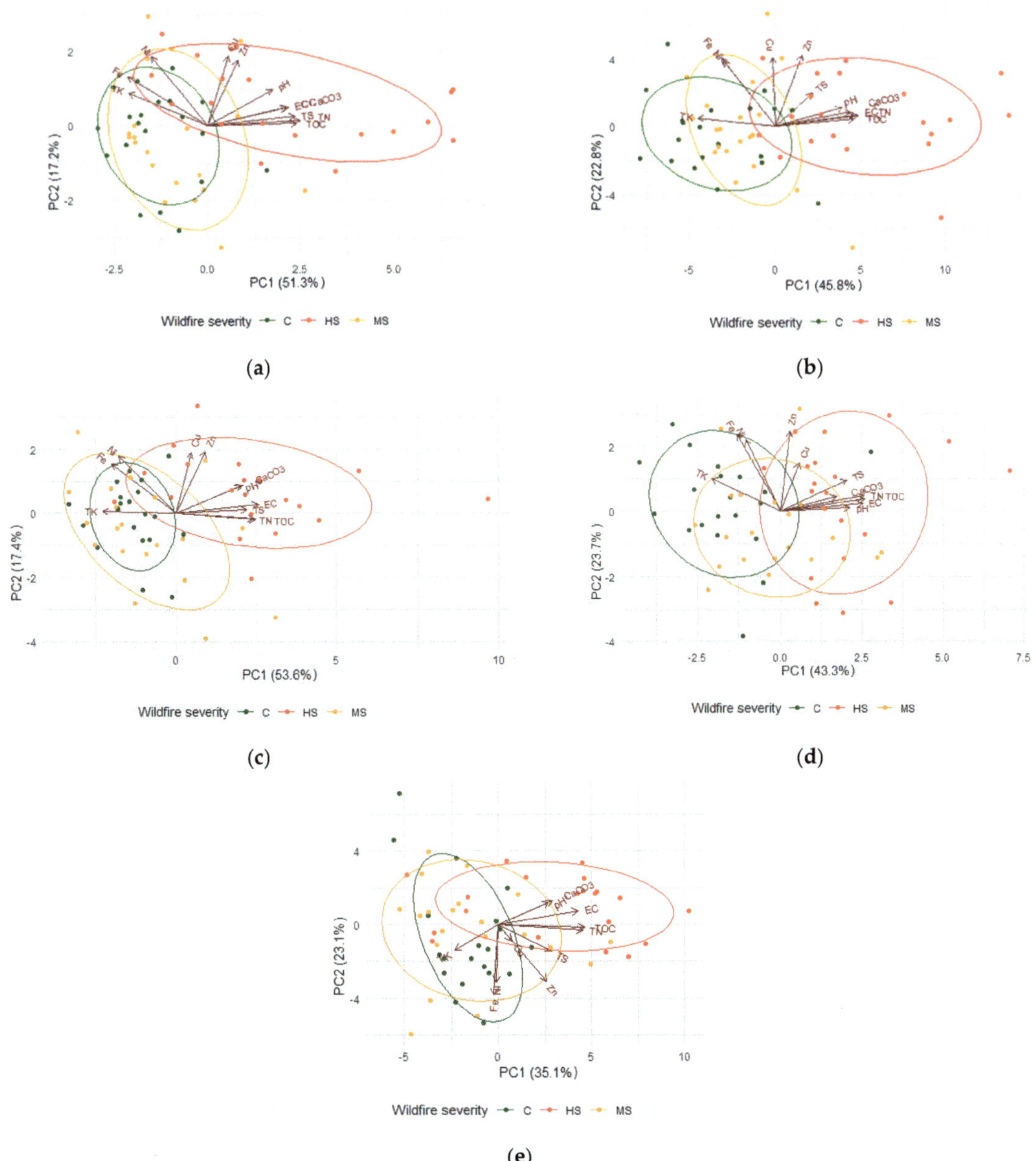

Figure 3. The relationship between PC1 and PC2: (**a**) 0 MAF; (**b**) 3 MAF; (**c**) 6 MAF; (**d**) 9 MAF; (**e**) 12 MAF (MAF—months after fire).

The relationships between PC1 and PC2 shown in the graphs reveal that pH, EC, $CaCO_3$, TN, TS, and TOC are positively correlated and are most responsible for separating the C, MS, and HS groups. In addition, the graph for 0 MAF shows that EC, $CaCO_3$, TN, TS, and TOC are inversely related to TK and Fe content, and while this relationship persists at 12 MAF, it is less pronounced for Fe.

4. Discussion

4.1. Basic Chemical Properties

Soil pH values were significantly higher on both MS and HS than on C taken under both vegetation species during the entire study period. The observed increase was due to the combustion of above-ground organic matter, which produced large quantities of ash containing base cations [28,29,34], and to the loss of hydroxyl groups (-OH) and organic acids during the oxidation process of burning [35]. At 12 MAF, the observed pH values were still significantly higher at MS and HS than at C, which was consistent with previous studies that reported high pH values one year [1,36] and up to two years post-fire [37,38]. Higher pH in *Juniperus c.* samples was consistent with a study by Huang et al. [39] that found a relationship between increasing SOM content and a decrease in pH due to the release of organic acids following SOM decomposition. In this study, *Juniperus c.* samples had lower TOC contents and consequently higher pH values.

Higher levels of EC and $CaCO_3$ content were also observed in burned areas. However, significantly higher values were only observed in HS, while MS was not significantly different from C. High values of EC and $CaCO_3$ in HS soil samples observed under both vegetation species and throughout the study period might have resulted from the burning of organic matter, which produced high levels of inorganic ions [40]. However, many authors reported increased EC to be ephemeral and recorded its recovery to pre-fire levels within a year, i.e., 8 months [41] and 12 months post-fire [37]. In this study, the increased EC values 12 MAF persisted in HS samples, although they began to show signs of decline, possibly due to the effects of rainfall, ion leaching [42], or nutrient uptake by plants [43]. High EC values 6 MAF could be related to the overall slower recovery of vegetation in HS compared to MS, considering that EC is a measure of soil salinity that negatively affects plant germination [44]. The absence of vegetation recovery in HS was observed in the first 6 months, and recovery began to be visible after 9 months (both vegetation types), although some sites remained without significant vegetation growth after this time.

The initial increase in $CaCO_3$ content, especially at 3 and 6 MAF, showed a declining trend at 12 MAF in HS. This could be attributed to the alkaline oxides in the ash reacting with atmospheric CO_2 and water vapour, resulting in the formation of soluble hydroxides and carbonates in the soil [45,46]. Moreover, Goforth et al. [46] observed that white ash originating from completely combusted organic matter contained more alkaline oxides available to form carbonates, supporting the significantly higher content found in this study in HS areas.

In this study, there was a significant increase in TOC content in the HS soil samples compared to MS and C, and no change in MS compared to the C samples throughout the study period. The higher TOC content in the samples from HS could be caused by residual ash and charred residue after the combustion of organic matter incorporated within the soil. Namely, the combustion of organic material leads to the formation of aromatic compounds that may be more recalcitrant to decomposition [47], which could explain the increase in TOC in HS. These findings coincide with a study by Muráňová and Šimanský [48], who observed a 24% increase in SOM in high-severity fires.

The more pronounced increase in TOC content in burned *Quercus p.* compared to burned *Juniperus c.* area could be related to the greater overall amount of biomass in *Quercus p.* vegetation (i.e., fuel) available for combustion and subsequent incorporation into the soil, as well as the different nutrient composition of the two vegetation species [49]. This assumption stems from the fact that the interaction between fire severity and vegetation factor was not significant, implying that further research is needed to determine the reasons for this discrepancy.

Nevertheless, the higher content of soil nutrients in the form of TOC and TN in MS -Q and HS -Q compared to MS -J and HS -J suggested that the wildfire caused an increase that was beneficial for the regrowth and development of *Quercus p.* seedlings that began to grow in the spring after the fire (9 MAF).

Similar to TOC and TN content, significantly higher TS content was found in the HS samples compared to the MS and C samples, and the increase was equally pronounced in both vegetation species. The TS content returned to pre-fire levels after 12 months, which is in agreement with a study by Kong et al. [50] that observed an initial increase in this soil nutrient and its rapid recovery to pre-fire levels. Although the increase was evident, the seasonality of TS content appeared to remain relatively unchanged during the monitoring period and was highest in 9 MAFs at both C and MS and HS, most likely due to accelerated mineralization at higher temperatures and plant uptake rather than the effects of a wildfire [51].

The content of TK was the only soil property that decreased 0 MAF in both MS and HS. However, this change appears to be temporary because 12 MAF content in MS and HS did not differ from that of unburned soils. Caon et al. [52], Litton and Santelices [53], Simelton [53], Úbeda et al. [54], and Xue et al. [36] reported a continuous decrease in total and exchangeable potassium in the time-since-fire period. The decrease observed in this study could have further been caused by the above-average rainfall in November and December 2019 (Figure 2), which resulted in consistent depletion of potassium through leaching and plant uptake [55].

4.2. Soil Trace Elements

After the wildfire, a decrease in Fe and Ni content was observed in MS and HS. In general, the main source of these elements is mineral soil [15,55]. However, under favourable pH conditions (between 6.5 and 7.5), Fe and Ni form complexes with organic acids in the SOM [56,57], and it is possible that these compounds volatilised as a result of the high wildfire temperatures. A similar observation was made by Delač et al. [58], who found significantly lower Fe content in the post-fire period in burned areas due to leaching and plant consumption. Furthermore, source rocks containing crystallised Fe oxides are also known to contain certain amounts of Ni [59], and this is why they are highly correlated throughout the study period (Figure 3). Santorufo et al. [60] and Fernández et al. [61] also observed a decrease in Fe and Ni during the post-fire period.

A significant increase in Cu and Zn content was observed in the soil samples from HS, while it was not significant in MS compared to C throughout the study period. These results could be explained by the fact that, in addition to mineral soil, biomass is an important source of Cu and Zn in the soil [15,56], so the ash incorporated into the soil after the combustion of vegetation most likely increased the content of these elements in the topsoil of HS. Notably, in the samples from HS, the entire canopy was burned, and no leaves were left on the branches of the burned trees, while in MS, some leaves remained untouched by the fire, which meant that quantitatively not as much ash reached the soil surface. Additionally, ash from lower severity fires contain a greater amount of incompletely combusted material and thus more organic compounds [62], which could have led to the formation of complex ligands that affect the bioavailability and mobility of Zn and Cu in the soil. This is most likely the reason why the effects of a medium wildfire severity are less pronounced throughout the study period. Mitic et al. [63] also found that Zn and Cu levels increased after a fire.

Moreover, at this stage, we can rule out the possibility that elevated concentrations of trace elements Cu, Ni, and Zn in the studied soils are an environmental problem. According to the Croatian legislation [64], the maximum permissible levels for pollutants in agricultural soils in Croatia are 150, 75, and 200 mg/kg for Cu, Ni, and Zn, respectively. Although this legislation concerns agricultural soils, it is the only available reference to determine whether the concentrations are likely to cause environmental damage. The established limits were not exceeded even in the immediate post-fire period. Only the Ni content was found to be borderline problematic at the beginning of the sampling campaign in August 2019. However, the elevated concentrations were found in both unburned and burned soil, suggesting that other external factors were the cause of this occurrence. Further studies are needed to confirm that other factors may have compromised soil health, as these metals can be involved in a number of complex chemical and biological interactions that are beyond the scope of this investigation.

4.3. Interrelations between Properties

PCA analysis revealed similar relationships between soil properties in PC1 and PC2 in all 5 sampling intervals (Figure 3) and identified pH, EC, $CaCO_3$, TOC, TN, and TS as the group of variables most affected by wildfire severity. These properties were inversely related to TK throughout the study period. This indicates that the former increased and the latter decreased after a wildfire. Additionally, Fe was also inversely related to the first group 0 MAF, suggesting that it was the trace element most affected by the fire, although this relationship decreased with time. Furthermore, PCA analysis identified the highly correlated Fe and Ni from parent rock and Cu and Zn from biomass as two separate groups of variables that were less affected by the wildfire. As expected, the greater impact of HS compared to MS is most evident at 0 MAF. At 9 MAF, the difference between HS and MS is least pronounced, likely due to intense vegetation recovery in the spring. The simultaneous recovery of vegetation and soil properties was also observed by Muñoz-Rojas et al. [42]. Previous studies also observed the correlated relationships between pH, EC, TN, and TOC in burned soils and concluded that soil system recovery could take years [12,58,65].

4.4. Implications for Soil Management

Mediterranean ecosystems are adapted to wildfire, so in many situations, there is no need for post-fire management activities [28]. In areas more susceptible to soil degradation, such as hillsides, post-fire management is desirable [66] but not often used. Most post-fire management practices include mulching to reduce soil runoff and erosion on slopes or adding organic amendments to help soil nutrient recovery [67]. Other practices include afforestation and seeding, salvage logging, erosion barriers, or soil preparation. None of the above practices has been tested under current conditions. However, given the soil conditions and visual observations, it is important to mention that the rapid growth of vegetation in the study area in spring 2020 (9 MAF) indicates continued ecosystem recovery. The recuperation could be further supported by the fact that the wildfire occurred in an area with well-developed Cambisols, which are characterised by favourable aggregate structure and high content of weatherable minerals [27] that provide the nutrients needed for resprouting. This suggests that this Mediterranean locality is well adapted to wildfires. Moreover, nutrient-rich soil and no risk of erosion by water support the "leave as-is" recommendation in this area. However, considering that forest managers typically conduct post-fire restoration activities such as salvage logging, burned tree removal, tillage, and reforestation, the vegetation used as a factor in this study could be useful for conducting future activities in a similar setting. The greater soil disturbance observed at HS compared to MS appears to depend on vegetation composition and should be considered in post-fire management. Although not directly measured, the authors visually observed vigorous new sprouting of *Quercus p.* in the spring after wildfire (9 MAF), while *Juniperus c.* exhibited low intensity of new sprouting, especially in HS. Therefore, in the case of reforestation of the study area, the results suggest that *Quercus p.* would be more suitable because the soil under *Quercus p.* had higher TOC and TN content, especially in HS, compared to *Juniperus c.* As both carbon and nitrogen are essential for plant growth and development, in the event of future fire occurrence, the higher soil TOC and TN content would provide the essential nutrients for faster vegetation recovery.

Vegetation was shown in this study to be an important factor controlling soil quality after a fire and should be taken into consideration for future management activities.

5. Conclusions

The studied wildfire had a disturbing impact on the soil, especially in the HS treatment. Significantly higher levels of soil pH, EC and $CaCO_3$, TOC, TN, Cu, and Zn content were observed throughout the study period, especially in HS, most likely due to the slow recovery of vegetation. TOC and TN content were higher under *Quercus pubescens* Willd. than under *Juniperus communis* L., especially during the first three months, which may be attributed to the overall higher initial fuel quantity of the former. Moreover, the increase in Cu and Zn content in HS was due to the complete canopy combustion.

Under *Juniperus c.*, TOC and TN content at HS returned to pre-fire levels 12 MAF, while the persistently higher levels at HS under *Quercus p.* may be due to the higher biomass and higher wildfire resistance of *Quercus p.* species. Elevated soil pH, EC, and $CaCO_3$ content persisted in both HS -J and HS -Q 1 year after a fire, indicating that more than 12 months are required for soil system recovery. Multivariate analysis showed a positive correlation between pH, EC, $CaCO_3$, TOC, TN, and TS content and identified them as most affected by wildfire severity. The difference between the HS and MS categories was less visible 9 and 12 MAF, indicating that recovery of the soil system was underway. The differences in soil disturbance under HS and MS depend on vegetation composition and suggest that this factor should be imperatively considered in post-fire management. Namely, the higher TOC and TN content observed in *Quercus p.*, especially in HS, suggests that this deciduous species is a more suitable choice for reforestation than the coniferous *Juniperus c.* However, the differences in the properties studied among vegetation types and their evolution over time suggest that further research on the recovery process of soil properties is recommended to understand the mechanisms of long-term soil recovery better.

Author Contributions: Conceptualization, I.H., I.B. and P.P.; methodology, I.B. and P.P.; software, I.H. and I.Š.; validation, I.H., I.B. and I.Š.; formal analysis, I.H., I.B. and D.D.; investigation, I.H., I.B., I.Š. and P.P.; resources, I.B. and I.Š.; writing—original draft preparation, I.H.; writing—review and editing, I.H., I.Š., I.B. and D.D.; visualization, I.H.; supervision, I.B., I.Š. and P.P.; project administration, I.B. and I.Š.; funding acquisition, I.B. All authors have read and agreed to the published version of the manuscript.

Funding: This research was funded by CROATIAN SCIENCE FOUNDATION through the project "Soil erosion and degradation in Croatia" (UIP-2017-05-7834) (SEDCRO).

Institutional Review Board Statement: Not applicable.

Informed Consent Statement: Not applicable.

Data Availability Statement: The data presented in this study are available on request from the corresponding author. The data are not publicly available due to their current usage for the preparation of another draft for publication.

Conflicts of Interest: The authors declare no conflict of interest. The funders had no role in the design of the study; in the collection, analyses, or interpretation of data; in the writing of the manuscript, or in the decision to publish the results.

References

1. Alcañiz, M.; Outeiro, L.; Francos, M.; Farguell, J.; Úbeda, X. Long-Term Dynamics of Soil Chemical Properties after a Prescribed Fire in a Mediterranean Forest (Montgrí Massif, Catalonia, Spain). *Sci. Total Environ.* **2016**, *572*, 1329–1335. [CrossRef] [PubMed]
2. Pausas, J.G.; Llovet, J.; Rodrigo, A.; Vallejo, R. Are Wildfires a Disaster in the Mediterranean Basin?-A Review. *Int. J. Wildland Fire* **2008**, *17*, 713. [CrossRef]
3. Halofsky, J.E.; Peterson, D.L.; Harvey, B.J. Changing Wildfire, Changing Forests: The Effects of Climate Change on Fire Regimes and Vegetation in the Pacific Northwest, USA. *Fire Ecol.* **2020**, *16*, 4. [CrossRef]
4. Pavlek, K.; Bišćević, F.; Furčić, P.; Grđan, A.; Gugić, V.; Malešić, N.; Moharić, P.; Vragović, V.; Fuerst-Bjeliš, B.; Cvitanović, M. Spatial Patterns and Drivers of Fire Occurrence in a Mediterranean Environment: A Case Study of Southern Croatia. *Geogr. Tidsskr. Dan. J. Geogr.* **2017**, *117*, 22–35. [CrossRef]
5. Westerling, A.L. Increasing Western US Forest Wildfire Activity: Sensitivity to Changes in the Timing of Spring. *Phil. Trans. R. Soc. B* **2016**, *371*, 20150178. [CrossRef] [PubMed]
6. IPCC. *Climate Change 2013: The Physical Science Basis. Contribution of Working Group I to the Fifth Assessment Report of the Intergovernmental Panel on Climate Change;* Stocker, T.F., Qin, D., Plattner, G.-K., Tignor, M., Allen, S.K., Boschung, J., Nauels, A., Xia, Y., Bex, V., Midgley, P.M., Eds.; Cambridge University Press: Cambridge, UK, 2013; Available online: https://www.ipcc.ch/report/ar5/wg1/ (accessed on 5 May 2022).
7. Pausas, J.G.; Keeley, J.E. Wildfires as an Ecosystem Service. *Front. Ecol. Environ.* **2019**, *17*, 289–295. [CrossRef]

8. Pereira, P.; Francos, M.; Brevik, E.C.; Ubeda, X.; Bogunovic, I. Post-Fire Soil Management. *Curr. Opin. Environ. Sci. Health* **2018**, *5*, 26–32. [CrossRef]

9. Moya, D.; González-De Vega, S.; Lozano, E.; García-Orenes, F.; Mataix-Solera, J.; Lucas-Borja, M.E.; de las Heras, J. The Burn Severity and Plant Recovery Relationship Affect the Biological and Chemical Soil Properties of *Pinus Halepensis* Mill. Stands in the Short and Mid-Terms after Wildfire. *J. Environ. Manag.* **2019**, *235*, 250–256. [CrossRef] [PubMed]

10. Dzwonko, Z.; Loster, S.; Gawroński, S. Impact of Fire Severity on Soil Properties and the Development of Tree and Shrub Species in a Scots Pine Moist Forest Site in Southern Poland. *For. Ecol. Manag.* **2015**, *342*, 56–63. [CrossRef]

11. Lombao, A.; Barreiro, A.; Carballas, T.; Fontúrbel, M.T.; Martín, A.; Vega, J.A.; Fernández, C.; Díaz-Raviña, M. Changes in Soil Properties after a Wildfire in Fragas Do Eume Natural Park (Galicia, NW Spain). *Catena* **2015**, *135*, 409–418. [CrossRef]

12. Francos, M.; Úbeda, X.; Pereira, P.; Alcañiz, M. Long-Term Impact of Wildfire on Soils Exposed to Different Fire Severities. A Case Study in Cadiretes Massif (NE Iberian Peninsula). *Sci. Total Environ.* **2018**, *615*, 664–671. [CrossRef] [PubMed]

13. Murphy, S.F.; McCleskey, R.B.; Martin, D.A.; Holloway, J.M.; Writer, J.H. Wildfire-Driven Changes in Hydrology Mobilize Arsenic and Metals from Legacy Mine Waste. *Sci. Total Environ.* **2020**, *743*, 140635. [CrossRef]

14. Harper, A.R.; Santin, C.; Doerr, S.H.; Froyd, C.A.; Albini, D.; Otero, X.L.; Viñas, L.; Pérez-Fernández, B. Chemical Composition of Wildfire Ash Produced in Contrasting Ecosystems and Its Toxicity to Daphnia Magna. *Int. J. Wildland Fire* **2019**, *28*, 726. [CrossRef]

15. Santín, C.; Doerr, S.H.; Otero, X.L.; Chafer, C.J. Quantity, Composition and Water Contamination Potential of Ash Produced under Different Wildfire Severities. *Environ. Res.* **2015**, *142*, 297–308. [CrossRef] [PubMed]

16. Fernández, S.; Cotos-Yáñez, T.; Roca-Pardiñas, J.; Ordóñez, C. Geographically Weighted Principal Components Analysis to Assess Diffuse Pollution Sources of Soil Heavy Metal: Application to Rough Mountain Areas in Northwest Spain. *Geoderma* **2018**, *311*, 120–129. [CrossRef]

17. Li, C.; Zhou, K.; Qin, W.; Tian, C.; Qi, M.; Yan, X.; Han, W. A Review on Heavy Metals Contamination in Soil: Effects, Sources, and Remediation Techniques. *Soil Sediment Contam.* **2019**, *28*, 380–394. [CrossRef]

18. Zald, H.S.J.; Dunn, C.J. Severe Fire Weather and Intensive Forest Management Increase Fire Severity in a Multi-Ownership Landscape. *Ecol. Appl.* **2018**, *28*, 1068–1080. [CrossRef]

19. Stavi, I. Wildfires in Grasslands and Shrublands: A Review of Impacts on Vegetation, Soil, Hydrology, and Geomorphology. *Water* **2019**, *11*, 1042. [CrossRef]

20. Bond, W. Fires, Ecological Effects of. In *Encyclopedia of Biodiversity*; Academic Press: Cambridge, MA, USA, 2001; Volume 2, pp. 745–753.

21. Pereira, P.; Bogunovic, I.; Zhao, W.; Barcelo, D. Short-Term Effect of Wildfires and Prescribed Fires on Ecosystem Services. *Curr. Opin. Environ. Sci. Health* **2021**, *22*, 100266. [CrossRef]

22. Birch, D.S.; Morgan, P.; Kolden, C.A.; Abatzoglou, J.T.; Dillon, G.K.; Hudak, A.T.; Smith, A.M.S. Vegetation, Topography and Daily Weather Influenced Burn Severity in Central Idaho and Western Montana Forests. *Ecosphere* **2015**, *6*, 1–23. [CrossRef]

23. Caldwell, P.V.; Elliott, K.J.; Liu, N.; Vose, J.M.; Zietlow, D.R.; Knoepp, J.D. Watershed-scale Vegetation, Water Quantity, and Water Quality Responses to Wildfire in the Southern Appalachian Mountain Region, United States. *Hydrol. Process* **2020**, *34*, 5188–5209. [CrossRef]

24. Fernández-Guisuraga, J.M.; Suárez-Seoane, S.; García-Llamas, P.; Calvo, L. Vegetation Structure Parameters Determine High Burn Severity Likelihood in Different Ecosystem Types: A Case Study in a Burned Mediterranean Landscape. *J. Environ. Manag.* **2021**, *288*, 112462. [CrossRef]

25. Kottek, M.; Grieser, J.; Beck, C.; Rudolf, B.; Rubel, F. World Map of the Köppen-Geiger Climate Classification Updated. *Metz* **2006**, *15*, 259–263. [CrossRef]

26. Zaninović, K.; Gajić Čapka, M.; Perčec Tadić, M.; Vučetić, M. *Klimatski Atlas Hrvatske Climate Atlas of Croatia 1961–1990. 1971–2000*; Državni hidrometeorološki zavod: Zagreb, Croatia, 2008; ISBN 978-953-7526-01-6.

27. IUSS Working Group WRB. *World Reference Base for Soil Resources 2014, Update 2015: International Soil Classification System for Naming Soils and Creating Legends for Soil Maps*; World Soil Resources Reports No. 106; FAO: Rome, Italy, 2015; p. 192.

28. Pereira, P.; Brevik, E.; Bogunović, I.; Ferran, E.-S. Ash and soils: A Close Relationship in Fire-Affected Areas. In *Fire Effects on Soil Properties*; Pereira, P., Mataix-Solera, J., Úbeda, X., Rein, G., Cerdà, A., Eds.; CSIRO Publishing: Clayton, Australia, 2019; ISBN 9781486308149.

29. Úbeda, X.; Pereira, P.; Outeiro, L.; Martin, D.A. Effects of Fire Temperature on the Physical and Chemical Characteristics of the Ash from Two Plots of Cork Oak (*Quercus Suber*). *Land Degrad. Dev.* **2009**, *20*, 589–608. [CrossRef]

30. Kannan, K.S.; Manoj, K.; Arumugam, S. Labeling methods for identifying outliers. *Int. J. Stat. Syst.* **2015**, *10*, 231–238.

31. R Core Team. R: A Language and Environment for Statistical Computing. Available online: https://www.r-project.org (accessed on 3 April 2021).

32. StatSoft, Inc. STATISTICA (Data Analysis Software System), Version 12.0. Available online: https://www.statsoft.de/en/home (accessed on 12 January 2022).

33. Plotly Technologies Inc. Collaborative Data Science. Available online: https://plot.ly (accessed on 26 April 2022).

34. Alcañiz, M.; Outeiro, L.; Francos, M.; Úbeda, X. Effects of Prescribed Fires on Soil Properties: A Review. *Sci. Total Environ.* **2018**, *613*, 944–957. [CrossRef]

35. Heydari, M.; Rostamy, A.; Najafi, F.; Dey, D.C. Effect of Fire Severity on Physical and Biochemical Soil Properties in Zagros Oak (*Quercus Brantii* Lindl.) Forests in Iran. *J. For. Res.* **2017**, *28*, 95–104. [CrossRef]

36. Xue, L.; Li, Q.; Chen, H. Effects of a Wildfire on Selected Physical, Chemical and Biochemical Soil Properties in a *Pinus Massoniana* Forest in South China. *Forests* **2014**, *5*, 2947–2966. [CrossRef]
37. Granged, A.J.P.; Zavala, L.M.; Jordán, A.; Bárcenas-Moreno, G. Post-Fire Evolution of Soil Properties and Vegetation Cover in a Mediterranean Heathland after Experimental Burning: A 3-Year Study. *Geoderma* **2011**, *164*, 85–94. [CrossRef]
38. Litton, C.M.; Santelices, R. Effect of wildfire on soil physical and chemical properties in a *Nothofagus glauca* forest, Chile. *Rev. Chil. De Hist. Nat.* **2003**, *76*, 529–542. [CrossRef]
39. Huang, Y.; Wu, L.; Yu, J. Relationships between soil organic matter content (SOM) and pH in topsoil of zonal soils in China. *Acta Pedol. Sin.* **2009**, *46*, 851–860.
40. Certini, G. Effects of Fire on Properties of Forest Soils: A Review. *Oecologia* **2005**, *143*, 1–10. [CrossRef]
41. Kutiel, P.; Inbar, M. Fire Impacts on Soil Nutrients and Soil Erosion in a Mediterranean Pine Forest Plantation. *Catena* **1993**, *20*, 129–139. [CrossRef]
42. Muñoz-Rojas, M.; Erickson, T.E.; Martini, D.; Dixon, K.W.; Merritt, D.J. Soil Physicochemical and Microbiological Indicators of Short, Medium and Long Term Post-Fire Recovery in Semi-Arid Ecosystems. *Ecol. Indic.* **2016**, *63*, 14–22. [CrossRef]
43. Pereira, P.; Cerda, A.; Martin, D.; Úbeda, X.; Depellegrin, D.; Novara, A.; Martínez-Murillo, J.F.; Brevik, E.C.; Menshov, O.; Comino, J.R.; et al. Short-Term Low-Severity Spring Grassland Fire Impacts on Soil Extractable Elements and Soil Ratios in Lithuania. *Sci. Total Environ.* **2017**, *578*, 469–475. [CrossRef]
44. Moreno-Casasola, P. Dunes. In *Encyclopedia of Ecology*; Jorgensen, S.E., Fath, B., Eds.; Elsevier: Amsterdam, The Netherlands, 2008.
45. Ulery, A.L.; Graham, R.C.; Amrhein, C. Wood-ash composition and soil pH following intense burning. *Soil Sci.* **1993**, *156*, 358–364. [CrossRef]
46. Goforth, B.R.; Graham, R.C.; Hubbert, K.R.; Zanner, C.W.; Minnich, R.A. Spatial Distribution and Properties of Ash and Thermally Altered Soils after High-Severity Forest Fire, Southern California. *Int. J. Wildland Fire* **2005**, *14*, 343. [CrossRef]
47. Santín, C.; Doerr, S.H. Carbon. In *Fire Effects on Soil Properties*; Pereira, P., Mataix-Solera, J., Úbeda, X., Rein, G., Cerdà, A., Eds.; CSIRO Publishing: Clayton, Australia, 2019; ISBN 9781486308149.
48. Muráňová, K.; Šimanský, V. The effect of different severity of fire on soil organic matter and aggregates stability. *Acta Fytotech. Zootech.* **2015**, *18*, 1–5. [CrossRef]
49. Iglesias, T.; Cala, V.; Gonzalez, J. Mineralogical and Chemical Modifications in Soils Affected by a Forest Fire in the Mediterranean Area. *Sci. Total Environ.* **1997**, *204*, 89–96. [CrossRef]
50. Kong, J.; Yang, J.; Bai, E. Long-Term Effects of Wildfire on Available Soil Nutrient Composition and Stoichiometry in a Chinese Boreal Forest. *Sci. Total Environ.* **2018**, *642*, 1353–1361. [CrossRef] [PubMed]
51. Williams, C. Seasonal Fluctuations in Mineral Sulphur under Subterranean Clover Pasture in Southern New South Wales. *Soil Res.* **1968**, *6*, 131. [CrossRef]
52. Caon, L.; Vallejo, V.R.; Ritsema, C.J.; Geissen, V. Effects of Wildfire on Soil Nutrients in Mediterranean Ecosystems. *Earth Sci. Rev.* **2014**, *139*, 47–58. [CrossRef]
53. Simelton, E. Texture and Nutrient Status in the Topsoil Six Years after Low and High Intensity Wildfires, NE Catalonia, Spain. Ph.D. Thesis, Earth Sciences Centre, Göteborg University, Gothenburg, Sweden, 2001.
54. Úbeda, X.; Bernia, S.; Simelton, E. Chapter 6 The Long-Term Effects on Soil Properties from a Forest Fire of Varying Intensity in a Mediterranean Environment. In *Developments in Earth Surface Processes*; Elsevier: Amsterdam, The Netherlands, 2005; Volume 7, pp. 87–102. ISBN 978-0-444-52084-5.
55. Lasanta, T.; Cerdà, A. Long-Term Erosional Responses after Fire in the Central Spanish Pyrenees. *Catena* **2005**, *60*, 81–100. [CrossRef]
56. Burton, C.A.; Hoefen, T.M.; Plumlee, G.S.; Baumberger, K.L.; Backlin, A.R.; Gallegos, E.; Fisher, R.N. Trace Elements in Stormflow, Ash, and Burned Soil Following the 2009 Station Fire in Southern California. *PLoS ONE* **2016**, *11*, e0153372. [CrossRef]
57. Mellis, E.V.; Cruz, M.C.P.d.; Casagrande, J.C. Nickel Adsorption by Soils in Relation to pH, Organic Matter, and Iron Oxides. *Sci. Agric.* **2004**, *61*, 190–195. [CrossRef]
58. Delač, D.; Kisić, I.; Bogunović, I.; Pereira, P. Temporal Impacts of Pile Burning on Vegetation Regrowth and Soil Properties in a Mediterranean Environment (Croatia). *Sci. Total Environ.* **2021**, *799*, 149318. [CrossRef]
59. Massoura, S.T.; Echevarria, G.; Becquer, T.; Ghanbaja, J.; Leclerc-Cessac, E.; Morel, J.-L. Control of Nickel Availability by Nickel Bearing Minerals in Natural and Anthropogenic Soils. *Geoderma* **2006**, *136*, 28–37. [CrossRef]
60. Santorufo, L.; Memoli, V.; Panico, S.C.; Santini, G.; Barile, R.; Di Natale, G.; Trifuoggi, M.; De Marco, A.; Maisto, G. Early Post-Fire Changes in Properties of Andosols within a Mediterranean Area. *Geoderma* **2021**, *394*, 115016. [CrossRef]
61. Fernández, I.; Cabaneiro, A.; Carballas, T. Organic Matter Changes Immediately after a Wildfire in an Atlantic Forest Soil and Comparison with Laboratory Soil Heating. *Soil Biol. Biochem.* **1997**, *29*, 1–11. [CrossRef]
62. Pereira, P.; Jordán, A.; Cerdà, A.; Martin, D. Editorial: The role of ash in fire-affected ecosystems. *Catena* **2014**, *135*, 337–339. [CrossRef]
63. Mitić, V.D.; Stankov Jovanović, V.P.; Ilić, M.D.; Nikolić Mandić, S.D. Impact of Wildfire on Soil Characteristics and Some Metal Content in Selected Plants Species of *Geraniaceae* Family. *Environ. Earth Sci.* **2015**, *73*, 4581–4594. [CrossRef]
64. Ministry of Agriculture. *Regulation on the Protection of the Agricultural Land from Pollution*; Official Gazette 71/2019; Narodne Novine: Zagreb, Croatia, 2019. Available online: https://narodne-novine.nn.hr/clanci/sluzbeni/2019_07_71_1507.html (accessed on 6 May 2022). (In Croa)

65. Francos, M.; Úbeda, X.; Pereira, P. Impact of Torrential Rainfall and Salvage Logging on Post-Wildfire Soil Properties in NE Iberian Peninsula. *Catena* **2019**, *177*, 210–218. [CrossRef]
66. Prats, S.; Malvar, M.; Martins, M.A.S.; Keizer, J.J. Post-fire soil erosion mitigation: A review of the last research and techniques developed in Portugal. *Cuadernos de Investigación Geográfica* **2014**, *40*, 403–428. [CrossRef]
67. Fontúrbel, M.T.; Barreiro, A.; Vega, J.A.; Martín, A.; Jiménez, E.; Carballas, T.; Fernández, C.; Díaz-Raviña, M. Effects of an Experimental Fire and Post-Fire Stabilization Treatments on Soil Microbial Communities. *Geoderma* **2012**, *191*, 51–60. [CrossRef]

Article

A Flood Mapping Method for Land Use Management in Small-Size Water Bodies: Validation of Spectral Indexes and a Machine Learning Technique

Lorena Lombana * and Antonio Martínez-Graña

Department of Geology, Faculty of Sciences, University of Salamanca, Plaza de la Merced s/n, 37008 Salamanca, Spain; amgranna@usal.es
* Correspondence: llombanag@usal.es

Abstract: The assessment of flood disasters is considered an essential factor in land use management, being necessary to understand and define the magnitude of past events. In this regard, several flood diagnoses have been developed using Sentinel-2 multispectral imagery, especially in large water bodies. However, one of the main challenges is still related to floods, where water surfaces have sizes similar to the spatial resolution of the analyzed satellite images, being difficult to detect and map. Therefore, the present study developed a combined methodology for flood mapping in small-sized water bodies using Sentinel-2 MSI imagery. The method consisted of evaluating the effectiveness of the application and combination of (a) a super-resolution algorithm to improve image resolution, (b) a set of seven spectral indices for highlighting water-covered areas, such as AWE indices, and (c) two methods for flood mapping, including a machine learning method based on unsupervised classification (EM cluster) and 14 thresholding methods for automatic determination. The processes were evaluated in the Carrión River, Palencia, Spain. It was determined that the approach with the best results in flood mapping was the one that combined AWE spectral indices with methods such as Huang and Wang, Li and Tam, Otsu, moment preservation, and EM cluster classification, showing global accuracy and Kappa coefficient values higher than 0.88 and 0.75, respectively, when applying the quantitative accuracy index.

Keywords: flood mapping; Sentinel-2; spectral indices; cluster analysis

Citation: Lombana, L.; Martínez-Graña, A. A Flood Mapping Method for Land Use Management in Small-Size Water Bodies: Validation of Spectral Indexes and a Machine Learning Technique. *Agronomy* **2022**, *12*, 1280. https://doi.org/10.3390/agronomy 12061280

Academic Editor: Francisco Manzano-Agugliaro

Received: 24 February 2022
Accepted: 25 May 2022
Published: 26 May 2022

Publisher's Note: MDPI stays neutral with regard to jurisdictional claims in published maps and institutional affiliations.

1. Introduction

Floods are among the most common and disastrous worldwide natural events, being the cause of many human life losses, heritage and environmental damages, and economic costs [1]. Those events are progressively more recurrent and catastrophic not only due to extreme weather-related to climate change but also owing to the increase in the number of exposed elements triggered by socio-economic activities that settle in fluvial spaces [2]. Thus, in order to implement flood risk management measures, it is necessary to understand the magnitude, severity, and frequency of past events [3]. Hence, flood disaster assessment must determine the flood event extension, which depends on timely and effective monitoring at a regional scale [4]. However, this diagnosis has difficulties due to the lack of coverage and density of meteorological and gauging stations [5].

Thus, taking advantage of the accelerated development of remote sensing techniques, massive satellite data have become an available practical source for flood monitoring [6]. Several studies have been carried out for this purpose, applying different methods [7,8]. Nevertheless, identifying the surfaces covered by water is the main task in flooding diagnoses, and hence it is necessary to implement remote sensing techniques that provide high spatial images with high acquisition frequency [4].

In this regard, the European Space Agency (ESA) has developed Earth observation missions under the Copernicus program for land monitoring, including the Sentinel-2 mission, which supports global satellite data with a wide width and multispectral band [9]. This mission provides freely available multispectral Images (MSI) with a 5-day revisit frequency and a 10 m spatial resolution for visible bands, making Sentinel-2 imagery an excellent resource for surface water mapping [7,10,11], becoming one of the main satellites used to detect flood extents [12,13].

In this aspect, during the last decades, different methodologies have been developed for the digital processing of satellite images, reducing costs, time, and efforts [11]. These methodologies involve several techniques, which can be classified into four main classes: single band, spectral index, machine learning, and spectral unmixing based methods [7].

Among the techniques applied, water spectral indexes are the most widely used methods to classify surfaces into water and non-water using multispectral images. These indices allow enhancing the contrast between pixels of these two categories, exploiting the information of different bands [8,14]. In this regard, the near-infrared (NIR) band is usually used to identify water and non-water pixels due to its high absorption and low reflectance rates [15,16].

The water indices have been improved to achieve good performances in the delineation of water bodies. In 1996, McFeeters developed the Normalized Difference Water Index (NDWI) [17], which uses green and NIR bands to obtain an index representing water with positive values and non-water with negative values. However, it was found that this index is not efficient in built-up suppressing, so the identified water surfaces can mix water and built-up land noises. Thus, the modified Normalized Difference Water Index (mNDWI) was implemented by using the shortwave-infrared (SWIR) band instead of the NIR band [18]. On the other hand, considering the classification issues caused by shadow noises, the AWE indexes (AWEIsh and AWEInsh) were developed [19]. These use different band combinations and weightings, including the blue, green, NIR, and SWIR bands, where water pixel values are above 0 and non-water pixels below 0. It was determined that AWEInsh is appropriate for images where clouds do not represent a problem, suppressing dark surfaces in urban areas. On the other hand, AWEIsh successfully removes shadow noises. Likewise, other studies have used vegetation indices to extract water features [20].

Although the spectral water indexes normally represent water and non-water pixels with values above or below 0, it is necessary to determine a suitable threshold value depending on the characteristics of the water surfaces in each multispectral image [11]. Different studies have established optimal threshold values by visual interpretation of frequency histograms; however, it is not a useful method, especially when analyzing large multispectral images [8]. Therefore, automatic histogram-based thresholding algorithms have been applied for this purpose, with Otsu's method being the most widely used algorithm to improve the classification accuracy [11,15]. This one operates with the maximum between-class variance to recognize objects and background [10,21].

On the other hand, surface water mapping methodologies based on machine learning have been applied [7,22], including the classification algorithms, which can be categorized into supervised and unsupervised methods depending on the human training needed [7]. The unsupervised methods have been demonstrated to be effective in differentiating between water from non-water pixels [12,23]. One of the methods used is the Clustering algorithm, where points with similar features are automatically grouped into the same cluster, such as K-means clustering, mean-shift clustering, and expectation–maximization (EM) clustering [7,12].

Although many methods have been developed to improve water surface mapping, there are still many challenges in this subject, including the issues related to the delineation accuracy considering the resolution of remotely sensed images and the water bodies' sizes. In this case, when water surfaces have sizes similar to the spatial resolution of the satellite images used, they are hard to detect and map [7,10,24].

In an effort to address part of these challenges, some studies have improved the spatial resolution of remotely sensed images by using sharpening algorithms; however, these have been applied to large water bodies, for instance, Poyang Lake in China, Aras River in Turkey, or Po River in Italy [6,10,25,26]. Some of them had used panchromatic bands with high resolution to sharper the bands with lower resolution and improved the delineation accuracy [25]. Nevertheless, while many satellite imagery includes a panchromatic band, some do not, such as the Sentinel-2 images. For these images, other sharpening algorithms have been applied [11,26]. These methods can be classified into three categories: pan-sharpening per band, inverting an explicit imaging model, and machine learning methods [27]. The pan-sharpening per band method increases the spatial resolution of each band independently, mixing information from a high-resolution band [11,25,28]. However, the complexity of these processes can restrict the efficiency of water body mapping [10]. On the other hand, the model-based methods obtain a high-resolution image by minimizing residual errors in a single optimization for all bands simultaneously [27]. Accordingly, Brodu [29] developed an algorithm for super-resolved multiresolution images, showing good results in agricultural lands with large uniform areas. It separates band-dependent spectral information from information that is common across all bands ("geometry of scene elements"), super-resolving the low-resolution bands while preserving their overall reflectance [29]. Although this method proved efficient in improving Sentinel-2 images' spatial resolution, it has not been directly tested to enhance flood mapping.

Considering the above, and taking into consideration the challenges in surface water mapping and the diversity of methods and techniques developed around this issue, flood mapping is considered an open research topic because no single method has been found suitable for all data sets and all conditions [7].

So, by combining the advantages of data held in Sentinel-2 multispectral images (MSI), image pre-processing techniques, spectral indexes, and machine learning methods, the present study developed a combined automated methodology for flood mapping for one of the biggest challenges: small-sized water bodies. Accordingly, this study involved four main phases (Figure 1): (1) Improve Sentinel-2 images resolution by applying and comparing resample, and a super-resolving algorithm; (2) Assess and compare the effectiveness of seven spectral indexes in highlight flood surfaces; (3) Apply and compare different flood extent mapping methods including 14 thresholding algorithms and a machine learning method for unsupervised classification; (4) Evaluate the performance of the flood mapping methods used; to finally define the most accurate combined method for flood mapping in a small-sized water body.

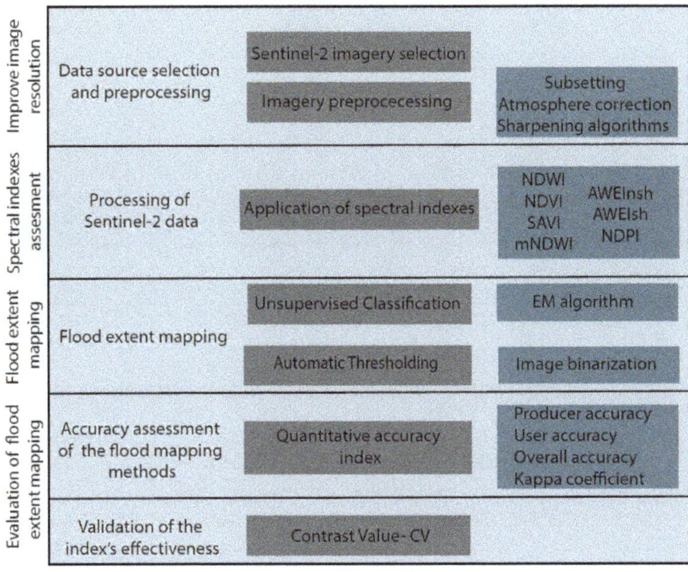

Figure 1. Methodology of the study.

The water body selected to validate the methodology framework was the Carrión River in the Duero basin, located in Palencia, Spain (Figure 2). It has a channel length of 197 km, with a drainage area of 3368 km^2, average values of width of 40 m, river level of 0.5 m, discharges of 12 m^3/s, and an annual maximum of 38 m^3/s [30]. This river presents a marked lateral migration with a wide alluvial plain, where an active and/or abandoned drainage network can be identified [31]. Throughout its alluvial plain, agricultural activities and urban land uses can be found, so different elements can be exposed to flood risks [32]. Thus, although the Carrión river can be classified as a narrow water body, it presents recurrent flood events due to its morphometric characteristics, fluvial dynamics, occupations, land cover, and land uses.

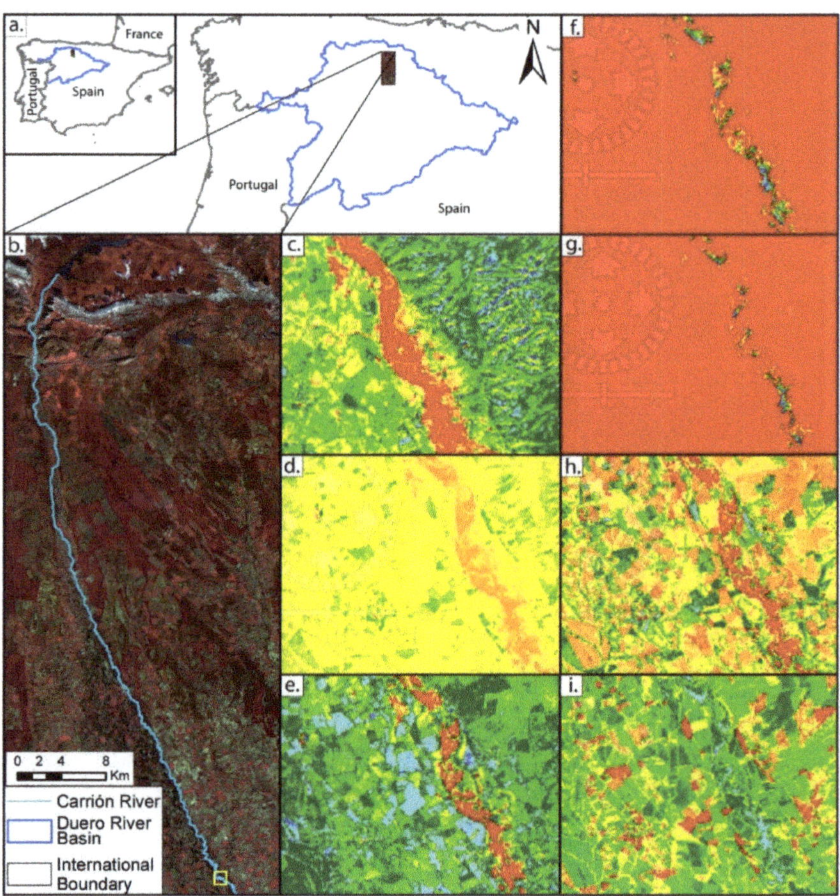

Figure 2. Location of the study area and Spectral indices' images obtained from Sentinel-2: (**a**) General localization. (**b**) Studied River. (**c**) AWEInsh. (**d**) AWEIsh. (**e**) MNDWI. (**f**) NDPI. (**g**) NDVI. (**h**) NDWI. (**i**) SAVI.

2. Methods

2.1. Data Source Selection and Pre-Processing

The imagery selected for the study was the high-resolution optical images provided by the Sentinel-2 satellite mission under the Copernicus Earth Observation program led by the European Commission and operated by the European Space Agency (ESA) (https://earth.esa.int/web/sentinel/home, accessed on 10 July 2021).

The time and areas of study in Carrion River were selected to assess post-flood events, considering diverse criteria. Regarding the selection of study dates, the hydrological year analyses were considered. The period selected corresponded to the 2019–2020 wet year, when the Duero basin presented rainfalls 105% higher than the 1981–2020 average value, with November and December being the most important months [33]. On the other hand, the study area selection in the basin took into consideration that the water and non-water surfaces should be diverse and include principal channels, reservoirs, vegetation, buildings, bare land, and fields. Accordingly, Level-2A S2 images were downloaded employing the Sentinel-2 Toolbox for ArcGIS 10.8, using a maximum cloud percentage of 20%, between 1 November 2019 and 31 December 2019. The detailed description of the image obtained is shown in Table 1.

Table 1. Sentinel-2A image features.

Feature	Value
Name	S2A_MSIL2A_20191228T111451_N0213_R137_T30TUN_20191228T122432
Sector	Superior
Cloud cover percentage	14.108462
Cloud shadow percentage	2.133657
Ingestion Date	2019-12-28T20:40:54.449Z
Orbit number (start)	23586
Pass direction	Descending
Vegetation percentage	29.200098
Water percentage	0.675628

The Level-2A images downloaded were pre-processed using the Sentinel Application Platform (SNAP), developed by ESA (https://earth.esa.int/eogateway/tools/snap, accessed on 10 July 2021). This process consisted of three different steps. First, to work over the areas belonging to the Carrion River, the image subsetting was carried out to obtain a raster of 3132 × 6540 pixels. Sentinel-2A data have multiple bands that include four 10 m visible and near-infrared bands, six 20 m vegetation red edge and short wave infrared bands, and three 60 m coastal aerosol, water vapor, and SWIR-Cirrus bands [34].

Considering that the bands have different resolutions, it is necessary to obtain images of high quality and the same pixel size. In order to obtain high spatial resolution data of 10 m-bands, two methods were applied and compared. Firstly, a resampling algorithm was used, implementing a tool available in Raster-Geometry subset tools of SNAP. It was used by taking as reference data the Blue band and a nearing algorithm. Secondly, a super-resolution method was applied using the Sen2Res tool in SNAP [29]. This method was developed for multispectral and multiresolution images, such as Sentinel-2, which obtains information from pixels that have the highest resolution and reproduce these details to all other bands; thus, an image with the best resolution can be obtained [29]. The method works for uniform areas such as agricultural lands. Finally, a land use band combination was composited (using the 11, 8A, and 4 bands) and the two methods applied were compared by visual interpretation.

2.2. Processing of Sentinel-2 Data: Spectral Indices

For the study, a set of 7 spectral indices were applied to highlight surfaces covered by water before applying flood extent mapping methods. This kind of indices used complex ratios of multiple bands, which are selected depending on the spectral characteristics of the target element studied [7] (a further description of the algorithms can be found in Table 2). In this case, to calculate the indexes, the Raster Math tool available in SNAP was implemented.

Table 2. Summary of indexes used in the study.

Index	Algorithms-Sentinel-2 *	Category
NDWI [17]	(B03 − B08)/(B03 + B08)	Water
NDVI [35]	(B08 − B04)/(B08 + B04)	Vegetation
SAVI [36]	(B08 − B04)/(B08 + B04 + 0.428) × (1.428)	Vegetation
mNDWI [18]	(B03 − B11)/(B03 + B11)	Water
AWEInsh [19]	4 × (B03 − B11) − (0.25 × B08 + 2.75 × B11)	Water
AWEIsh [19]	B02 + 2.5 × B03 − 1.5 × (B08 + B11) − 0.25 × B12	Water
NDPI [37]	(B12 − B03)/(B12 + B03)	Water

* B: Band.

2.3. Flood Extent Mapping

The flood extent mapping consists of the surface water detection and delineation using remote sensing images; therefore, multiple studies have been carried out on this subject [7,8,14,38]. In this study, from the indices calculated, two different methods were evaluated, the unsupervised classification and the Automatic Threshold Determination.

The unsupervised classification was performed using the expectation–maximization (EM) algorithm, which considers the number of clusters that must be established as well as their center points which are randomly initialized [7]. In the study, eight classes were performed according to the cover land. For that, the EM cluster analysis tool supported in SNAP was applied to each spectral index.

On the other hand, the Automatic Threshold Determination, a single band method, consisted of the application of image binarization algorithms to classify pixels into different categories, in this case, water and non-water surfaces [7]. A total of 14 automatic thresholding methods were evaluated for each spectral index applied using the ImageJ software [39]. The algorithms used are presented in Table 3.

Table 3. Thresholding methods applied in the study.

No	Threshold Automatic Algorithm	No	Threshold Automatic Algorithm
1	Huang and Wang's [40]	8	Moment-preserving [41]
2	Intermode [42]	9	Otsu's thresholding [21]
3	Isodata [43]	10	Percentile (p-tile) [44]
4	Li and Tam's [45]	11	Renyi's entropy [46]
5	Maximum entropy [46]	12	Shanbhang's [47]
6	Mean [48]	13	Triangle [49]
7	Minimum [42]	14	Yen's [50]

2.4. Accuracy Assessment of the Flood Mapping Methods

Regarding the assessment of the extraction of the flooded areas, the quantitative accuracy index was applied [10]. This widely used method involves the generation of random water and non-water sample points where water extraction results are verified. Thus, for the study, this process included four steps:

a. Generation of 200 water and non-water sample points over the study area by applying the Create Random Points tool in ArcGIS.

b. Once the points were generated, high-resolution images were employed to verify and adjust the random sample points. In this case, the images were obtained from the Spanish National Geographic Institute, IGN, provided by National Aerial Orthophotography Plan (PNOA 2017). Additionally, scatter plots were plotted to verify the reflectance of the water sample points, which is expected to be low if the Red and NIR-1 bands are contrasted.

c. Subsequently, it was identified whether each of the random points corresponded to surfaces classified as water or non-water according to each flood mapping method applied, so the total number of points correctly identified was calculated.

d. The accuracy of each method was evaluated by applying four assessment indices. For this purpose, a confusion matrix-based approach was developed (Table 4). Thus, by comparing the extracted water and non-water points with the reference data, four types of pixels were obtained: true positive (TP), the number of correctly extracted water pixels; false negative (FN), the number of undetected water pixels; false positive (FP), the number of incorrectly extracted water pixels; and true negative (TN), the number of correctly rejected non-water pixels [25]. Based on the four types of pixels, the producer accuracy, user accuracy, overall accuracy, and Kappa coefficient were applied according to the Equations (1)–(4), respectively [10].

$$Producer's\,accuracy = \frac{TP}{TP + FN} \tag{1}$$

$$User's\,accuracy = \frac{TP}{TP + FP} \tag{2}$$

$$Overall\,accuracy = \frac{TP + TN}{T} \tag{3}$$

$$Kappa\,coefficient = \frac{T(TP + TN) - ((TP + FP)(TP + FN) + (FN + TN)(FP + TN))}{T^2 - ((TP + FP)(TP + FN) + (FN + TN)(FP + TN))} \tag{4}$$

Table 4. Confusion matrix-based applied in the study.

		Reference Data		User
		Water	Non-Water	
Classified data	Water	TP	FP	TP + FP
	Non-Water	FN	TN	FN + TN
Producer		TP + FN	FP + TN	T = TP + FP + FN + TN

2.5. Validation of the Index's Effectiveness

By comparing the mean value of the index values in water and non-water points, the ability of each index to distinguish flood surfaces can be assessed [10], so the formula of the Contrast Value (CV) was applied (Equation (5)).

$$CV = E - K \tag{5}$$

where, E and K denote the mean value of each index in terms of the water and non-water surface, respectively.

3. Results

3.1. Image Processing

After pre-processing Sentinel-2 images, an atmosphere corrected subset image was obtained. By applying the resample and super-resolution methods, two different images were produced; thus, the 11 and 12 bands were improved to a 10 m-resolution (Figure 3). As can be seen, although two images with the same pixel size were obtained, the image produced by the super-resolution method shows better results than the resampling method. This is because the resampling method resamples each band independently, while super-resolution obtains information from other bands that have higher resolution [29]. In this sense, this method allowed obtaining high-resolution bands that are necessary to calculate spectral indexes, in this case, the shortwave-infrared bands (11-band and 12-band), going from 20 m to 10 m resolution. Consequently, the following steps in the proposed methodology were applied to the super-resolution image obtained.

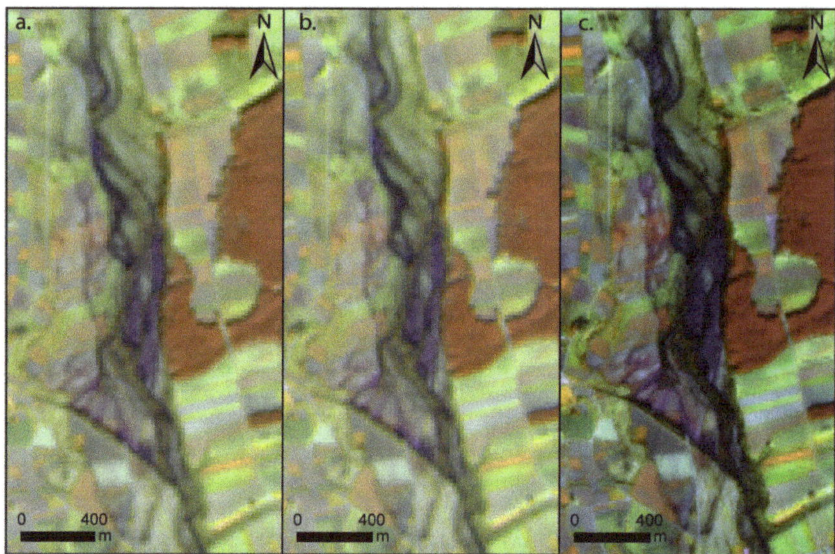

Figure 3. Sentinel-2 images using composite bands: (**a**) Raw image. (**b**) Resampled image. (**c**) Super-resolution image.

3.2. Application of the Spectral Indices and Effectiveness Validation

A total of seven indices were effectively applied to the Sentinel-2 image to identify flooded areas, as shown in Figure 4. By comparing the results with a false-color composite image created to enhance water surfaces (Figure 4b) through a visual interpretation, it can be indicated that the AWEInsh and AWEIsh indices show accurate results in reflecting flooded areas, but in any case, with diffuse boundaries. It seems that, although the MNDWI index does not fully bright flooded areas, it manages to delimit the zones with sharper boundaries, as does the NDWI index, although this shows lesser precision. The NDPI, NDVI, and SAVI indices do not seem to represent good results in flooded areas reflecting. However, the effectiveness of each index and the application of flood extent mapping methods will be evaluated later through quantitative methods.

As a product, histograms of each spectral index were also obtained (Figure 5). As shown, AWEInsh and AWEIsh index show a marked unimodal distribution, while MNDWI, NDWI, and NDPI tend to represent a bimodal distribution. The other indices present different peaks, so they show multimodal distributions.

In order to assess the effectiveness of each spectral index used, the mean and contrast values were calculated for the water and non-water random points previously generated. Accordingly, boxplots were made to compare the distribution of the data (Figure 6). As shown, in terms of the water points, the mean values of all indexes are closer to zero (between -0.23 and 0.05) than the values obtained for non-Water points (which are between -1.08 and 0.37). The AWEIsh, MNDWI, and NDWI indices showed similar mean values for both water and non-water points. The same case occurred with the NDVI and SAVI indices. Therefore, the highest contrast value was calculated for the AWEInsh index, followed by the MNDWI and NDPI indices. At first, it suggests that AWEInsh, MNDWI, and NDPI spectral indices are more effective in highlighting water surfaces in Sentinel-2 images.

3.3. Flood Extent Mapping Performance

As could be interpreted from visual interpretation and contrast values, AWEInsh highlights water surfaces brighter than the other indexes. However, it is necessary to generate flood maps by different methods to develop spatial and quantitative analysis, as was previously detailed. Firstly, the threshold values were calculated for each index, as shown in Table 5. As can be seen, different thresholding methods provide the same results, such as the AWEInsh index applying the Huang's, Li, and Otsu algorithms, and the AWEish index using the Huang's, Isodata, and Percentile methods. In these cases, the flood maps produced are the same.

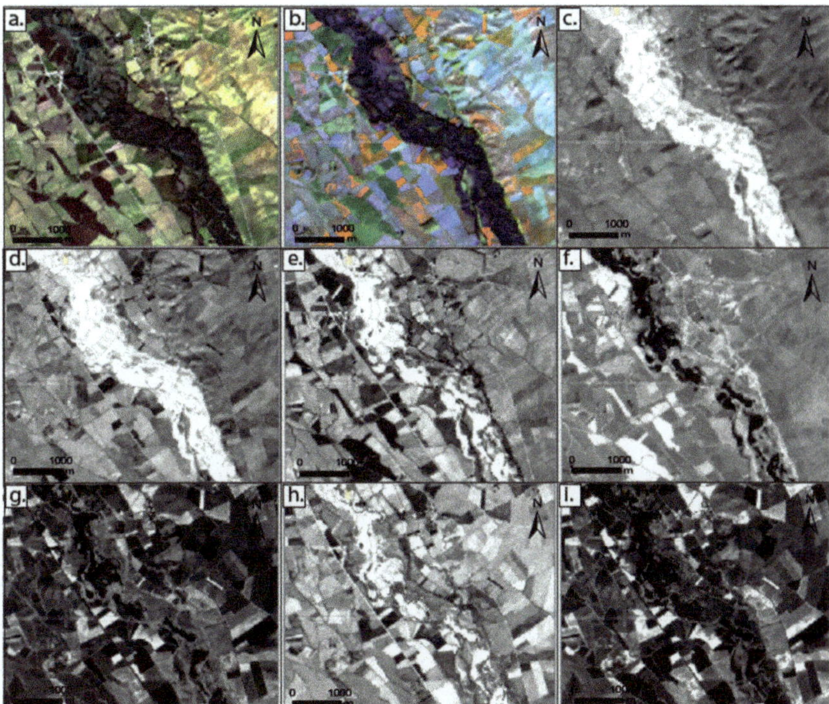

Figure 4. Flooded areas detected by each index used in the study area. (**a**) Real color image. (**b**) Composite image (R: 8, G: 11, B: 4). (**c**) AWEInsh. (**d**) AWEIsh. (**e**) MNDWI. (**f**) NDPI. (**g**) NDVI. (**h**) NDWI. (**i**) SAVI.

Each threshold value was used to classify pixels into the water and non-water surfaces; therefore, thematic maps from a single-band image were created by using discrete binary colors. Figure 7 shows the performances of each spectral index by applying the different thresholding methods and the unsupervised classification. It should have been previously indicated that Minimum, Yen, and Shanbhang thresholding methods did not work to classify pixels for any of the indices used.

As seen in Figure 7, mean and percentile algorithms extracted water bodies worse than the other methods for all indices. In addition, maximum entropy and intermodel methods only worked for the MNDWI and NDPI indexes. In contrast, the unsupervised classification method (EM Cluster) illustrated the water surfaces better than the thresholding methods in most cases.

Furthermore, the methods of Huang, Isodata, and Li and Tam only worked for the AWEInsh index. Thus, it seems that the best performances are obtained by classifying the AWEInsh index pixels with different methods.

Although the accuracy of each method can be explained by visual interpretation, it is important to analyze the results of the assessment methods implemented for this purpose. Thus, the overall accuracy and the Kappa coefficient values are summarized in Figure 8 and plotted in Figure 9 (the detailed results obtained by applying the producer accuracy, user accuracy, overall accuracy, and Kappa coefficient can be found in the Appendix A).

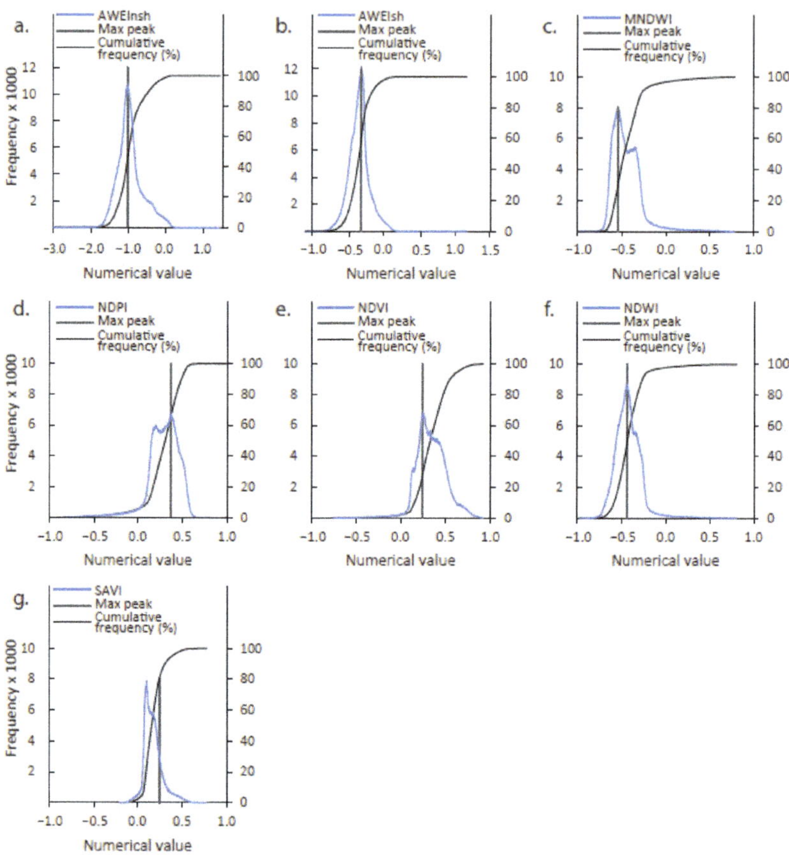

Figure 5. Histograms obtained from different spectral indices. (**a**) AWEInsh. (**b**) AWEIsh. (**c**) MNDWI. (**d**) NDPI. (**e**) NDVI. (**f**) NDWI. (**g**) SAVI.

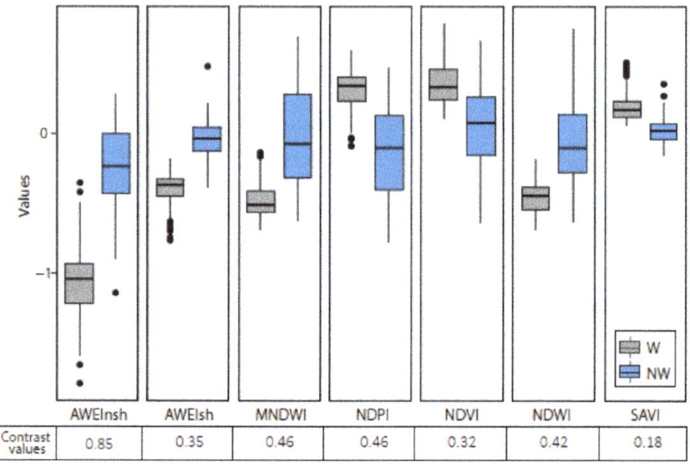

Contrast values	0.85	0.35	0.46	0.46	0.32	0.42	0.18
	AWEInsh	AWEIsh	MNDWI	NDPI	NDVI	NDWI	SAVI

Figure 6. Box plots classified into water (W) and non-water (NW) categories and contrast values obtained for each index.

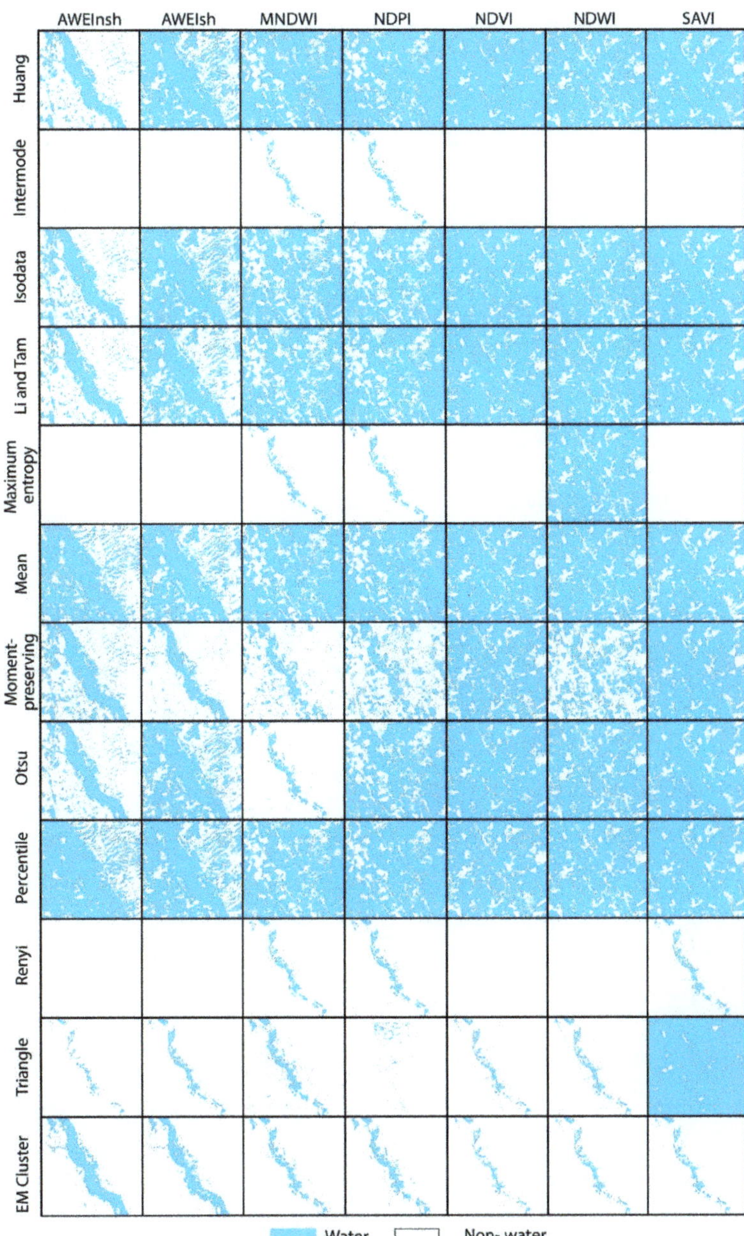

Figure 7. Flood-areas extraction performances generated using all indexes by each mapping method (refer to Table 5).

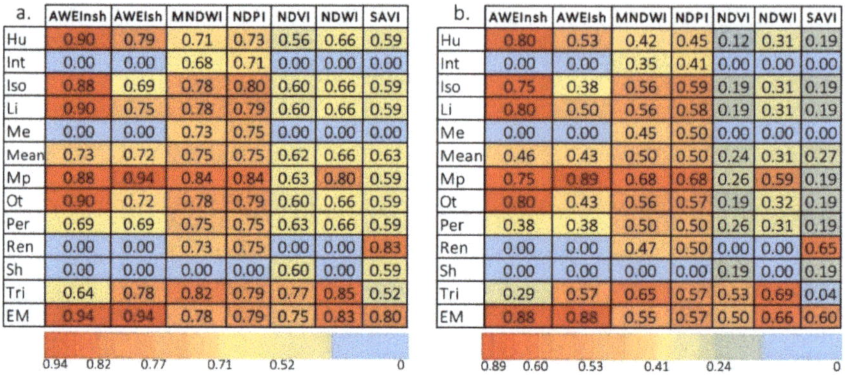

a.

	AWEInsh	AWEIsh	MNDWI	NDPI	NDVI	NDWI	SAVI
Hu	0.90	0.79	0.71	0.73	0.56	0.66	0.59
Int	0.00	0.00	0.68	0.71	0.00	0.00	0.00
Iso	0.88	0.69	0.78	0.80	0.60	0.66	0.59
Li	0.90	0.75	0.78	0.79	0.60	0.66	0.59
Me	0.00	0.00	0.73	0.75	0.00	0.00	0.00
Mean	0.73	0.72	0.75	0.75	0.62	0.66	0.63
Mp	0.88	0.94	0.84	0.84	0.63	0.80	0.59
Ot	0.90	0.72	0.78	0.79	0.60	0.66	0.59
Per	0.69	0.69	0.75	0.75	0.63	0.66	0.59
Ren	0.00	0.00	0.73	0.75	0.00	0.00	0.83
Sh	0.00	0.00	0.00	0.00	0.60	0.00	0.59
Tri	0.64	0.78	0.82	0.79	0.77	0.85	0.52
EM	0.94	0.94	0.78	0.79	0.75	0.83	0.80

0.94   0.82   0.77   0.71   0.52   0

b.

	AWEInsh	AWEIsh	MNDWI	NDPI	NDVI	NDWI	SAVI
Hu	0.80	0.53	0.42	0.45	0.12	0.31	0.19
Int	0.00	0.00	0.35	0.41	0.00	0.00	0.00
Iso	0.75	0.38	0.56	0.59	0.19	0.31	0.19
Li	0.80	0.50	0.56	0.58	0.19	0.31	0.19
Me	0.00	0.00	0.45	0.50	0.00	0.00	0.00
Mean	0.46	0.43	0.50	0.50	0.24	0.31	0.27
Mp	0.75	0.89	0.68	0.68	0.26	0.59	0.19
Ot	0.80	0.43	0.56	0.57	0.19	0.32	0.19
Per	0.38	0.38	0.50	0.50	0.26	0.31	0.19
Ren	0.00	0.00	0.47	0.50	0.00	0.00	0.65
Sh	0.00	0.00	0.00	0.00	0.19	0.00	0.19
Tri	0.29	0.57	0.65	0.57	0.53	0.69	0.04
EM	0.88	0.88	0.55	0.57	0.50	0.66	0.60

0.89   0.60   0.53   0.41   0.24   0

Figure 8. Accuracy values obtained for each index and mapping method: (**a**) Overall accuracy. (**b**) Kappa coefficient.

Table 5. Thresholding values obtained from each method.

Thresholding Methods	Threshold Value						
	AWEInsh	AWEIsh	MNDWI	NDPI	NDVI	NDWI	SAVI
Huang and Wang's (Hu)	−0.88	−0.44	−0.53	0.33	0.57	−0.52	0.26
Intermode (Int)	−3.92	−1.37	0.1	−0.21	−0.05	0.05	0.05
Isodata (Iso)	−0.93	−0.44	−0.48	0.27	0.5	−0.52	0.26
Li and Tam (Li)	−0.88	−0.39	−0.48	0.29	0.48	−0.52	0.25
Maximum entropy (Me)	0.91	0.38	−0.03	−0.11	0.05	−0.11	0.04
Mean	−1.08	−0.41	−0.5	0.31	0.45	−0.52	0.24
Minimum (Min)	−6.31	−2.26	0.39	−0.47	−0.25	0.36	−0
Moment-preserving (Mp)	−0.93	−0.27	−0.33	0.17	0.43	−0.4	0.25
Otsu (Ot)	−0.88	−0.41	−0.19	0.29	0.5	−0.5	0.26
Percentile (p-tile) (Per)	−1.13	−0.44	−0.52	0.32	0.43	−0.52	0.23
Renyi's entropy (Ren)	0.61	0.38	−0.05	−0.11	0.07	−0.12	0.05
Shanbhag's (Sh)	2.66	0.08	0.46	−0.4	0.49	0.01	0.26
Triangle (Tri)	−0.03	−0.09	−0.26	0.02	0.09	−0.23	0.46
Yen' (Y)	0.96	0.48	0	−0.14	0.03	−0.1	0.03

As can be seen, the AWE indexes showed the highest accuracy in water and non-water pixels categorization. In these cases, the most effective methods used were the EM unsupervised classification (which represents an overall accuracy of 0.94 and a Kappa coefficient of 0.88 in both cases) and the moment-preserving thresholding method (reaching overall and Kappa values of 0.94 and 0.89, respectively).

However, thresholding methods such as Huang and Wang, Li and Tam, Isodata and Otsu also illustrated optimal results for the AWEInsh index, with overall and Kappa values over 0.88 and 0.75, respectively.

The NDVI, NDWI, and SAVI indices presented low accurate results in most of the cases, showing the NDVI the worst performance. Nevertheless, the NDWI index reaches an overall accuracy of 0.85 and a kappa coefficient of 0.69 by using the Triangle thresholding method.

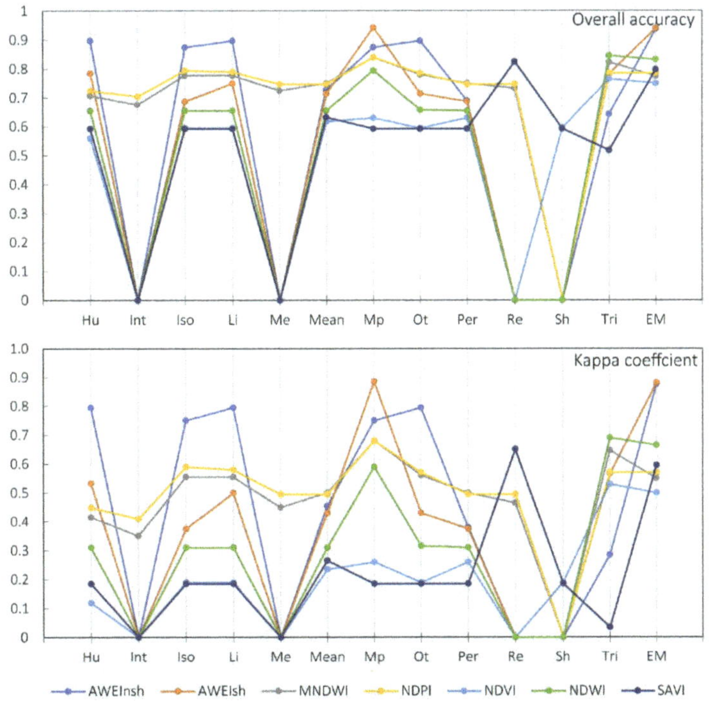

Figure 9. Accuracy values plotted for each index and mapping method by Overall and Kappa algorithms.

4. Discussion

The super-resolve method developed by Brodu [29] showed good results in improving the resolution of Sentinel-2 images in the area of interest, confirming its efficiency when applied to uniform areas representative of agricultural lands. This algorithm simulates a panchromatic sharpening, which can extract geometric information and use it to separate the low-resolution pixels while preserving their overall reflectance [29], so it is effective for super-resolved bands needed to calculate water spectral indexes, especially the SWIR bands [7,11,19,26].

Once the Sentinel-2 bands had the same high resolution (10 m), it was possible to apply the seven spectral indexes. Regarding the histograms obtained for each index (Figure 5), previous studies have shown similar results for MNDWI, NDWI, and AWEIsh histograms, where peak values usually represent land or water surfaces [15]. In this regard, different authors have demonstrated that index histograms with bimodal and multimodal distributions are appropriate for implementing thresholding methods [8]; however, this study was successful in using automatic thresholding algorithms in index histograms with a unimodal distribution, such as the case of the AWE indices. Additionally, it is important to emphasize that in all histograms, except for the SAVI, the maximum peaks were close to the index mean values calculated for non-water points.

By analyzing the contrast values, the AWEInsh, MNDWI, and NDPI represented the maximum values, suggesting that they may be more effective in highlighting water surfaces. However, as verified in the flood mapping performances, the AWE indexes showed the most accurate results, considering the overall and Kappa coefficients; on the contrary, the MNDWI and NDPI did not present satisfactory results, obtaining accuracy values below 0.84. On the other hand, spectral indexes, such as NDWI and mNDWI, usually used in water surface identification, were not useful for detecting flood areas in the analyzed small-size water body [10]. Consequently, as revealed, the AWE indexes were the most suitable for extracting flood areas in the study zone. This type of spectral indices has been structured to identify water features, so Ish was designed to remove shadow pixels and Insh for areas with an urban background [19]. Although the method was built for Landsat images, it works effectively for the Sentinel-2 image used in this study, as reported by other authors [14,15].

As for flood mapping methods, among the automatic thresholding algorithms employed, the moment-preserving thresholding method showed the most accurate results, followed by the Huang and Wang, Li and Tam, and Otsu algorithms, which also presented good outputs. However, the machine learning method for unsupervised classification (the EM cluster) showed the best results in all cases, only reached by the Moment-preserving thresholding algorithm. Consequently, according to the visual interpretation and the accuracy values obtained, it can be said that the flood mapping methods with Overall and Kappa values over 0.88 and 0.75, respectively, represent effective methods for the identification of flooded areas.

In this regard, the recommended combination of spectral indexes, thresholding algorithms, and/or unsupervised methods for flood mapping in small rivers is as follows: (a) Super-resolved Sentinel-2 images by applying the super-resolution method developed by Brodu [29], and obtain 10-m-resolution bands, including SWIR bands. (b) Calculate the AWEInsh index and apply the Huang and Wang, Li and Tam, Otsu, and EM cluster methods. (c) Calculate the AWEIsh index and use the moment-preserving algorithm and EM cluster method for pixel categorization. (d) Assess the accuracy of the flood mapping methods by using validation points and a derived confusion matrix.

The method proposed involves the use of simple and available access data and tools such as Sentinel-2 MSI, the SNAP (and its Sen2Res tool, Raster Math, cluster analysis packages), and ImageJ software. It holds the advantages of free access to information and an unsupervised algorithm that does not depend on training data efficiency [7], which makes the methodology a practical process for mapping floods in narrow rivers. However, further studies could evaluate the use of other machine learning techniques that can keep the pros of unsupervised classification [23].

Furthermore, the accuracy assessment results suggest that the method developed in this study is not just simple but effective in narrow rivers. Nevertheless, one limitation is related to the analysis of the effect of the presence of mixed spectra in the Sentinel-2 image pixels, so further studies should include this kind of evaluation [7]. In this regard, these analyses should include other accuracy assessment methods that allow measuring the results obtained by comparing flood maps with a Very High Resolution (VHR) imagery of the exact date of the flood event. It may involve new methods to quantitatively evaluate the extraction accuracy taking into account the spatial position [15]. On the other hand, it can be recommended that further investigations evaluate the methodology proposed using SAR images.

5. Conclusions

The combined automated method for water mapping developed in the present study contributes to improving the detection of surfaces covered by water during flood events in narrow rivers. This methodology exploits the advantages of data held in Sentinel-2 MSI by applying image pre-processing techniques and simple methods such as single band techniques, which include thresholding algorithms, combined with spectral indexes and unsupervised machine learning techniques. This can show accurate results in highlighting flood areas in small-size waterbodies whose width is similar to the size pixel.

It can be emphasized the effectiveness of AWE indexes in water surface detection in narrow rivers by combining mapping methods such as Huang and Wang, Li and Tam, and Otsu, moment-preserving thresholding algorithms, and EM cluster classification. In this order, the super-resolution method developed by Brodu was efficient in SWIR band super-resolving, enhancing the water spectral indexes application.

The method such as the proposed becomes a practical procedure for mapping floods in small-size water bodies due to it involves the use of free access data and software, such as Sentinel-2 MSI and SNAP, holding the advantages of unsupervised algorithms.

Future studies can evaluate this combined methodology in other situations in addition to agricultural lands, for instance, agricultural areas plus settlements and mixed water plus other vegetation conditions, among others. Another analysis could be performed, considering spectral mixture in pixels, different machine learning techniques based on unsupervised classification, and the application of other quantitative methods to accuracy assessment in flood mapping.

Author Contributions: Conceptualization, L.L.; methodology L.L. and A.M.-G.; software, L.L.; validation, L.L. and A.M.-G.; formal analysis, L.L.; investigation, L.L. and A.M.-G.; resources, L.L. and A.M.-G.; data curation, L.L. and A.M.-G.; writing—original draft preparation, L.L.; writing— review and editing, L.L.; visualization, L.L.; supervision, L.L. and A.M.-G.; project administration, A.M.-G.; funding acquisition, A.M.-G. All authors have read and agreed to the published version of the manuscript.

Funding: This research received no external funding.

Acknowledgments: This work has been supported by the Regional Government of Castilla y León 2014–2020, European Social Fund Operational Programme, and the GEAPAGE research group (Environmental Geomorphology and Geological Heritage) of the University of Salamanca.

Conflicts of Interest: The authors declare no conflict of interest.

Appendix A

Results obtained in the accuracy assessment of the flood mapping methods by applying the producer accuracy, user accuracy, overall accuracy, and Kappa coefficient.

References

1. European Court of Auditors—ECA. *Floods Directive: Progress in Assessing Risks, While Planning and Implementation Need to Improve;* Special Report No 25, 2018; ECA: Luxembourg, 2019. [CrossRef]
2. Alfieri, L.; Burek, P.; Feyen, L.; Forzieri, G. Global warming increases the frequency of river floods in Europe. *Hydrol. Earth Syst. Sci.* **2015**, *19*, 2247–2260. [CrossRef]
3. Sofia, G.; Nikolopoulos, E.I. Floods and rivers: A circular causality perspective. *Sci. Rep.* **2020**, *10*, 5175. [CrossRef]
4. Dong, Z.; Wang, G.; Amankwah, S.O.Y.; Wei, X.; Hu, Y.; Feng, A. Monitoring the summer flooding in the Poyang Lake area of China in 2020 based on Sentinel-1 data and multiple convolutional neural networks. *Int. J. Appl. Earth Obs. Geoinf.* **2021**, *102*, 102400. [CrossRef]
5. Caballero, I.; Ruiz, J.; Navarro, G. Sentinel-2 satellites provide near-real time evaluation of catastrophic floods in the West Mediterranean. *Water* **2019**, *11*, 2499. [CrossRef]
6. Goffi, A.; Stroppiana, D.; Brivio, P.A.; Bordogna, G.; Boschetti, M. Towards an automated approach to map flooded areas from Sentinel-2 MSI data and soft integration of water spectral features. *Int. J. Appl. Earth Obs. Geoinf.* **2020**, *84*, 101951. [CrossRef]
7. Bijeesh, T.V.; Narasimhamurthy, K.N. Surface water detection and delineation using remote sensing images: A review of methods and algorithms. *Sustain. Water Resour. Manag.* **2020**, *6*, 68. [CrossRef]
8. Sekertekin, A. Potential of global thresholding methods for the identification of surface water resources using Sentinel-2 satellite imagery and normalized difference water index. *J. Appl. Remote Sens.* **2019**, *13*, 044507. [CrossRef]
9. European Space Agency. *SENTINEL-2 User Handbook;* European Space Agency: Paris, France, 2015; Available online: https://sentinel.esa.int/documents/247904/685211/Sentinel-2_User_Handbook (accessed on 30 July 2021).
10. Jiang, W.; Ni, Y.; Pang, Z.; Li, X.; Ju, H.; He, G.; Lv, J.; Yang, K.; Fu, J.; Qin, X. An Effective Water Body Extraction Method with New Water Index for Sentinel-2 Imagery. *Water* **2021**, *13*, 1647. [CrossRef]
11. Du, Y.; Zhang, Y.; Ling, F.; Wang, Q.; Li, W.; Li, X. Water Bodies' Mapping from Sentinel-2 Imagery with Modified Normalized Difference Water Index at 10-m Spatial Resolution Produced by Sharpening the SWIR Band. *Remote Sens.* **2016**, *8*, 354. [CrossRef]
12. Landuyt, L.; Verhoest, N.E.C.; Van Coillie, F.M.B. Flood Mapping in Vegetated Areas Using an Unsupervised Clustering Approach on Sentinel-1 and -2 Imagery. *Remote Sens.* **2020**, *12*, 3611. [CrossRef]
13. Hu, P.; Zhang, Q.; Shi, P.; Chen, B.; Fang, J. Flood-induced mortality across the globe: Spatiotemporal pattern and influencing factors. *Sci. Total Environ.* **2018**, *643*, 171–182. [CrossRef]
14. Zhou, Y.; Dong, J.; Xiao, X.; Xiao, T.; Yang, Z.; Zhao, G.; Zou, Z.; Qin, Y. Open Surface Water Mapping Algorithms: A Comparison of Water-Related Spectral Indices and Sensors. *Water* **2017**, *9*, 256. [CrossRef]

15. Yue, H.; Li, Y.; Qian, J.; Liu, Y. A new accuracy evaluation method for water body extraction. *Int. J. Remote Sens.* **2020**, *41*, 7311–7342. [CrossRef]
16. Wang, H.; Chu, Y.; Huang, Z.; Hwang, C.; Chao, N. Robust, Long-term Lake Level Change from Multiple Satellite Altimeters in Tibet: Observing the Rapid Rise of Ngangzi Co over a New Wetland. *Remote Sens.* **2019**, *11*, 558. [CrossRef]
17. McFeeters, S.K. The use of the Normalized Difference Water Index (NDWI) in the delineation of open water features. *Int. J. Remote Sens.* **1996**, *17*, 1425–1432. [CrossRef]
18. Xu, H. Modification of normalised difference water index (NDWI) to enhance open water features in remotely sensed imagery. *Int. J. Remote Sens.* **2006**, *27*, 3025–3033. [CrossRef]
19. Feyisa, G.L.; Meilby, H.; Fensholt, R.; Proud, S.R. Automated Water Extraction Index: A new technique for surface water mapping using Landsat imagery. *Remote Sens. Environ.* **2014**, *140*, 23–35. [CrossRef]
20. Rokni, K.; Ahmad, A.; Selamat, A.; Hazini, S. Water Feature Extraction and Change Detection Using Multitemporal Landsat Imagery. *Remote Sens.* **2014**, *6*, 4173–4189. [CrossRef]
21. Otsu, N. Threshold selection method from gray-level histograms. *IEEE Trans. Syst. Man Cybern. SMC* **1979**, *9*, 62–66. [CrossRef]
22. Huang, X.; Xie, C.; Fang, X.; Zhang, L. Combining Pixel-and Object-Based Machine Learning for Identification of Water-Body Types from Urban High-Resolution Remote-Sensing Imagery. *IEEE J. Sel. Top. Appl. Earth Obs. Remote Sens.* **2015**, *8*, 2097–2110. [CrossRef]
23. Zhang, Q.; Zhang, P.; Hu, X. Unsupervised GRNN flood mapping approach combined with uncertainty analysis using bi-temporal Sentinel-2 MSI imageries. *Int. J. Digit. Earth* **2021**, *14*, 1561–1581. [CrossRef]
24. Bhangale, U.; More, S.; Shaikh, T.; Patil, S.; More, N. Analysis of Surface Water Resources Using Sentinel-2 Imagery. *Procedia Comput. Sci.* **2020**, *171*, 2645–2654. [CrossRef]
25. Acharya, T.D.; Subedi, A.; Lee, D.H. Evaluation of Water Indices for Surface Water Extraction in a Landsat 8 Scene of Nepal. *Sensors* **2018**, *18*, 2580. [CrossRef] [PubMed]
26. Yang, X.; Zhao, S.; Qin, X.; Zhao, N.; Liang, L.; Kuenzer, C.; Mishra, D.R.; Huang, W.; Thenkabail, P.S. Mapping of Urban Surface Water Bodies from Sentinel-2 MSI Imagery at 10 m Resolution via NDWI-Based Image Sharpening. *Remote Sens.* **2017**, *9*, 596. [CrossRef]
27. Lanaras, C.; Bioucas-Dias, J.; Galliani, S.; Baltsavias, E.; Schindler, K. Super-resolution of Sentinel-2 images: Learning a globally applicable deep neural network. *ISPRS J. Photogramm. Remote Sens.* **2018**, *146*, 305–319. [CrossRef]
28. Gašparović, M.; Jogun, T. The effect of fusing Sentinel-2 bands on land-cover classification. *Int. J. Remote Sens.* **2017**, *39*, 822–841. [CrossRef]
29. Brodu, N. Super-Resolving Multiresolution Images with Band-Independent Geometry of Multispectral Pixels. *IEEE Trans. Geosci. Remote Sens.* **2017**, *55*, 4610–4617. [CrossRef]
30. Confederación Hidrográfica del Duero—CHD. Demarcación Hidrográfica del Duero—Revisión y Actualización de la Evaluación Preliminar del Riesgo de Inundación 2o Ciclo. Available online: https://www.chduero.es/documents/20126/7047 46/0_REVISION_EPRI_DUERO_MEMORIA.pdf/d42faee0-dd95-f352-f820-b150904a7131?t=1566469961152 (accessed on 20 August 2021).
31. Lombana, L.; Martínez-Graña, A. Hydrogeomorphological analysis for hydraulic public domain definition: Case study in Carrión River (Palencia, Spain). *Environ. Earth Sci.* **2021**, *80*, 193. [CrossRef]
32. Lombana, L.; Martínez-Graña, A. Multiscale Hydrogeomorphometric Analysis for Fluvial Risk Management. Application in the Carrión River, Spain. *Remote Sens.* **2021**, *13*, 2955. [CrossRef]
33. Agencia Estatal de Meteorología—AEMET. Año Hidrológico 2019–2020. Available online: https://www.miteco.gob.es (accessed on 10 July 2021).
34. Yan, L.; Roy, D.P.; Zhang, H.; Li, J.; Huang, H. An Automated Approach for Sub-Pixel Registration of Landsat-8 Operational Land Imager (OLI) and Sentinel-2 Multi Spectral Instrument (MSI) Imagery. *Remote Sens.* **2016**, *8*, 520. [CrossRef]
35. Rouse, J.; Haas, R.; Schell, J.; Deering, D. Monitoring Vegetation Systems in the Great Plains with Erts. *NASA Spec. Publ.* **1974**, *351*, 309.
36. Huete, A.R. A soil-adjusted vegetation index (SAVI). *Remote Sens. Environ.* **1988**, *25*, 295–309. [CrossRef]
37. Lacaux, J.P.; Tourre, Y.M.; Vignolles, C.; Ndione, J.A.; Lafaye, M. Classification of ponds from high-spatial resolution remote sensing: Application to Rift Valley Fever epidemics in Senegal. *Remote Sens. Environ.* **2007**, *106*, 66–74. [CrossRef]
38. Rouibah, K.; Belabbas, M. Applying Multi-Index Approach from Sentinel-2 Imagery to Extract Urban Areas in Dry Season (Semi-Arid Land in North East Algeria). *Rev. Teledetec.* **2020**, *56*, 89–101. [CrossRef]
39. Landini, G. Auto Threshold. Available online: https://imagej.net/plugins/auto-threshold (accessed on 8 January 2022).
40. Huang, L.K.; Wang, M.J.J. Image thresholding by minimizing the measures of fuzziness. *Pattern Recognit.* **1995**, *28*, 41–51. [CrossRef]
41. Tsai, W.H. Moment-preserving thesolding: A new approach. *Comput. Vis. Graph. Image Process.* **1985**, *29*, 377–393. [CrossRef]
42. Prewitt, J.M.S.; Mendelsohn, M.L. The Analysis of Cell Images. *Ann. N. Y. Acad. Sci.* **1966**, *128*, 1035–1053. [CrossRef]
43. Ridler, T.W.; Calvard, S. Picture Thresholding Using an Iterative Selection Method. *IEEE Trans. Syst. Man Cybern.* **1978**, *8*, 630–632. [CrossRef]
44. Doyle, W. Operations Useful for Similarity-Invariant Pattern Recognition. *J. ACM (JACM)* **1962**, *9*, 259–267. [CrossRef]

45. Li, C.H.; Tam, P.K.S. An iterative algorithm for minimum cross entropy thresholding. *Pattern Recognit. Lett.* **1988**, *19*, 771–776. [CrossRef]
46. Kapur, J.N.; Sahoo, P.K.; Wong, A.K.C. A new method for gray-level picture thresholding using the entropy of the histogram. *Comput. Vis. Graph. Image Process.* **1985**, *29*, 273–285. [CrossRef]
47. Shanbhag, A.G. Utilization of Information Measure as a Means of Image Thresholding. *CVGIP Graph. Models Image Process.* **1994**, *56*, 414–419. [CrossRef]
48. Glasbey, C.A. An Analysis of Histogram-Based Thresholding Algorithms. *CVGIP Graph. Models Image Process.* **1993**, *55*, 532–537. [CrossRef]
49. Zack, G.W.; Rogers, W.E.; Latt, S.A. Automatic measurement of sister chromatid exchange frequency. *J. Histochem. Cytochem.* **1977**, *25*, 741–753. [CrossRef]
50. Yen, J.C.; Chang, F.J.; Chang, S. A New Criterion for Automatic Multilevel Thresholding. *IEEE Trans. Image Process.* **1995**, *4*, 370–378. [CrossRef]

 **agronomy**

Article

Food and Medicinal Uses of Ancestral Andean Grains in the Districts of Quinua and Acos Vinchos (Ayacucho-Peru)

Roberta Brita Anaya [1], Eusebio De La Cruz [2], Luz María Muñoz-Centeno [3,*], Reynán Cóndor [1], Roxana León [4] and Roxana Carhuaz [1]

1   Faculty of Biological Sciences, National University of San Cristobal de Huamanga, Ayacucho 05001, Peru; roberta.anaya@unsch.edu.pe (R.B.A.); reynan.condor@unsch.edu.pe (R.C.); roxana.carhuaz@unsch.edu.pe (R.C.)
2   Chemical Engineering and Metallurgy, National University of San Cristobal de Huamanga, Ayacucho 05001, Peru; eusebio.delacruz@unsch.edu.pe
3   Department of Botany and Plant Physiology, University of Salamanca, 37008 Salamanca, Spain
4   Faculty of Health Sciences, National University of San Cristobal de Huamanga, Ayacucho 05001, Peru; roxana.leon@unsch.edu.pe
*   Correspondence: luzma@usal.es

Abstract: Andean grains are key elements in the construction of family production systems. These seeds speak of the history of a people, their customs and ancestral knowledge. The general objective of the work was to evaluate the food use, crop management and traditional knowledge about the medicinal use of ancestral Andean grains among the inhabitants of the districts of Quinua and Acos Vinchos (Ayacucho-Peru). Basic descriptive research, carried out by means of convenience sampling, the sample size determined by the Law of Diminishing Returns, after signing an informed consent form. Semi-structured individual interviews were applied to 96 informants. A total of 96.9% of the informants reported that they obtained *quinoa* grain from their own crops, and 24.0% obtained *achita* grain that they sowed directly on their land; no *cañihua* was cultivated. A total of 58.3% use *quinoa* and *achita* in their diet. The variability of the food use of ancestral grains, specifically *quinoa* and *achita*, constitute a natural source of vegetable protein of high nutritional value, which represents one of the main foods of the inhabitants of Quinua and Acos Vinchos. Traditional medicine derived from the ancestral knowledge of Andean grains is barely preserved, but this is not the case for other medicinal plants in the area, as this knowledge is still preserved.

Keywords: *Chenopodium quinoa* Willd (quinoa); *Amaranthus caudatus* L. (achita); *Chenopodium pallidicaule* Aellen (cañihua); ethnobotany; Andean grains; food uses; medicinal uses; edaphic resilience and crops

Citation: Anaya, R.B.; De La Cruz, E.; Muñoz-Centeno, L.M.; Cóndor, R.; León, R.; Carhuaz, R. Food and Medicinal Uses of Ancestral Andean Grains in the Districts of Quinua and Acos Vinchos (Ayacucho-Peru). *Agronomy* **2022**, *12*, 1014. https://doi.org/10.3390/agronomy12051014

Academic Editor: Jirui Wang

Received: 21 February 2022
Accepted: 21 April 2022
Published: 23 April 2022

Publisher's Note: MDPI stays neutral with regard to jurisdictional claims in published maps and institutional affiliations.

1. Introduction

Andean grains have a high value, as they are key elements in the construction of Andean family production systems. These seeds speak of the history of a people, their habits and resistance. As Arias states [1], defending Andean grains is like protecting real possibilities of an independence that defies the market and money. Currently, the relationship created so many years ago between the inhabitants of this area and their grains is being lost, resulting in the loss of this genetic material and the ancestral knowledge related to its management and uses. Plants are present in all areas of human activity, and ethnobotany combines the application of various disciplines to understand the relationship between a culture and the plant world that surrounds it [2]. It is a science that links botany and anthropology and also draws on other disciplines such as ecology, pharmacognosy, medicine, nutrition, agronomy, sociology, linguistics and history. The interaction between humans and plants is one of the aspects that indicates how a culture relates to the natural environment, and ethnobotany is therefore situated within ethnoecology. Ethnoecology approaches the study of traditional cultures not as obsolete systems but as a fraction of society that possesses valuable resilient ecological wisdom. According to Aceituno's review, a distinction is made within ethnobotany between the cognitive stream, concerned with how humans perceive and classify plants, and the utilitarian stream, concerned with how they use and manage them. The former uses methods from the social sciences, while the latter uses an approach from the natural sciences [2].

The FAO, the Food and Agriculture Organization of the United Nations, considers quinoa (*Chenopodium quinoa* Willd.) to be one of the most promising foods for humanity, along with achita (*Amaranthus caudatus* L.) and cañihua (*Chenopodium pallidicaule* Aellen), not only because of its great beneficial properties and multiple uses, but also because it is considered an alternative to solve serious human nutrition problems, especially because of its ability to grow in adverse environments [3,4]. Food security has been affected by many factors, including the growing demand for basic agricultural commodities for production, the disappearance of traditional varieties that are highly resilient to climate change, the increase in the world's hungry population, inadequate food and nutrition policies in the country, and the misallocation of domestic and foreign aid, all of which have contributed to malnutrition among the most vulnerable populations and, consequently, to deficiencies in the integral development of the human being. The recent COVID-19 pandemic has exacerbated this situation. Therefore, if the world is to cope with pandemics without stopping food production and distribution, the cultivation of staple food crops, which are basically grain crops, will have to be increased. Crops that are drought, pest and disease-resistant, nutritionally superior, and health-promoting will have to be selected. These include *quinoa, achita* and *cañihua*. In addition, future grain crops must be produced in a profitable, sustainable and environmentally friendly way [5].

This ethnobotanical study has focused on the ancestral medicinal and nutritional knowledge of Andean grains, due to their great benefits as a protein and functional food that reduces the risk of various diseases. *Quinoa*, in particular, is considered one of the most important grains of the 21st century, sometimes referred to as a "superfood" It is high in protein, carbohydrates, polyunsaturated fatty acids, vitamins, fiber and minerals and is gluten-free. It also has high concentrations of bioactive compounds such as phenolic acids, flavonoids, bioactive peptides, saponins, phytosteroids and phytosterols. [6–9]. These bioactive compounds provide quinoa with anti-diabetic, antioxidant, anti-inflammatory, immunoregulatory, antimicrobial, anti-obesity, anti-cancer and heart-healthy properties [10,11].

The Andean region comprises one of the eight largest centers of cultivated plant domestication in the world, giving rise to one of the most genetically diverse and sustainable agricultural systems in the world. *Quinoa* is a good example of this, showing a great diversity of genotypes and wild progenitors in Peru and Bolivia; this crop groups around 3000 varieties. Currently, Peru is the world's leading producer of quinoa, surpassing Bolivia [6,12–14].

With the results of this research, we intend to highlight the great importance of these natural resources and their associated knowledge in order to recover their mass consumption in all social strata, especially in those most in need, and thus reduce the high levels of anemia and malnutrition in which a large part of the population of this area is immersed. Sustainable cultivation based on the local farmers' own grains and therefore on traditional varieties would provide alternative crops for self-sufficiency and trade that would make the population less dependent on introduced products with low nutritional levels, which a priori are cheaper, but which in the long run produce irreparable consequences such as chronic malnutrition in these Andean peoples [13,15].

Our research hypothesis is to study whether traditional knowledge on the use and management of Andean grains is still valid or has been lost due to globalization. The main aim of this research is to recover, evaluate and enhance local knowledge about ancestral Andean grains in the districts of Quinua and Acos Vinchos (Peru). To this end, the specific objectives of this study are (a) to collect data on the socio-economic characteristics of the informants; (b) to evaluate the information on the nutritional use of these grains; (c) to determine the type of management of these crops; (d) to collect traditional knowledge on the medicinal use of these grains and assess its validity.

2. Methodology

2.1. Study Area

The study was carried out in the districts of Quinua (3280 m asl) and Acos Vinchos (2840 m asl) in the province of Huamanga, department of Ayacucho-Peru (Figure 1). The province of Huamanga is located in the Peruvian Andes and presents a very diverse physiography with a particular flora and vegetation. Pulgar [16] defined and described the existence of eight natural regions in Peru based on geographical location and natural indicator vegetation. Tapia [17] has added agronomic variables to this classification into natural regions and proposes a classification into agro-ecological zones, which he complements with local peasant knowledge, crops, varieties and cultivation practices. According to this zoning of agro-ecological zones in the Peruvian Andes, the study districts are in the Quechua region (between 2300 and 3500 m above sea level). The climate in this region is temperate, varying according to latitude. Temperatures can fluctuate between an average annual temperature of 11 to 16 °C, with maximums between 22 and 29 °C and minimums between 7 and 4 °C during the winter (May to August); humidity rates are between 500 to 1200 mm of precipitation, increasing from south to north. These conditions make it possible to differentiate the Quechua region into three zones: arid, semi-arid and semi-humid. In 2019, the year of sampling, the average rainfall in the study area was 893 mm; the rainy season covered about 7 months (October–April), with an average total accumulation of 74 mm [14].

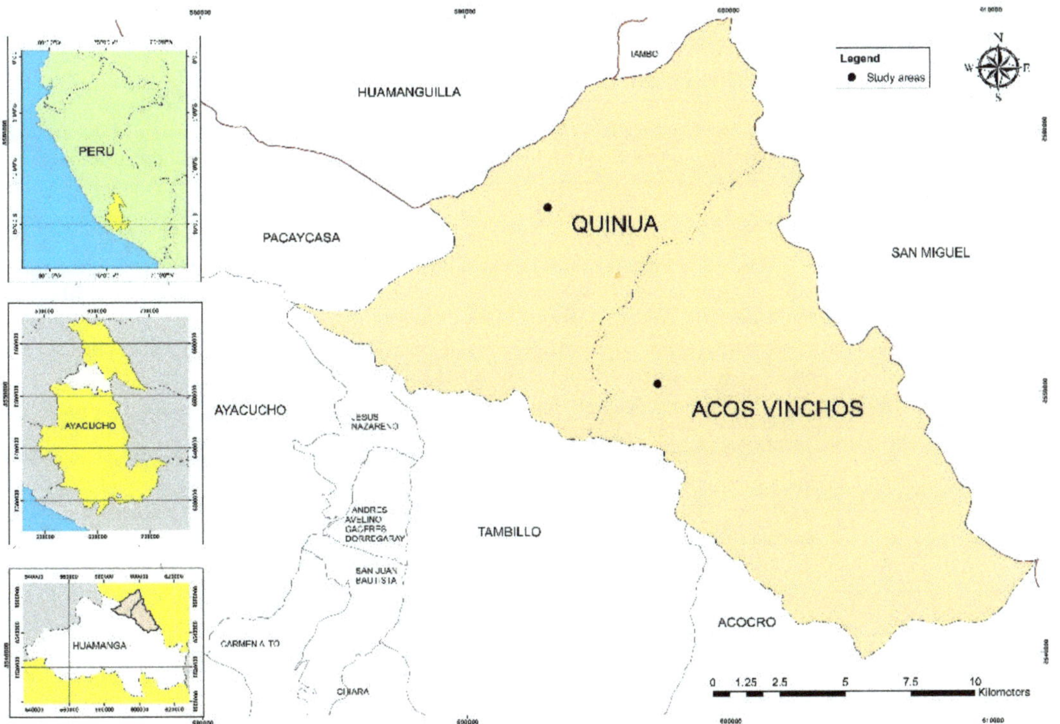

Figure 1. Location map of the study areas: Quinua and Acos Vinchos, province of Huamanga, Region Ayacucho-Peru.

2.2. Ethnobotanical Data Collection

Ethnobotanical fieldwork was conducted between May and June 2019 in the urban and surrounding areas of the districts of Quinua and Acos Vinchos. We interviewed 96 informants (40 women and 56 men) selected purposively, looking for "experts" within the local population. [18,19]. By "local population" we mean people who have lived in the area since birth or who migrated more than 20 years ago. By "experts" we mean people who have kept part of the cultural richness related to plants in their memories or customs [20]. First of all, older people were interviewed because they are the ones who have the most ancestral knowledge about the use of plants in their area [21]. Informant age ranged from 27 to 75 years old, with a majority being between 45 and 60 years old.

The sample size is determined by the Law of Diminishing Returns and is therefore not defined in advance [22]. The surveys were conducted in Spanish and in Quechua (the native language), using previously validated questionnaires in which the same questions were always asked about these Andean grains. The basic interview was a one-to-one meeting.

After direct and participatory observation, information was collected through semi-structured individual interviews and field interviews. In the semi-structured interviews, the informants were conversed within a relaxed manner so that they felt comfortable and could transmit their knowledge in a simple and sincere way. Beneath this apparent simplicity, it is necessary to control how the interview is conducted, how the questions are constructed and presented, and how responses are recorded [23]. For this purpose, the previously validated questionnaire was used. Emphasis was placed on recording whether the uses, practices or beliefs are traditional, whether the informant has ever put them into practice or has only heard about them, and whether the use of these plants is still in use today, in order to record ancestral knowledge and study changes in their use and management. In addition to interviews in the urban area, interviews were also carried out in the houses around the fields and in the countryside.

Plants and seeds were collected in situ. The vouchers collected during the walks were dried and stored in the Herbarium Huamangensis of the Faculty of Biological Sciences of the National University of San Cristóbal de Huamanga (Ayacucho-Perú) and the seeds in the Biochemistry Laboratory of the National University of San Cristóbal de Huamanga [22,24]. The species were determined by Blga. Laura Aucasime Medina.

2.3. Data Analysis

The ethnobotanical information obtained in the surveys carried out with the inhabitants of Quinua and Acos Vinchos was organized in a Microsoft Office Excel 2019 database. The data analysis was carried out by obtaining descriptive statistics. Frequencies and percentages were used to elaborate statistical tables and graphs. Statistical processing and analysis were carried out in R version 3.6.3 [25].

3. Results

The results of the interviews were grouped into six frequency tables based on the questions asked about socio-economic profile of the surveyed villagers (Tables 1 and 2), food use of Andean grains (Table 3), agronomic management of Andean grains (Table 4), medicinal use of Andean grains (Table 5) and health care (Table 6).

With regard to the socio-economic profile (Table 1), the surveyed inhabitants of Quinua cover an age range between 38 and 65 years, with a mean of 53 years, SD of 6.74; most of them correspond to the age group between 45 and 60 years with 71.4%, and those who participated in greater proportion were men with a 20% difference. Regarding the level of education, 71% are between those with no education and those with only primary education; however, 57% practice agriculture; 21.4% are also housewives. A total of 97.6% are natives and live in Quinua, 83.4% of whom are between 45 and 65 years old, i.e., adults and older people who still preserve their ancestral knowledge.

Table 1. Socio-economic profile of the surveyed villagers in *Quinoa*.

Variable	Frequency (*n* = 42)	%
Age (min = 38, max = 65, mean = 52.93, SD = 6.74)		
<45 years	5	11.9
45–60 years	30	71.4
>60 years	7	16.7
Gender		
Female	17	40.5
Male	25	59.5
Education level		
Higher Education	1	2.4
Secondary school	8	19.0
Basic education	14	33.3
No schooling	16	38.1
No answer	3	7.1
Present occupation		
Merchant	3	7.1
Farmer	24	57.1
Housewife	2	4.8
Civil servant and farmer	1	2.4
Farmer and housewife	9	21.4
No answer	3	7.1
Birth place		
Lima	1	2.4
Quinua	41	97.6
Living region		
Quinua	41	97.6
Quinua-Chihuanpampa	1	2.4
Years of residence (min = 15, máx = 65, mean = 51.21, SD = 9.89)		
<45 years	7	16.7
45–60 years	28	66.7
>60 years	7	16.7

Table 2 shows the socio-economic profile of the surveyed inhabitants of the district of Acos Vinchos, with an age range between 27 and 75 years, with an average age of 55 years, SD of 9.25, the largest age group being between 45 and 60 years with 63%, followed by older adults with 24.1%, and those who participated in the highest proportion were men with 59.3%, followed by women with 37%. Regarding the level of education, 90.8% are among those who have no education (51.9%) and the difference is that they have only primary education (38.9%); 61.1% of them are farmers, and 27.8% of them are housewives and traders. A total of 90.7% are locals from the area, and almost 100% live in Acos Vinchos and Huinchupata (district town). As shown in Table 2, it is the older adults who participated most in the interviews.

Regarding the food use of Andean grains (Table 3), 58.3% use *quinoa* and achita, and only 41.7% use *quinoa*. Paradoxically, in Acos Vinchos, they use more *quinoa* (66.7%) than in the district of Quinua (9.5%). The vast majority (87.5%) of *quinoa* and *achita* grains are used for food, and only 12.5% also use the tender leaves of *quinoa*. In both districts, these grains are consumed almost entirely (96.9%) at breakfast, as soups and seconds. A total of 66.7% cultivate only for consumption and 31.3% for commercialization. A total of 65.6% say they know which plants are more nutritious, compared to 28.1% who say they do not know.

Based on the results and in situ verification of Andean grain crops, most of the inhabitants grow *quinoa* for their own consumption and trade, and to a lesser extent *achita*. No cultivation of *cañihua* was found, confirming that this species thrives at higher altitudes than the areas visited, as it is typical of high altitudes such as Puno, where it reaches altitudes of around 3800 m above sea level [4,16].

Table 2. Socio-economic profile of the inhabitants surveyed in Acos Vinchos.

Variable	Frequency (*n* = 54)	%
Age (min = 27, max = 75, mean = 53.49, SD = 9.25)		
<45 years	6	11.1
45–60 years	34	63.0
>60 years	13	24.1
No answer	1	1.9
Gender		
Female	20	37.0
Male	32	59.3
No answer	2	3.7
Education level		
Secondary school	3	5.6
Basic education	21	38.9
No schooling	28	51.9
No answer	2	3.7
Present occupation		
Merchant	1	1.9
Farmer	33	61.1
Housewife	4	7.4
Farmer and housewife	12	22.2
Merchant and farmer	3	5.6
No answer	1	1.9
Birth place		
Acosvinchos	49	90.7
Ayacucho	1	1.9
Chaupirara	1	1.9
Chaupirara-Acosvinchos	1	1.9
Huinchupata	1	1.9
Lima	1	1.9
Living region		
Acosvinchos	53	98.1
Huinchupata	1	1.9
Years of residence (min = 47, max = 70.5, mean = 52.31, SD = 10.67)		
<45 years	7	13.0
45–60 years	32	59.3
>60 years	10	18.5
No answer	5	9.3

The results on the crop management of Andean grains by the inhabitants of Quinua and Acos Vinchos are shown in Table 4. A total of 96.9% of the informants' report obtaining *quinoa* grain from their own crops and 24.0% from the *achita* grain that they sow directly on their land. *Cañihua* is not cultivated in these districts because of the altitude required for its optimal development, given that it is a species native to the circum-lacustrine zone of Lake Titicaca, shared between Bolivia and Peru in the altiplano region, mainly at altitudes above 3800 m [16]. In terms of production, 39.6% of the informants say that the yield of the *quinoa* crop is good, and 56.3% that it is fair; in the case of *achita*, 6.3% say it is good and 18.8% say it is fair. A total of 93.8% of the informants use their own seeds for replanting.

Concerning the products used for the treatment and prevention of plant diseases caused by fungi and other pathogens, 85.4% of informants use chemical products and 10.4% use homemade products, mainly based on a combination of cupric sulphate and hydrated lime, dissolved separately in water in nonmetallic containers and without heating the ingredients.

Table 3. Food use and production of ancestral Andean grains in the districts of Quinua and Acos Vinchos. Ayacucho-Peru.

		Quinua		Acos Vinchos	
		Frequency	**%**	**Frequency**	**%**
Types of ancient Andean grains used for food	*Quinoa*	4	9.5	36	66.7
	Quinoa y *Achita*	38	90.5	18	33.3
	Total	42	100.0	54	100.0
Parts of the plant used for food	Grain	36	85.7	48	88.9
	Grain and young leaves	6	14.3	6	11.1
	Total	42	100.0	54	100.0
Form of consumption	Do not know	1	2.4	2	3.7
	Breakfast, main course and soups	41	97.6	52	96.3
	Total	42	100.0	54	100.0
Production for consumption only	Yes	28	66.7	36	66.7
	No	11	26.2	17	31.5
	Do not know	3	7.1	1	1.9
	Total	42	100.0	54	100.0
Production for trade	Yes	11	26.2	19	35.2
	No	27	64.3	35	64.8
	Do not know	4	9.5	0	0
	Total	42	100.0	54	100
They know which plants are the most nutritious	Yes	34	81.0	29	53.7
	No	6	14.3	21	38.9
	Do not know	2	4.8	4	7.4
	Total	42	100.0	54	100.0

Table 4. Crop management of ancestral Andean grains in the districts of Quinua and Acos Vinchos. Ayacucho-Peru.

		Quinua		Acos Vinchos	
		Frequency	**%**	**Frequency**	**%**
Informants who crop quinoa	Sowed	39	92.9	54	100.0
	No answer	3	7.1	0	0.0
	Total	42	100.0	54	100
Informants who crop achita	Sowed	8	19.0	15	27.8
	Do not know	34	81.0	39	72.2
	Total	42	100	54	100
Quinoa production	Good	18	42.9	20	37.0
	Fair	20	47.6	34	63.0
	Do not know	4	9.5	0	0.0
	Total	42	100.0	54	100.0
Achita production	Good	4	9.5	2	3.7
	Fair	5	11.9	13	24.1
	Do not know	33	78.6	39	72.2
	Total	42	100.0	54	100.0
Origin of seeds for cultivation	Do not know	2	4.8	1	1.9
	Purchased	2	4.8	1	1.9
	Home-grown	38	90.5	52	96.3
	Total	42	100.0	54	100.0
Substances used to treat plant diseases	Chemicals	37	88.1	45	83.3
	Homemade products	2	4.8	8	14.8
	Do not know	3	7.1	1	1.9
	Total	42	100.0	54	100.0

Dietary consumption of these Andean grains reduces the risk of suffering from certain diseases, thanks to their nutritional and functional characteristics. For example, quinoa has anti-diabetic, antioxidant, anti-inflammatory, immunoregulatory and anticarcinogenic properties, among others [10]. This is the reason why we wanted to complete our research by compiling the traditional knowledge of the medicinal uses of Andean grains of the inhabitants of the area (Table 5). The results show that only 13.5% recall any medicinal use of *quinoa*. A total of 76.9% use *quinoa* as a purgative and 23% for colic. The parts of the plant used for medicinal purposes are the grains (46.2%), the leaves (30.8%) and a mixture of leaves and grains (23.1%), which are harvested when the plant is ripe. It is administered as a wash (38.5%), as an infusion (30.8%), and a mixture of both, infusion and wash (23.1%). The interviewees do not remember how these infusions and washes are prepared and administered, nor do they provide information on the proportions (quantity of leaves or grains) to be included in the preparation, the cooking or maceration time, the dosage and frequency of administration. A total of 84.6% indicated that mixing *quinoa* with other plants could cause adverse effects, and 84.6% also stated that at high doses it could be toxic, without mentioning the reasons. Only 15.4% indicated that the medicinal use of *quinoa* is maintained, while 46.2% confirmed that it is disappearing.

Table 5. Medicinal use of ancestral Andean grains in the districts of Quinua and Acos Vinchos. Ayacucho-Peru.

		Quinua		Acos Vinchos	
		Frequency	%	Frequency	%
Informants who use quinoa and achita as medicinal plants	Yes	7	16.7	6	11.1
	No	35	83.3	48	88.9
	Total	42	100.0	54	100.0
Ailments for which quinoa is used	Purgante	7	100.0	3	50.0
	Colics-purgante	0	0.0	3	50.0
	Total	7	100	6	100.0
Part of the quinoa plant used	Grain	3	42.9	3	50.0
	Leaves	3	42.9	1	16.7
	Leaves and grains	1	14.3	2	33.3
	Total	7	100.0	6	100.0
Forms of administration of quinoa	Infusion	3	42.9	1	16.7
	Infusion and wash	1	14.3	2	33.3
	Wash	3	42.9	2	33.3
	Do not know	0	0.0	1	16.7
	Total	7	100.0	6	100.0
Adverse effects of quinoa mixed with other plants	Yes	5	71.4	6	100.0
	No	2	28.6	0	0.0
	Total	7	100.0	6	100
Quinoa toxicity in high doses	Yes	6	85.7	5	83.3
	No	1	14.3	1	16.7
	Total	7	100.0	6	100.0
Frequency of medicinal use of quinoa	Common use	1	14.3	1	16.7
	Disappearing	5	71.4	1	16.7
	Do not know	1	14.3	4	66.7
	Total	7	100.0	6	100.0

Table 6. Health care and use of medicinal plants in the districts of Quinua and Acos Vinchos. Ayacucho-Peru.

		Quinua		Acos Vinchos	
		Frequency	%	Frequency	%
Places and practices carried out for health care	Health center	5	11.9	5	9.3
	Healer and use medicinal plants	9	21.4	2	3.1
	Health centre, healer and use medicinal plants	28	66.7	47	87.0
	Total	42	100	54	100
Places where medicinal plants are obtained	Purchased	3	7.1	3	5.6
	Collected from the field	15	35.7	20	37.0
	Grown in the orchard	11	26.2	8	14.8
	Collected from the field and orchard	13	31.0	22	40.7
	Do not use	0	0.0	1	1.9
	Total	42	100.0	54	100.0
The use of medicinal plants is common in the family	Yes	39	92.9	49	90.7
	Does not use	3	7.1	5	9.3
	Total	42	100.0	54	100.0
The use of medicinal plants is recommended	Yes	39	92.9	54	100.0
	Do not know	3	7.1	0	0.0
	Total	42	100.0	54	100
Informants who know healers in their community	Yes	22	52.4	15	27.8
	No	17	40.5	39	72.2
	Do not know	3	7.1	0	0
	Total	42	100.0	54	100.0

Figure 2 shows that those interviewed between 45 and 60 years of age have the most knowledge about the medicinal use of these grains. A total of 11.9% of those were interviewed in Quinoa, and 7.5% in Acos Vinchos. This traditional knowledge is scarcely preserved among adults over 45 years of age, but it has been alarmingly lost among those under 45 years of age.

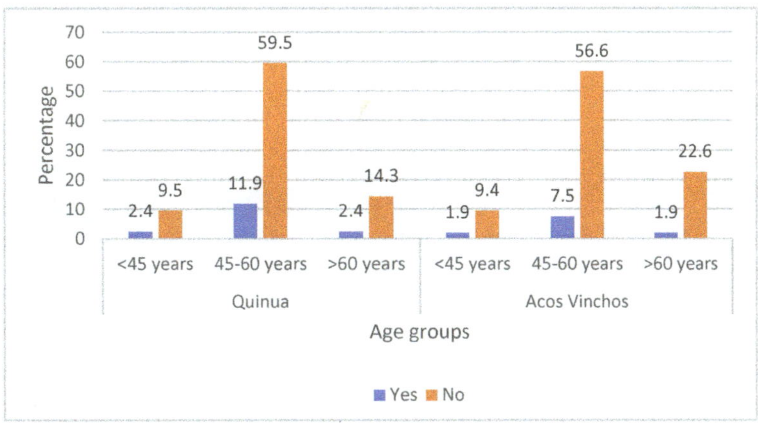

Figure 2. Response to the use of quinoa and/or cañihua plants as medicinal plant according to age groups.

While we asked about the traditional medicinal uses of Andean grains, we were also interested in the use of other medicinal plants and health care (Table 6). A total of 66.7% of the villagers interviewed from Quinua go to health center, use medicinal plants and visit the healer, and that number is 87.0% in Acos Vinchos. A total of 52.4% of those interviewed in Quinua know a healer, and in Acos Vinchos, that number is 27.8%. They recognize the healers as having traditional knowledge about the medicinal use of plants.

Figure 3 shows that informants in both districts go to health centers but also use medicinal plants and visit healers to satisfy their primary health care needs. It is noteworthy that in the district of Quinua, there are still informants older than 45 years who do not go to the health center and instead take care of their health by using medicinal plants and visiting the healer; in Acosvinchos, the proportion of informants is much lower, and it is women older than 60 years who carry out these practices (use of medicinal plants and visiting the healers).

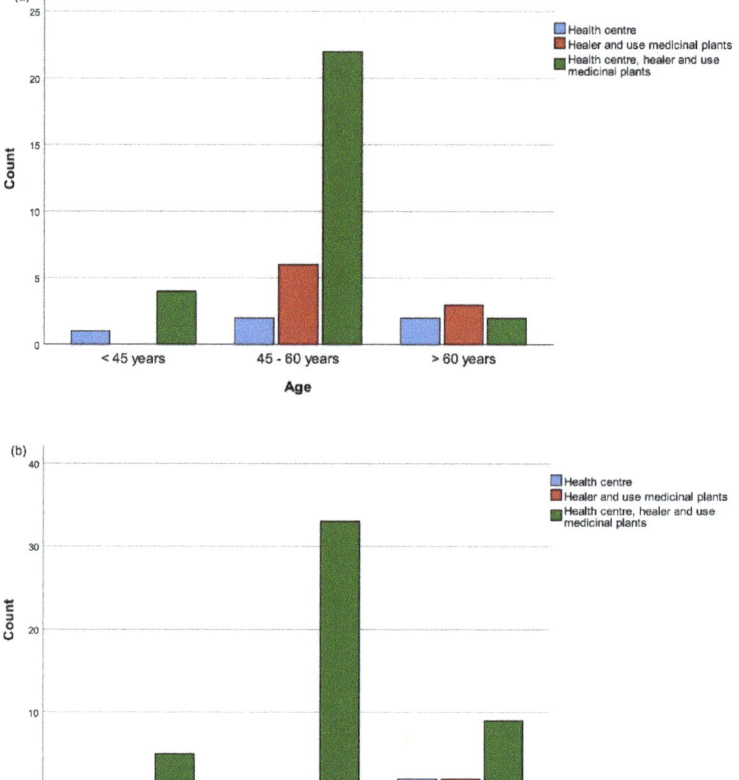

Figure 3. Places where respondents of Quinua (**a**) and Acos Vinchos (**b**) go when they become sick, according to age groups.

Figure 4 shows how the inhabitants of the districts studied obtain medicinal plants, according to gender. Men in Quinua (Figure 4a) mostly report that they collect them from the field, as well as from the field and the orchard. Very few people buy the plants to be used for medicinal purposes. In the case of women, the trend is similar, but to a lesser extent in all cases, except for purchasing. In Acos Vinchos (Figure 4b), the majority of men report that they collect them from the field and the orchard; a smaller proportion collect only from the field, and few grow them in the orchard. More women go to the fields to collect medicinal plants, although they also grow them in the orchard and buy them.

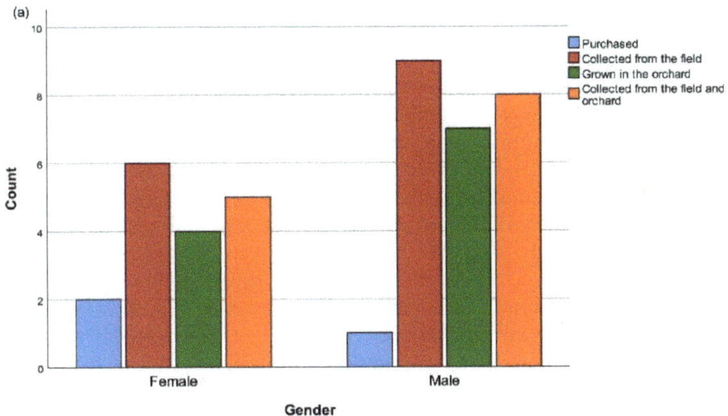

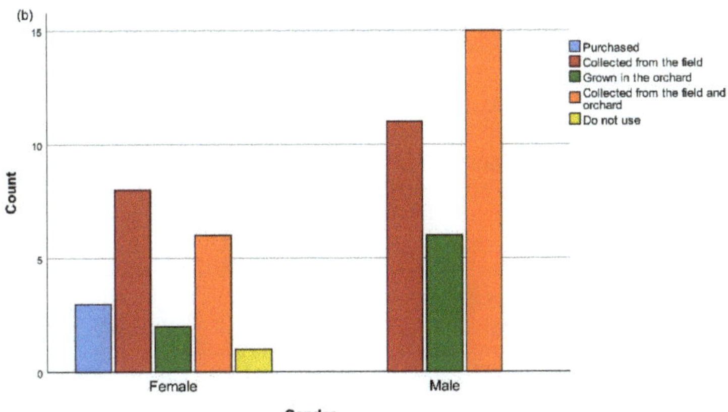

Figure 4. How the inhabitants of Quinua (**a**) and Acos Vinchos (**b**) obtain medicinal or aromatic plants according to gender.

4. Discussion

Peru is the world's largest producer of Quinoa, one of the most promising crops of the future [5,14]. It is important to know, evaluate and recover the knowledge that the villagers have about these ancestral Andean grains because this will indicate their degree of information about the benefits and advantages of cultivating these grains. This study has evaluated this knowledge with the collaboration of 96 informants, 40 women and 56 men, between 27 and 75 years of age, of different socio-economic characteristics and who were very participative. According to the information collected in Tables 1 and 2 and looking at the socio-economic profile of the two districts studied, it should be noted that, despite the low levels of education, the villagers have the practical management of their crops and have provided us with the required information about the use and management of these grains. Thanks are due to all participants for their generosity in sharing their knowledge with us.

It is important to remember that the Andes are the center of domestication of many cultivated species and that indigenous communities maintain native varieties. The characteristic crop of this agro-ecological region is maize, accompanied by cucurbits (*calabaza, caihua, zapallo*) and Andean grains (*quinoa, achita and cañihua*). *Quinoa* cultivation in the inter-Andean valleys of the Peruvian highlands was mostly carried out in a traditional system, characterized by the sowing of quinoa in very small plots, interspersed with other species and on the edge of fields of other crops. Cultivation was mainly for self-consumption with native varieties, traditional soil preparation techniques, use of organic matter and their own pest control products, which gave them independence from external inputs. From the 2000s onwards, the growing demand in the national and international market and the high prices of quinoa had an impact on quinoa cultivation, resulting in an increase in the cultivation area and a change in the management of these crops [26–29]. The results obtained (Table 3) indicate that the most commonly used Andean grains in the districts studied are *quinoa* and *achita*, with more than half of the population interviewed, specifically 58.3%, consuming *quinoa* or *achita*. These grains are consumed in different ways and form part of the daily diet, a fact that helps to reduce the nutritional problems of the rural population, as these products are characterized by their high biological value proteins, as well as carbohydrates, polyunsaturated fatty acids, fiber, vitamins and minerals with antioxidant properties. It is considered a nutraceutical or functional food, with a high protein content (15.7% to 18.8%) and a significant proportion of essential amino acids, including lysine (7.1%), an amino acid that is scarce in plant-based foods and forms part of the human brain [6–9,30–32].

A total of 66.7% of the informants cultivate quinoa for their own consumption and, to a lesser extent, 31.3% for commercialization. Many farmers do not have adequate technological support, in addition to the problems of pests and drought, but the cultivation of *quinoa* and *achita*, even with these problems, is a profitable activity and is considered a good alternative for cultivation in the region [33]. The sowing of *quinoa* and *achita* has increased throughout the region, which is in line with Ministry of Agriculture (MINAGRI) information; thus, at the national level, the *quinoa* sowing area for the 2019–2020 agricultural campaign was 67,777 ha, slightly higher than the 2018–2019 campaign of 66,388 ha; in the Ayacucho region, 14,439 ha were sown in the 2018–2019 agricultural campaign, slightly lower than the 2017–2018 agricultural campaign in which 14,617 ha were sown [14]. Quinoa production is of economic and social importance for many Peruvian regions above 3000 m above sea level because there are not many other cultivation options. One example is the region of Puno, where this crop is the livelihood of approximately 100,000 rural families. This is one of the poorest regions in Peru with a poverty rate of 24.2% [34].

Quinoa has been selected as one of the crops destined to offer food security in the 21st century, due to its high capacity to withstand extreme environmental conditions and its qualities as a functional food. In addition, *quinoa* has gained space in international consumer markets in the last decade, which opens up economic opportunities for Andean producers [3,5,33,35–37].

The orchards are the best preserved agro-ecosystem in Peru, representing a biological and cultural refuge. Before 1960, the diversification of land use, cultivated varieties and the use of wild plants was a survival strategy. Cultural identity is one of the reasons for maintaining certain traditional uses, such as the collection of wild plants or the cultivation of orchards and traditional varieties [2].

In this perspective, our study shows a high percentage of informants (93.8%) who use their own seeds for replanting, thus becoming independent from commercially available seeds and maintaining local varieties, which is very positive. Regarding the adapted local varieties of *quinoa*, the most common and one of the most commercialized in Peru is the Blanca Junín, which has two types, Blanca and Rosada, and is typical of the central region. Other varieties are as follows: Hualhuas, obtained by the Universidad Nacional del Centro de Huancayo (white grains); Amarilla Marangani, a variety from Cusco (Peru), with the characteristic of having a high saponin content; Roja Pasankalla, from the district of Ácora, province of Puno (red grains); Negra, INIA (2013) composed of 13 accessions, commonly known as "Quytu jiwras", which have their origin in the accessions that were collected from the localities of Caritamay, district of Ácora, province of Puno. The latter variety is also resistant to mildew [38,39]. In the Ayacucho Region, different varieties of quinoa (Amarilla Marangani, Illpa INIA, INIA Salcedo, Altiplano, Hualhuas, Rosada Junín, Huancayo, INIA 433, Pasankalla, Negra Collana, Amarilla Sacaca and Blanca Junín) from different areas were evaluated and adapted in the farming community of San Antonio de Manallasacc in the district of Chiara, province of Huamanga and Ayacucho Region, Peru. The INIA 433 variety achieved the highest yield with 4.72 mt per hectare and 14.3% protein. The highest protein variety was Amarilla Marangani with 16% protein. Therefore, INIA 433 and Blanca Junín are the recommended varieties for this area because they have the best yields per hectare: 4.72 mt and 4.62 mt respectively [40].

With regard to the *achita* crop, the most common and most commercial is the variety of the farmer Oscar Blanco, (Ayacucho-INIA), with white grains. It responds to the agro-ecological conditions of the Ayacucho region in terms of health, production and grain quality. The *achita* production areas in Peru are mainly located in inter-Andean valleys with large climatic fluctuations that put the food security of small producers at risk. Furthermore, production technology is still traditional, sowing, field work and harvesting is manual, and mechanization is only used during field preparation and threshing, which is carried out with stationary threshing machines [4].

The use of agrochemicals is widespread in the care of *quinoa* and *achita* crops in the municipalities studied. A total of 85.4% of the interviewed villagers use them, with a small percentage (10.4%) using home-made products. These homemade products are used for the prevention and control of diseases caused by fungi and other pathogens and are composed of a combination of cupric sulphate and hydrated lime, dissolved separately in water at room temperature and always in nonmetallic containers; they recommend using one part of broth plus one part of water (50:50) and using it immediately. The broth should not be applied to very small, newly germinated or flowering seedlings [41,42]. It is important to note that most *quinoa* production in Peru is carried out under the conventional production system that promotes intensive soil use, the planting of commercial varieties, inorganic fertilizers and the control of diseases and pests with synthetic pesticides [29]. This is because short-term profitability is the main concern, but the indiscriminate use of agrochemicals deteriorates crop fields and damages farmers' health [32]. Evidently, the green revolution, which focused on germplasm breeding and the use of agrochemicals, contributed to the increase in food plant production. However, the ecological and social damage of an agriculture dependent on external inputs, particularly imported products, raises concerns. Most of the technical advice comes through the agrochemical stores and sellers, who, thanks to their entrepreneurial capacity, have taken advantage of the situation to promote agriculture dependent on marketable products such as chemical fertilizers and pesticides, which now account for half of the total cost of production. This has contributed to an overuse of toxic chemicals that are damaging ecosystems, which in turn has resulted in new outbreaks of pests and diseases, soils with diminishing biotic life and polluted waters. In addition, pests become more resistant to agrochemicals, which has led to a vicious cycle in which farmers need to apply more and more toxic chemicals each year, while the productivity of their land continues to decline [36,37].

Regarding the commercialization of *quinoa* and *achita* in Peru, there are companies that grow and stockpile these Andean grains for export. These companies pay a low price to the small farmer, which makes a big difference compared to the price of packaged products in shopping centers. In recent years, *quinoa* and *achita* exports have increased, but this development opportunity must be framed within a framework of fair and sustainable trade for the Andean population. Peru has signed and ratified the Nagoya Protocol on access to genetic resources and fair and equitable sharing of the benefits arising from their utilization; therefore, in accordance with this international treaty, fair and equitable sharing of the benefits arising from the utilization of these genetic resources must be achieved, and payment for these harvests must be adequate [43].

The medicinal uses of *quinoa* and *achita* have been known to Andean people since ancient times. In traditional indigenous medicine, the leaves, stems and grains were used as healing, anti-inflammatory and analgesic agents, and they were therefore frequently used to treat bone fractures, bruises and internal hemorrhages. They were also used as antiseptics for the urinary tract and to treat blennorrhoea and urinary complaints [44]. Its lithium content reduces melancholy and sadness, phytoestrogens (daidzein and kinestein), can prevent uterine cancer and reduce menopausal problems as well as increase milk secretion. Its high calcium content, which is easily absorbed, prevents osteoporosis, and flavonoids promote blood circulation and reduce the risk of thrombosis. Anti-diabetic, antioxidant, anti-inflammatory, immunoregulatory, antimicrobial, anti-obesity, and heart-healthy properties are also being studied [4,10,11]. This scientific evidence supports the medicinal uses of *quinoa* and reveals that the use of its grains is not only good nutritionally but can also be used to improve health. This ancestral knowledge about the medicinal uses of *quinoa* and *achita* has been greatly eroded, as the results show that only 13.5% of the interviewed villagers remember any medicinal use of quinoa. The most common uses reported in our interviews are as a purgative (76.9%) and to relieve colic (23%). The use of quinoa as a *purgative* is not reflected in the works cited and, due to the high fiber content of these grains and leaves, is perfectly justified [45]. Although traditional knowledge about the medicinal use of Andean grains has been lost, Table 6 shows that the informants still have traditional knowledge about other medicinal plants in these districts; according to the data obtained, 91.7% of the informants use medicinal plants.

Traditional knowledge resides mainly in older people, and despite the efforts of many countries to preserve their ancestral knowledge, there has been a significant loss of traditional knowledge passed down from parents to children about the uses of the plants around them, mainly due to globalization, migration for education and work, and a reduced interest of young people in this cultural heritage [45,46]. This aspect is related to what is shown in Figures 2 and 3 of the research because it is adults over 45 years of age who still retain traditional knowledge about the medicinal plants and use them to take care of their health. Furthermore, various socio-cultural aspects of the relationship between human beings and ancestral Andean crops in the districts of Quinua and Acos Vinchos, whose economy is traditionally based on agriculture, handicrafts and little livestock, indicate that traditional medicine derived from ancestral knowledge is still maintained. Figures 3 and 4 show that the use, wild collection and cultivation of medicinal plants is very present among the inhabitants of both Quinua and Acos Vinchos. This indicates that knowledge about the medicinal use of plants, their location in the field and how to cultivate them in the garden still survives, especially among informants over 45 years of age. In Acos Vinchos, it is mainly women over 60 years of age who still maintain the traditional knowledge of their medicinal use and play the role of seed guardians. These results are in line with the studies carried out by Alberti-Manzanares y Luzuriaga-Quichimbo [23,47].

Most of these medicinal plants are collected in the field (36.4%) and from the orchard (20.5%), a very low percentage of which are bought (6.3%). A total of 91.7% report that their family uses medicinal plants, and 96.9% indicate that they would recommend their use to others. These results coincide with those reported by the WHO. Medicinal plants constitute a valuable resource in the health systems of developing countries, although there are no accurate data to assess the extent of global use of these plants, WHO has estimated that more than 80% of the world's population routinely uses traditional medicine to meet their primary health care needs and has developed the strategy on traditional medicine in 2014–2023 to support member states to harness their potential contribution to people's health, well-being and health care and to promote their safe and effective use through regulation and research [45,48]. As shown in Table 6, 78.1% of the villagers interviewed in Quinua and Acos Vinchos go to a health center, use medicinal plants and visit healer to improve their ailments. Although access to synthetic drugs is becoming easier, price can be a problem.

Random screening methods continue to be preferred in the search for active compounds by the pharmaceutical industry, but in recent years, special attention has been given to the use of ethnobotanical information for plant screening in the search for bioactive active compounds [45,49]. Indigenous Andean quinoa crops have excellent potential as a source of health-promoting bioactive compounds such as phenolic acids, saponins, phytosteroids and phytosterols, and in particular flavonoids; their content in *quinoa* grains is exceptionally high. *Bilberry* berries have been considered an excellent source of flavonoids, especially quercetin and myricetin, but when compared, the levels are 5 to 10 times lower than those found in quinoa grains. *Quinoa* seeds can be considered a very good source of flavonoids [50]. Efforts must be made to prevent the definitive loss of traditional knowledge on medicinal plants, not only to preserve this cultural heritage, but also to record information on certain useful species, which could be relevant for the development of new sources of medicines and other benefits for humanity while contributing to the protection of biodiversity [45,51].

5. Conclusions

The research shows that traditional knowledge about the medicinal use and crop management of these Andean grains has been lost, although they are still important in their diet. Two of the main foods consumed by the inhabitants of Quinua and Acos Vinchos are *quinoa* and *achita*. Some 58.3% of the informants consume them in their diet, and 66.7% grow them for their own consumption. The consumption of these grains in their daily diet helps to reduce the nutritional problems of the rural population because they contain proteins of high biological value, as well as carbohydrates, polyunsaturated fatty acids, fiber, vitamins and minerals. In the districts studied, traditional knowledge of the medicinal use of these Andean grains has been lost, but knowledge of other medicinal plants in the area is still preserved.

The cultivation of *quinoa* and *achita* is a common activity and is considered a good alternative in the region, despite pests and drought and little technological support. This study shows that a very high percentage of the farmers interviewed (93.8%) use their own seeds for replanting, thus becoming independent from commercial seeds and maintaining local varieties. The loss of traditional techniques of soil preparation, use of organic matter and preparation of pest control products indicate a change in crop management. A total of 85.4% use agrochemicals, on which they depend, thus decreasing the productivity of their land and edaphic resilience, causing various types of negative impacts on the environment and, in some cases, harming their own health. The use of agrochemicals does not result in higher production for most farmers, as only 39.6% of the informants' report that their production is good.

Finally, we suggest a framework of intervention to reward and promote the use of seeds of local varieties, in order to inform about the negative consequences of the use of agrochemicals and to offer them alternatives that promote responsible and sustainable management of the resources used in production, starting with soil management: use of organic matter, application of good agricultural practices and use of biocides to control diseases and pests. All of this will help us to take up the knowledge that reconciles us with our natural environment with the help of today's technology.

Author Contributions: Conceptualization, R.B.A. and L.M.M.-C.; methodology, L.M.M.-C. and R.C.; software, R.C. (Reynán Cóndor) and E.D.L.C.; validation, L.M.M.-C. and R.C. (Reynán Cóndor); formal analysis, R.C. (Reynán Cóndor) and L.M.M.-C.; Investigation: R.B.A., R.C. (Roxana Carhuaz) and R.L.; resources, R.B.A. and R.C. (Reynán Cóndor); data curation, R.C. (Reynán Cóndor) and L.M.M.-C.; writing—original draft preparation, R.B.A., L.M.M.-C., E.D.L.C. and R.C. (Reynán Cóndor); writing—review and editing, R.B.A. and L.M.M.-C.; visualization, L.M.M.-C. and R.B.A.; supervision, R.B.A. and L.M.M.-C.; project administration, R.B.A. and L.M.M.-C.; funding Acquisition, R.B.A. All authors have read and agreed to the published version of the manuscript.

Funding: Part of the research work has been funded by the Vice-rectorate of Research of the National University of San Cristobal de Huamanga.

Institutional Review Board Statement: Not applicable.

Informed Consent Statement: Informed consent was obtained from all subjects involved in the study.

Data Availability Statement: Not applicable.

Acknowledgments: To Ranulfo Cavero Carrasco in the Research of the UNSCH for the financial support, to the inhabitants of the intervened districts, to Manuel Pardo for validating the survey, to Blga. Laura Aucasime Medina, responsible for the Herbarium Huamangensis of the Faculty of Biological Sciences of the National University of San Cristóbal de Huamanga (Ayacucho-Peru) and to the Bachelor in Biological Sciences Carolina Fernández for collaborating in the application of the surveys.

Conflicts of Interest: The authors declare that they have no conflict of interest.

References

1. Arias, M.E.; Aguirre, M.G.; Luque, A.C. Caracterización Anatómica de Tallos de *Chenopodium* (Chenopodiaceae). Aportes al Estudio de Restos Arqueológicos. *Intersecc. Antropol.* **2014**, *15*, 265–276.
2. Aceituno-Mata, L. Estudio Etnobotánico y Agroecológico de La Sierra Norte de Madrid. Ph.D. Dissertation, Universidad Autónoma de Madrid, Madrid, Spain, 2010.
3. *FAO Dietary Protein Quality Evaluation in Human Nutrition*; Report of an FAO Expert Consultation; Food and Agriculture Organization of the United Nations Rome: Auckland, New Zealand, 2013; ISBN 978-92-5-107417-6.
4. Mujica, Á.; Moscoso, G.; Zavaleta, A.; Canahua, A.; Juarez, R.; Chambi, W.; Vignale, D. Usos Medicinales y Conocimientos Nutracéuticos Ancestrales de Granos Andinos: Quinua (*Chenopodium quinoa* Willd.), Kañihua (*Chenopodium pallidicaule* Aellen), Tubérculos Andinos: Izaño (*Tropaeolum tuberosum* R. y P), Olluco (*Ullucus tuberosus* Loz.), Oca (*Oxalis tuberosa* Mol.) y Parientes Silvestres En El Altiplano Peruano. *Agro Enfoque. Ago* **2015**, *29*, 10–16.
5. Bvenura, C.; Kambizi, L. Chapter 5—Future Grain Crops. In *Future Foods*; Bhat, R., Ed.; Academic Press: Cambridge, MA, USA, 2022; pp. 81–105, ISBN 978-0-323-91001-9.
6. Olivera, L.; Best, I.; Paredes, P.; Perez, N.; Chong, L.; Marzano, A. *Nutritional Value, Methods for Extraction and Bioactive Compounds of Quinoa*; IntechOpen: London, UK, 2022; ISBN 978-1-80355-181-4.
7. Villacrés, E.; Quelal, M.; Galarza, S.; Iza, D.; Silva, E. Nutritional Value and Bioactive Compounds of Leaves and Grains from Quinoa (*Chenopodium quinoa* Willd.). *Plants* **2022**, *11*, 213. [CrossRef]
8. Pereira, E.; Encina-Zelada, C.; Barros, L.; Gonzales-Barron, U.; Cadavez, V.; Ferreira, I.C.F.R. Chemical and Nutritional Characterization of *Chenopodium quinoa* Willd (Quinoa) Grains: A Good Alternative to Nutritious Food. *Food Chem.* **2019**, *280*, 110–114. [CrossRef]
9. Ponce de León Saavedra, P.; Valdez-Arana, J.d.C.; Ponce de León Saavedra, P.; Valdez-Arana, J.d.C. Evaluación Nutricional y Funcional de 17 Accesiones de Quinua (*Chenopodium quinoa* Willd) Cultivadas En La Zona Andina Del Perú. *Sci. Agropecuaria* **2021**, *12*, 15–23. [CrossRef]
10. Amit, S.; Irfan, A.M.; Gaurav, S.; Nalini, T.; Sarmad, M. A Review on Medicinal and Pharmaceutical Importance of Quinoa (*Chenopodium quinoa*). *Res. J. Pharm. Technol.* **2021**, *14*, 1779–1784. [CrossRef]
11. Pathan, S.; Siddiqui, R.A. Nutritional Composition and Bioactive Components in Quinoa (*Chenopodium quinoa* Willd.) Greens: A Review. *Nutrients* **2022**, *14*, 558. [CrossRef]
12. León, J. *Plantas Alimenticias Andinas*; Instituto Ineteramericano de Ciencias Agrícolas Zona Andina: Lima, Peru, 1964.
13. MINAGRI. *Quinua Peruana—Situación Actual y Perspectivas en el Mercado Nacional e Internacional al 2015*; Ministerio de Agricultura y Riego: Lima, Peru, 2015.
14. MINAGRI. *Encuesta Nacional de Intenciones de Siembra 2019 Campaña Agrícola Agosto 2019—Julio 2020*; Dirección General de Seguimiento y Evaluación de Políticas; Dirección de Estadística Agraria: Lima, Perú, 2019.
15. Cárdenas, D.; Bermúdez, C.; Echeverri, S.; Pérez, A.; Puentes, M.; López, L.; Correia, M.I.T.; Ochoa, J.B.; Ferreira, A.M.; Texeira, M.A.; et al. Declaración de Cartagena. Declaración Internacional sobre el Derecho al Cuidado Nutricional y la Lucha contra la Malnutrición. *Rev. Cuba. De Aliment. Y Nutr.* **2020**, *30*, 10–22.
16. Pulgar Vidal, J.P. Las ocho regiones naturales del Perú. Terra Brasilis (Nova Série). Revista da Rede Brasileira de História da Geografia e Geografia Histórica. *Terra Bras. (New Ser.)* **2014**, *3*, 1–21. [CrossRef]
17. Tapia Núñez, M.E. *Ecodesarrollo en los Andes Altos*; Fundación Friedrich Ebert: Bonn, Germany, 1996.
18. de Almeida, C.D.F.C.B.R.; Ramos, M.A.; Silva, R.R.V.; de Melo, J.G.; Medeiros, M.F.T.; Araújo, T.A.D.S.; de Almeida, A.L.S.; de Amorim, E.L.C.; Alves, R.R.D.N.; de Albuquerque, U.P. Intracultural Variation in the Knowledge of Medicinal Plants in an Urban-Rural Community in the Atlantic Forest from Northeastern Brazil. *Evid.-Based Complement. Altern. Med.* **2011**, *2012*, e679373. [CrossRef]
19. Corroto, F.; Macía, M.J. What Is the Most Efficient Methodology for Gathering Ethnobotanical Data and for Participant Selection? Medicinal Plants as a Case Study in the Peruvian Andes. *Econ. Bot.* **2021**, *75*, 63–75. [CrossRef]
20. Berkes, F.; Colding, J.; Folke, C. Rediscovery of Traditional Ecological Knowledge as Adaptive Management. *Ecol. Appl.* **2000**, *10*, 1251–1262. [CrossRef]
21. Macía, M.J.; García, E.; Vidaurre, P.J. An Ethnobotanical Survey of Medicinal Plants Commercialized in the Markets of La Paz and El Alto, Bolivia. *J. Ethnopharmacol.* **2005**, *97*, 337–350. [CrossRef] [PubMed]
22. Martin, G.J. Ethnobotany, Conservation and Community Development. In *Ethnobotany: A Methods Manual*; Martin, G.J., Ed.; Springer: Boston, MA, USA, 1995; pp. 223–251, ISBN 978-1-4615-2496-0.
23. Alberti-Manzanares, P. Los Aportes de Las Mujeres Rurales al Conocimiento de Plantas Medicinales En México: Análisis de Género. *Agric. Soc. Desarro.* **2006**, *3*, 139–153.

24. Alexiades, M.N. Collecting Ethnobotanical Data: An Introduction to Basic Concepts and Techniques. *Adv. Econ. Bot.* **1996**, *10*, 53–94.

25. R Core Team. European Environment Agency. 2020. Available online: https://www.eea.europa.eu/data-and-maps/indicators/oxygen-consuming-substances-in-rivers/r-development-core-team-2006 (accessed on 10 December 2021).

26. Tapia Núñez, M.E.; Fries, A.M.; Mazar, I.; Rosell, C. *Guía de Campo de los Cultivos Andinos*, 1st ed.; Asociación Nacional de Productores Ecológicos del Perú/Organización de Las Naciones Unidas para la Agricultura y la Alimentación: Lima, Roma, 2007; ISBN 978-92-5-305682-8.

27. Tapia, M.E. Ecodesarrollo Agropecuario en la Sierra del Perú. *Zonas Áridas* **1990**, *6*, 7. [CrossRef]

28. Benavides, M.; Vásquez Caicedo, G.; Casafranca, J. *La Pequeña Agroindustria En El Perú: Situación Actual y Perspectivas*; IICA Biblioteca Venezuela: San José, Costa Rica, 1996.

29. Pinedo-Taco, R.; Gómez-Pando, L.; Julca-Otiniano, A.; Pinedo-Taco, R.; Gómez-Pando, L.; Julca-Otiniano, A. Sostenibilidad de sistemas de producción de quinua (*Chenopodium quinoa Willd.*). *Ecosistemas Recur. Agropecu.* **2018**, *5*, 399–409. [CrossRef]

30. Diaz-Condori, J.G.; Florez-López, N.E. Evaluación Sensorial y Calidad Nutricional de Una Galleta a Base de Tarwi, Cañihua e Hígado de Pollo En Escolares de Una Institución Educativa de Cerro Colorado En El Año 2017. Ph.D. Dissertation, Universidad Nacional de San Agustín de Arequipa, Arequipa, Perú, 2017.

31. Blanco, T.B. *Alimentación y Nutrición*; Universidad Peruana de Ciencias Aplicadas (UPC): Lima, Perú, 2011.

32. Rodríguez, J.P.; Rahman, H.; Thushar, S.; Singh, R.K. Healthy and Resilient Cereals and Pseudo-Cereals for Marginal Agriculture: Molecular Advances for Improving Nutrient Bioavailability. *Front. Genet.* **2020**, *11*, 49. [CrossRef]

33. Guerrero, P.; Salazar, A.H.; Aguirre, N.C. Estudio Técnico y Económico de Cuatro Variedades de Quinua En La Región Andina Central Colombiana. *Rev. Luna Azul (On Line)* **2018**, *46*, 167–180.

34. Laurente Blanco, L.F.; Mamani Huanacuni, A. Modelamiento de La Producción de Quinua Aplicando ARIMA En Puno-Perú. *Fides Ratio-Rev. Difusión Cult. Científica Univ. Salle Boliv.* **2020**, *19*, 205–230.

35. FAO. *La Quinua, Cultivo Milenario Para Contribuir a La Seguridad Alimentaria Mundial*; Oficina Regional para América Latina y el Caribe: Lima, Perú, 2011.

36. Jacobsen, S.-E.; Sherwood, S. *Cultivo de Granos Andinos En Ecuador: Informe Sobre Los Rubros Quinua, Chocho y Amaranto*; Ediciones Abya Yala: Quito, Ecuador, 2002.

37. Pinedo-Taco, R.E.; Gómez-Pando, L.R.; Anderson-Berens, D. Production sustainability index of organic quinoa (Chenopodium Quinoa Willd.) in the interandean valleys of Peru. *Trop. Subtrop. Agroecosyst.* **2022**, *25*, 1–14.

38. Apaza Mamani, V.; Cáceres Sanizo, G.; Estrada Zuñiga, R.; Pinedo Taco, R.E. *Catálogo de Variedades Comerciales de Quinua en el Perú*; Instituto Nacional de Innovación Agraria: La Molina, Peru, 2013.

39. *MINAGRI Dirección General de Políticas Agrarias*; MINAGRI: Lima, Perú, 2017; pp. 1–8.

40. Soto, M.; Allende, R.A.; Romero, V.L. Estudio Comparativo En Rendimiento y Calidad de 12 Variedades de Quinua Orgánica En La Comunidad Campesina de San Antonio de Manallasac, Ayacucho. *Rev. Campus* **2020**, *25*, 57–65. [CrossRef]

41. Apaza, V.; Delgado, P. *Manejo y Mejoramiento de Quinua Orgánica*; Serie Manual; Instituto Nacional de Investigación y Extensión: Puno, Perú, 2005; 150p.

42. Restrepo, J. *El ABC de la Agricultura Orgánica, Fosfitos y Panes de Piedra: Manual Práctico*; Feriva: Cali, Colombia, 2007; ISBN 978-958-44-1261-4.

43. Protocolo de Nagoya. Available online: https://www.cbd.int/abs/doc/protocol/nagoya-protocol-es.pdf (accessed on 9 April 2022).

44. Scruzzi, G.; Cebreiro, C.; Pou, S.; Rodríguez Junyent, C. Salud Escolar: Una Intervención Educativa en Nutrición desde un Enfoque Integral. *Cuadernos. Info* **2014**, *35*, 39–53. [CrossRef]

45. Bermúdez, A.; Oliveira-Miranda, M.A.; Velázquez, D. La Investigación Etnobotánica Sobre Plantas Medicinales: Una Revisión de Sus Objetivos y Enfoques Actuales. *Interciencia* **2005**, *30*, 453–459.

46. Weckmüller, H.; Barriocanal, C.; Maneja, R.; Boada, M. Factors Affecting Traditional Medicinal Plant Knowledge of the Waorani, Ecuador. *Sustainability* **2019**, *11*, 4460. [CrossRef]

47. Luzuriaga-Quichimbo, C.X.; Hernández del Barco, M.; Blanco-Salas, J.; Cerón-Martínez, C.E.; Ruiz-Téllez, T. Plant Biodiversity Knowledge Varies by Gender in Sustainable Amazonian Agricultural Systems Called Chacras. *Sustainability* **2019**, *11*, 4211. [CrossRef]

48. Sanabria-Diago, O.L. La etnobotánica y su contribución a la conservación de los recursos naturales y el conocimiento tradicional. In *Manual de Herramientas Etnobotánicas Relativas a la Conservación y el Uso Sostenible de los Recursos Vegetales Una Contribución de la Red Latinoamericana de Botánica a la Implementación de la Estrategia Global para la Conservación de Las Especies Vegetales Hacia el Logro de Las Metas 13 y 15*; Red Latinoamericana de Botánica (RLB): Santiago, Chile, 2011; pp. 37–60, ISBN 978-956-9073-01-04.

49. Bussmann, R.W.; Sharon, D. Traditional Medicinal Plant Use in Northern Peru: Tracking Two Thousand Years of Healing Culture. *J. Ethnobiol. Ethnomed.* **2006**, *2*, 47. [CrossRef]

50. Repo-Carrasco-Valencia, R.; Hellström, J.K.; Pihlava, J.-M.; Mattila, P.H. Flavonoids and Other Phenolic Compounds in Andean Indigenous Grains: Quinoa (*Chenopodium quinoa*), Kañiwa (*Chenopodium pallidicaule*) and Kiwicha (*Amaranthus caudatus*). *Food Chem.* **2010**, *120*, 128–133. [CrossRef]

51. Cortés-Rodríguez, E.A.; Venegas-Cardoso, F.R. Conocimiento Tradicional y la Conservación de la Flora Medicinal en la Comunidad Indígena de Santa Catarina, BC, México. *Ra Ximhai Rev. Científica Sociedad Cultura Desarrollo Sostenible* **2011**, *7*, 117–122.

Article

Assessing Soil Organic Carbon, Soil Nutrients and Soil Erodibility under Terraced Paddy Fields and Upland Rice in Northern Thailand

Noppol Arunrat [1,*], Sukanya Sereenonchai [1], Praeploy Kongsurakan [2] and Ryusuke Hatano [3]

1 Faculty of Environment and Resource Studies, Mahidol University, Nakhon Pathom 73170, Thailand; sukanya.ser@mahidol.ac.th

2 Laboratory of Animal Ecology, Graduate School of Fisheries and Environmental Sciences, Nagasaki University, Nagasaki 852-8521, Japan; praeploy.kong@hotmail.com

3 Laboratory of Soil Science, Graduate School of Agriculture, Hokkaido University, Sapporo 060-8589, Japan; hatano@chem.agr.hokudai.ac.jp

* Correspondence: noppol.aru@mahidol.ac.th

Abstract: Terracing is the oldest technique for water and soil conservation on natural hilly slopes. In Northern Thailand, terraced paddy fields were constructed long ago, but scientific questions remain on how terraced paddy fields and upland rice (non-terraced) differ for soil organic carbon (SOC) stocks, soil nutrients and soil erodibility. Therefore, this study aims to evaluate and compare SOC stocks, soil nutrients and soil erodibility between terraced paddy fields and upland rice at Ban Pa Bong Piang, Chiang Mai Province, Thailand. Topsoil (0–10 cm) was collected from terraced paddies and upland rice fields after harvest. Results showed that SOC stocks were 21.84 and 21.61 Mg·C·ha^{-1} in terraced paddy and upland rice fields, respectively. There was no significant difference in soil erodibility between terraced paddies (range 0.2261–0.2893 t·h·MJ^{-1}·mm^{-1}) and upland rice (range 0.2238–0.2681 t·h·MJ^{-1}·mm^{-1}). Most soil nutrients (NH$_4$-N, NO$_3$-N, available K, available Ca and available Mg) in the terraced paddy field were lower than those in the upland rice field. It was hypothesized that the continuous water flows from plot-to-plot until lowermost plot caused dissolved nutrients to be washed and removed from the flat surface, leading to a short period for accumulating nutrients into the soil. An increase in soil erodibility was associated with decreasing SOC stock at lower toposequence points. This study suggested that increasing SOC stock is the best strategy to minimize soil erodibility of both cropping systems, while proper water management is crucial for maintaining soil nutrients in the terraced paddy field.

Keywords: soil organic carbon; soil erodibility; terraced paddy field; upland rice; Thailand

Citation: Arunrat, N.; Sereenonchai, S.; Kongsurakan, P.; Hatano, R. Assessing Soil Organic Carbon, Soil Nutrients and Soil Erodibility under Terraced Paddy Fields and Upland Rice in Northern Thailand. *Agronomy* **2022**, *12*, 537. https://doi.org/10.3390/agronomy12020537

Academic Editor: Antonio Miguel Martínez-Graña

Received: 24 November 2021
Accepted: 16 February 2022
Published: 21 February 2022

Publisher's Note: MDPI stays neutral with regard to jurisdictional claims in published maps and institutional affiliations.

1. Introduction

Globally, the soil organic carbon (SOC) pool has been a key challenging subject since it could generate either positive or negative feedback on atmospheric CO$_2$ changes [1–3]. As soil stores three-times more carbon than the atmosphere, many land uses have endangered soil degradation and also the related SOC sequestration potential [4,5], especially under agricultural systems [6]. In agricultural land, the SOC in soils is often depleted but also has the potential to sequester carbon (C) under agricultural practices [7,8]. In addition, SOC plays a role indicator of soil quality, as it contributes to soil biochemical and physical functions that are essential for plants and microorganisms [9]. Hence, agricultural management has become increasingly highlighted as a high-potential tool for mitigating climate changes and adapting to changing climate [10,11]. Concerning agriculture in mountainous areas, the slope gradient is the key factor affecting SOC dynamics and soil quality [12–14], while soil losses and soil erosion are highly correlated with slope steepness [15]. Budry and Curtis [16] and Tadele et al. [17] indicated that soil erosion causes land degradation that is the most serious problem for the farmer in tropical highlands, resulting in decreasing soil fertility and crop yield losses. Therefore, it is challenging to investigate the variation of SOC, soil nutrients and soil erodibility under different land management techniques in hilly and mountainous areas.

The most effective engineering measure for slope management is terracing, which, when well designed and appropriately maintained, has been reported to reduce soil and water loss [18–21]. Terracing is the oldest technique for water and soil conservation on natural hilly slopes [22,23], and some terraces were constructed in Southeast Asia more than 5000 years ago [24–27]. Under the terracing system, mountainous terrains are managed into narrow graduated steps to enable crop growing and management practices [28]. The terracing system plays a vital role in soil and water conservation not only by directly creating many microtopographical and specific hydrological pathways [29–31] but also by reducing slope gradient and connectivity of overland runoff [32,33]. In terraces, divided sections of cropland with gradient slopes also help farmers to alleviate flooding and erosion, while improving other ecosystem services, including food security [34], recreation [35] and C sequestration [36]. Previous studies, however, have found varying levels of SOC in different terraced sites. Xu et al. [37] revealed that mean SOC densities at the 0–60 cm soil depth in terraces were the highest when compared to forestland, grassland and sloping cropland. Meanwhile, Zhang et al. [38] presented the idea that soil erosion and cropping contributed to variations of SOC and TN losses along the sloping terrace. Inappropriate terraces, often randomly designed by local farmers, were also marked as having a higher risk of soil erosion and severe landslides caused by the unstable structure [39,40].

In Thailand, terraced paddy fields have been widely distributed in Northern Thailand under contribution by the Royal Development Project since 2003 [41]. Even though terraced paddy fields have been determined to be a highland sustainable agriculture approach [42], the surface runoff and soil and nutrient losses need to be examined [43]. However, the investigation of SOC stock, soil nutrients and soil erodibility under rice terraces is still lacking, indicating that it would be desirable to find out how terraced paddy field and upland rice differ in SOC stock, soil nutrients and soil erodibility. To fill these gaps, our study aims to evaluate and compare (1) SOC stock and soil nutrients and (2) soil erodibility between terraced paddy fields and upland rice as well as the different potentials of rice terraces among toposequences. This study could contribute to a clearer understanding of the relevant issues of hilly and mountainous areas.

2. Material and Methods

2.1. Study Area and Site Selection

The study area was carried out at Ban Pa Bong Piang, Chang Khoeng Subdistrict, Mae Chaem District, Chiang Mai Province, Northern Thailand (Figure 1a). The rainy season of Chiang Mai Province is from May to October. The average annual temperature ranged between 20 and 35 °C, while the average annual rainfall ranged between 600 and >1000 mm. The topography of Ban Pa Bong Piang is relatively steep with an elevation of 800 to 1400 m above sea level (m a.s.l.). Most of the farmers are Karen people who mainly cultivate upland rice and terraced paddy fields once a year. Cultivation begins in June for land preparation, and then rice planting is in July. Harvest occurs around mid-October to the beginning of November, and then the land is left fallow from December to May.

In this study, the similarities of original geography, soil formation and microclimate conditions were well considered for site selection. A terraced paddy field located upstream of the natural water canal for cultivation was selected (18°32′02.8″ N, 98°26′48.8″ E). This terraced paddy field (level terrace type) was constructed since 1982 (40 years ago), and the terrace risers or walls were built of soil. Selecting this terraced paddy field avoided nutrient and sediment discharges from other terraced upland rice fields. Its elevation ranged between 975 and 1014 m a.s.l. with slopes of 5–17% (Figure 1b). The widths of the flat section of the terrace ranged from 3 to 5 m, and the heights of the terrace riser or wall were between 0.5 and 0.8 m.

Meanwhile, an upland rice field (non-terraced) was selected (18°32′22.2″ N, 98°26′34.7″ E) as the comparison field, with an elevation of 908–935 m a.s.l. and 19–26% slopes. Upland rice is as traditional rice cultivation in this area, which has been cultivated for more than 70 years. This field has only grown upland rice once a year, whereas most fields comprise the upland rice–maize system (Figure 1c).

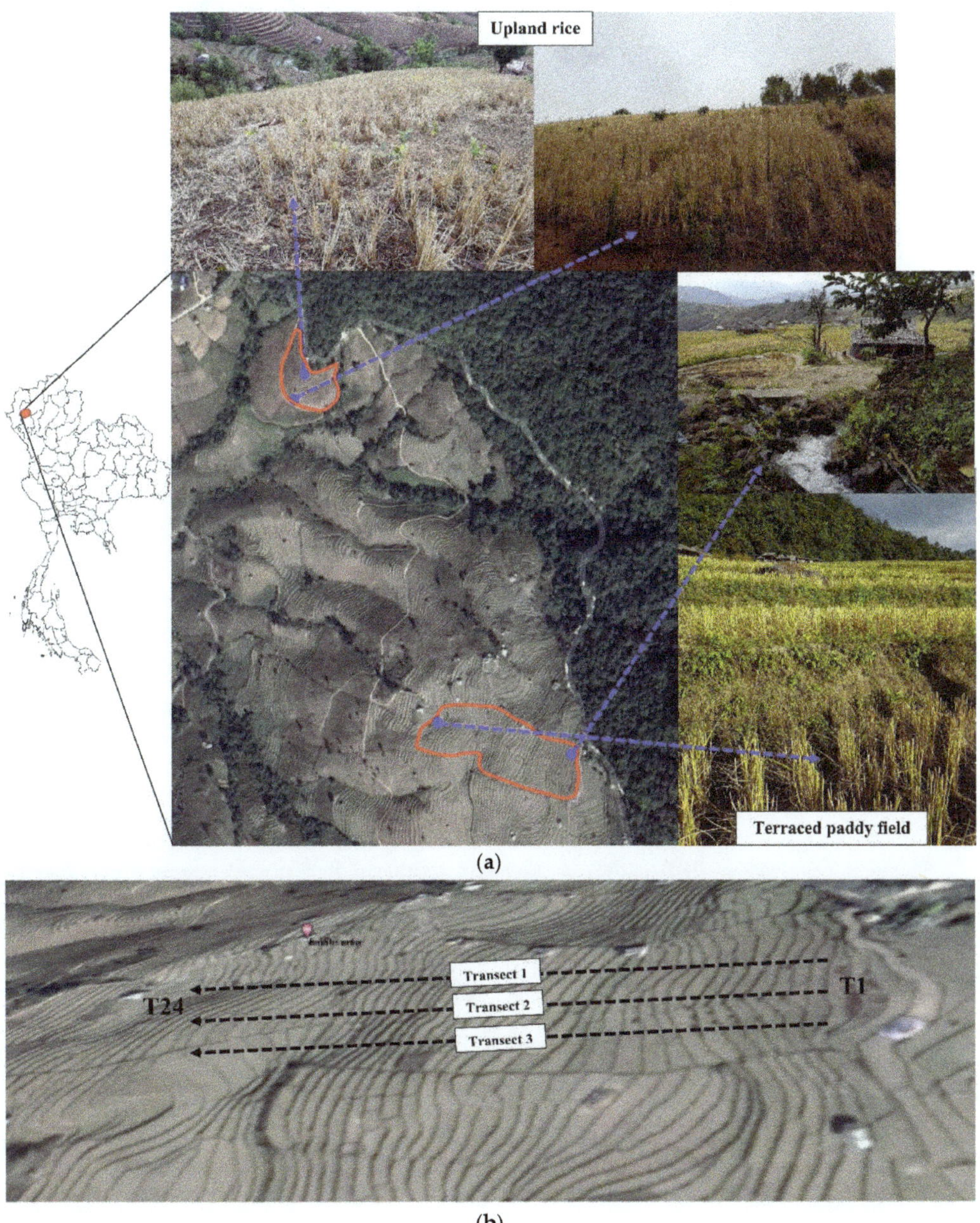

Figure 1. *Cont.*

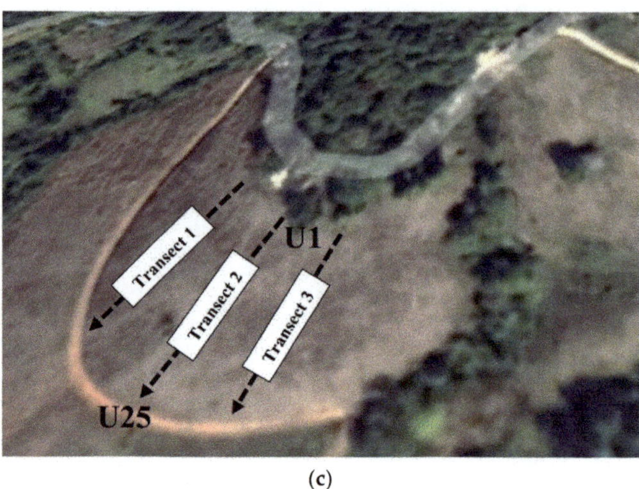

(c)

Figure 1. Study area. (**a**) Overall study area, (**b**) terraced paddy field and (**c**) upland rice. The aerial images were taken from Google maps on 25 February 2020. The photos were taken on 14 November 2020 by Noppol Arunrat.

2.2. Management Practices

Farm management practices were recorded from farm owners in November 2020. At the terraced paddy field, the transplanting method was used for rice cultivation. The nursery field was prepared in May. Rice seeds were placed into bags and soaked in water for 24 h, and then rice seeds in bags were drained and dried for 24 h in a shady area until small roots appeared at the end of the seeds. Then, pregerminating rice seeds were sown by hand in the nursery field. In June, puddling was prepared by a 15-horsepower tractor for all terraced paddy fields. The rice plants were removed from the nursery field and transplanted in terraced paddy fields by hand. The N, P_2O_5 and K_2O chemical fertilizers were applied using 46-0-0 (93.8 kg·ha^{-1}), 16-16-8 (156.3 kg·ha^{-1}), and 16-20-0 (156.3 kg·ha^{-1}). Harvesting was performed in October by hand. All rice residues were left in the field without burning.

Upland rice cultivation began in May with land preparation including removing weeds and vegetation. The no-tillage method was used due to the difficulty of using any machines on the hillslopes, whereas the drilling method was usually applied for planting. A hoe was used to dig the soil at a 10 cm depth, and then rice seeds were dropped by hand. The N, P_2O_5 and K_2O chemical fertilizers were applied, consisting of 16-20-0 (156.3 kg·ha^{-1}) and 16-16-8 (156.3 kg·ha^{-1}). Harvesting was performed in November by hand, and all rice residues were left in the field without burning.

2.3. Soil Sample Collection and Laboratory Analysis

Topsoil was collected at depth 0–10 cm on terraced paddy and upland rice fields after harvest in November 2020. At each field, three transects were designed with a distance of 7–8 m from one transect to another; then, the soil samples were collected from each point of each transect. At the terraced paddy field, the uppermost toposequence point was called T1, followed by T2, T3, . . . , and T24 (lowermost toposequence point), respectively (Figure 1b). At each point of each transect, the soil sample was collected using a mini shovel, then 1 kg of soil sample was packed into a plastic bag. Thus, 72 soil samples were obtained from the terraced paddy field. At the upland rice field, the uppermost toposequence point was called U1, followed by U2, U3 . . . , and U25 (lowermost toposequence point), respectively (Figure 1c). A total of 75 soil samples were gathered from the upland rice field. A compass was used to identify the slope gradient of each soil sample point. This is because the slope gradient influences the rate of runoff on the soil's surface, soil erosion and the movement of nutrients. Steel soil cores (5.0 cm width × 5.5 cm length) were used to collect soil samples from both fields for soil bulk density analysis.

At the laboratory, soil bulk density was measured by the dry weight per volume of soil core after drying in an oven at 105 °C for 24 h. All soil samples were air-dried at room temperature for 7–14 days, and then the dry soils were crushed and passed through a 2 mm sieve. Soil texture was determined by a hydrometer. Electrical conductivity (ECe) in saturation paste extracts was measured by an EC meter [44]. Soil pH was determined in a 1:2.5 soil to water mixture by a pH meter [45]. The cation exchange capacity (CEC) was determined by the NH_4OAc pH 7.0 method. Organic carbon (OC) was determined following the method described by Walkley and Black [46] using potassium dichromate ($K_2Cr_2O_7$) in sulfuric acid and reported as organic matter (OM) by multiplying with 1.724. Ammonium nitrogen (NH_4-N) and nitrate–nitrogen (NO_3-N) were measured by the KCL extraction method. Available phosphorus was determined based on the molybdate blue method (Bray II extraction) [47]. Available potassium, calcium and magnesium were extracted by NH_4OAc pH 7.0 and measured by atomic absorption spectrometry [48].

2.4. Soil Organic Carbon Calculation

The SOC stock was calculated using the following equation:

$$\text{SOC stock} = OC \times BD \times L, \tag{1}$$

where SOC is the soil organic carbon stock (Mg·C·ha^{-1}), *OC* is organic carbon (%), *BD* is bulk density (g·cm^{-3}) of soil and *L* is soil thickness (10 cm for this study).

2.5. Soil Erodibility Calculation

Among the estimators of the soil erodibility (*K*-value), the nomograph by Wischmeier et al. [49], the *K*-value in the Environmental Policy Integrated Climate (EPIC) model [50] and a formula proposed by Shirazi and Boersma [51] were widely used. Although there is no study in Thailand, several studies in China (e.g., Du et al. [52]; Wu et al. [53]; Chen et al. [54]; and Liu et al. [55]) achieved scientific results and revealed that the method of the EPIC model is reasonable to calculate *K*-value. Therefore, the *K*-value in our study area was calculated by following the EPIC model equation developed by Williams et al. [50] and Sharpley and Williams [56]:

$$K = \{0.2 + 0.3\exp[-0.256\text{SA}(1 - 0.01\text{SA})]\}\left(\frac{\text{SI}}{\text{SI}+\text{CL}}\right)^{0.3}\left[1 - \frac{0.25\text{C}}{\text{C}+\exp(3.72-2.95\text{C})}\right]$$
$$\times\left\{1 - \frac{0.7\left(\frac{1-\text{SA}}{100}\right)}{\left(\frac{1-\text{SA}}{100}\right)+\exp\left[-5.51+22.9\left(\frac{1-\text{SA}}{100}\right)\right]}\right\} \times 0.1317 \tag{2}$$

where SA is the fraction of sand (%), SI is the fraction of silt (%), CL is the fraction of clay (%), C refers to the organic carbon content (%) and 0.1317 is the conversion factor for United States business units (t acre h/100 acre/ft/tanf/in) to the international system of units. Thus, the unit of *K* value is t·h·MJ^{-1}·mm^{-1} (metric ton (t) × hour (h)/megajoule (MJ) × millimeter (mm)).

2.6. Statistical Analysis

The differences in soil properties between the terraced paddy and upland rice fields were compared by using independent *t*-tests. One-way ANOVA was employed to test for the differences in soil parameters within the group. The correlation between soil properties in terraced paddy samples and upland rice samples was separately investigated by Pearson's correlation coefficient. Meanwhile, principal components analysis (PCA) was applied to analyze the factors that influence the SOC stocks and soil erodibility under the terrace and upland rice paddies. The variables with high collinearity, correlation coefficients ǀrǀ > 0.7, were cut off in Pearson's correlation analysis. These statistical analyses and visualization were performed using R environment (v.4.0.2) with packages such as 'agricolae' [57], 'ggplot2' [58] and 'factoextra' [59]. Moreover, trend analysis was performed to analyze the variation of soil properties along the slope gradient using SPSS (v. 20.0).

3. Results and Discussion

3.1. Variation of Soil Properties under Terraced Paddy Field and Upland Rice

The proportion of topsoil textural composition of each sample of terraced paddy field and upland rice is shown in Figure 2. In the terraced paddy field, topsoils predominantly contained a high proportion of sand particles (49.3–67.1%), which varied significantly among different toposequences ($p < 0.01$, Table 1) from sandy loam to loam. Meanwhile, 29.1–47.9% of the sand proportions of topsoils varied from sandy clay loam to clayey loam in the upland rice field. Similarly, silt and clay contents were significantly different between the terraced paddy and upland rice fields ($p < 0.01$, Table 1), and higher average values were found under upland rice compared to terraced paddy fields (Table 1). In both cropping systems, there was a large difference in the particle size distribution depending on toposequence, as shown in Figure 3.

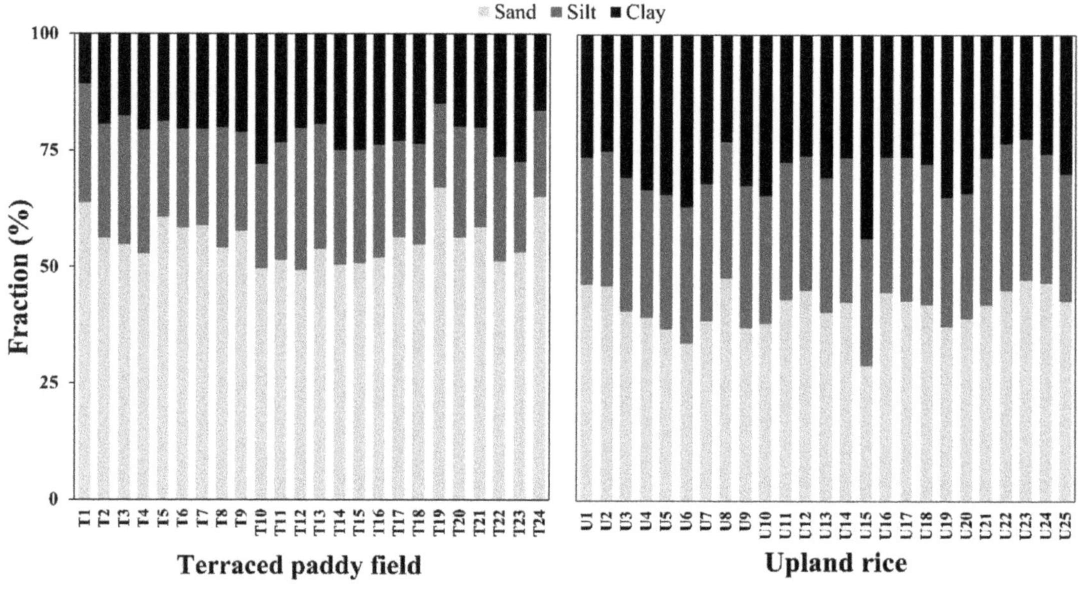

Figure 2. Particle size distribution for terraced paddy field and upland rice areas.

From the independent *t*-test in Table 1, the terrace paddy soils showed significantly higher values ($p < 0.05$) of OM, EC_e, available P and sand fraction than the upland rice soils. In addition, there were significantly lower ($p < 0.01$) values for bulk density, CEC, NH_4-N, NO_3-N, available K, available Ca, available Mg, silt fraction and clay fraction in terraced paddy than upland rice soils. However, there was no significant difference in pH, SOC and soil erodibility between these two cropping systems ($p > 0.05$).

In the same cropping system, more variation in soil properties for each toposequence was mostly found in terraced paddy soils rather than upland rice soils. Soil pH, OM, SOC, EC_e, CEC, available P, available Ca, available Mg, silt and clay contents as well as soil erodibility significantly fluctuated in different toposequence patches ($p < 0.05$) under terraced paddy cultivation (Table 1 and Figure 3). Meanwhile, significant differences ($p < 0.05$) of bulk density, NH_4-N, NO_3-N, available P, available K and available Mg were detected under the upland rice cultivation area (Table 1 and Figure 3). It is noteworthy that the available soil base elements responsively fluctuated under both upland rice and terrace paddy fields in different toposequences, especially available Ca (varied in ranges of 590.9–2302.2 mg·kg^{-1} and 198.7–984.2 mg·kg^{-1}, respectively, Figure 3).

Table 1. Independent *t*-test comparing soil properties between upland rice (n = 75) and terraced paddy field samples (n = 72).

Soil Properties	Upland Rice				Terraced Paddy Field				Between Group	
	Mean	Std	F	*sig.*	Mean	Std	F	*sig.*	*t* Value	*p* Value
pH	5.36	0.29	1.45	0.23	5.31	0.20	5.06	0.03	−1.15	0.25
Bulk density (g·cm^{-3})	1.46	0.03	4.60	0.04	1.35	0.01	3.43	0.07	−32.24	<0.01
Organic matter (%)	2.55	0.33	0.33	0.57	2.79	0.85	128.70	<0.01	2.29	0.02
Soil organic carbon (Mg·C·ha^{-1})	21.61	2.72	0.79	0.38	21.84	6.52	132.20	<0.01	0.28	0.78
Electrical conductivity (dS·m^{-1})	0.24	0.05	0.06	0.81	0.29	0.06	6.08	0.02	5.32	<0.01
Cation exchange capacity (meq·100 g^{-1})	13.87	1.75	0.07	0.79	8.50	1.45	145.70	<0.01	−20.21	<0.01
NH$_4$-N (mg·kg^{-1})	29.05	7.36	59.87	<0.01	22.85	6.68	0.59	0.45	−5.34	<0.01
NO$_3$-N (mg·kg^{-1})	17.44	4.52	14.16	<0.01	14.81	4.03	1.24	0.27	−3.72	<0.01
Available P (mg·kg^{-1})	2.57	0.97	8.83	<0.01	95.83	37.49	91.58	<0.01	21.54	<0.01
Available K (mg·kg^{-1})	215.25	48.84	23.97	<0.01	48.00	15.74	4.50	0.04	−27.71	<0.01
Available Ca (mg·kg^{-1})	1063.32	341.00	2.99	0.09	652.63	188.83	46.37	<0.01	−8.98	<0.01
Available Mg (mg·kg^{-1})	141.62	27.97	28.59	<0.01	28.28	7.52	5.85	0.02	−33.25	<0.01
%Sand	41.50	4.52	2.35	0.13	55.70	4.77	0.004	0.95	18.53	<0.01
%Silt	28.96	1.41	2.08	0.15	23.20	3.00	18.86	<0.01	−14.99	<0.01
%Clay	29.54	5.04	3.22	0.08	21.10	3.90	10.65	<0.01	−11.33	<0.01
Soil erodibility (t·h·MJ^{-1}·mm^{-1})	0.2402	0.0136	0.00	0.99	0.2351	0.0104	35.08	<0.01	0.17	0.86

The relative fluctuations of soil physical and chemical properties were reflected in the impact of patches heterogeneity and soil erodibility as well as practice management in these two areas. The fluctuation of soil physical and chemical properties in the terraced paddy field indicated the high variation of soil characteristics among plots from the uppermost to lowermost toposequence plots. Cui et al. [60] also found that the available K and available P were low in the fallow period compared with the tillage period, whereas OM was higher in the fallow period than that in the tillage period.

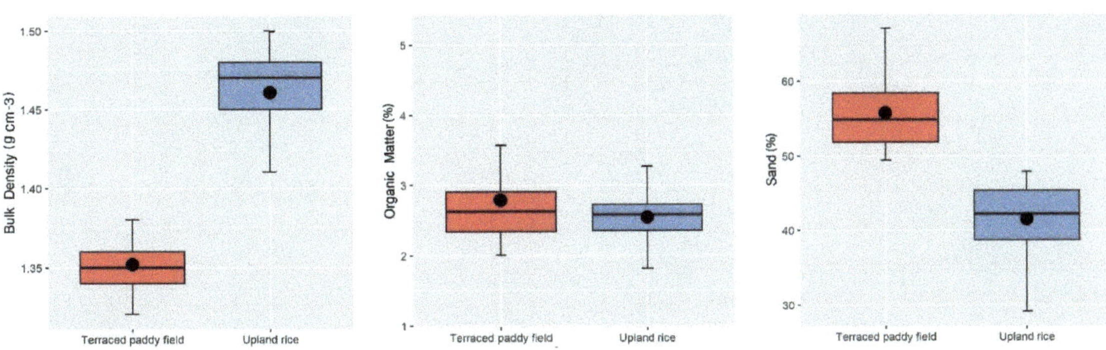

Figure 3. *Cont.*

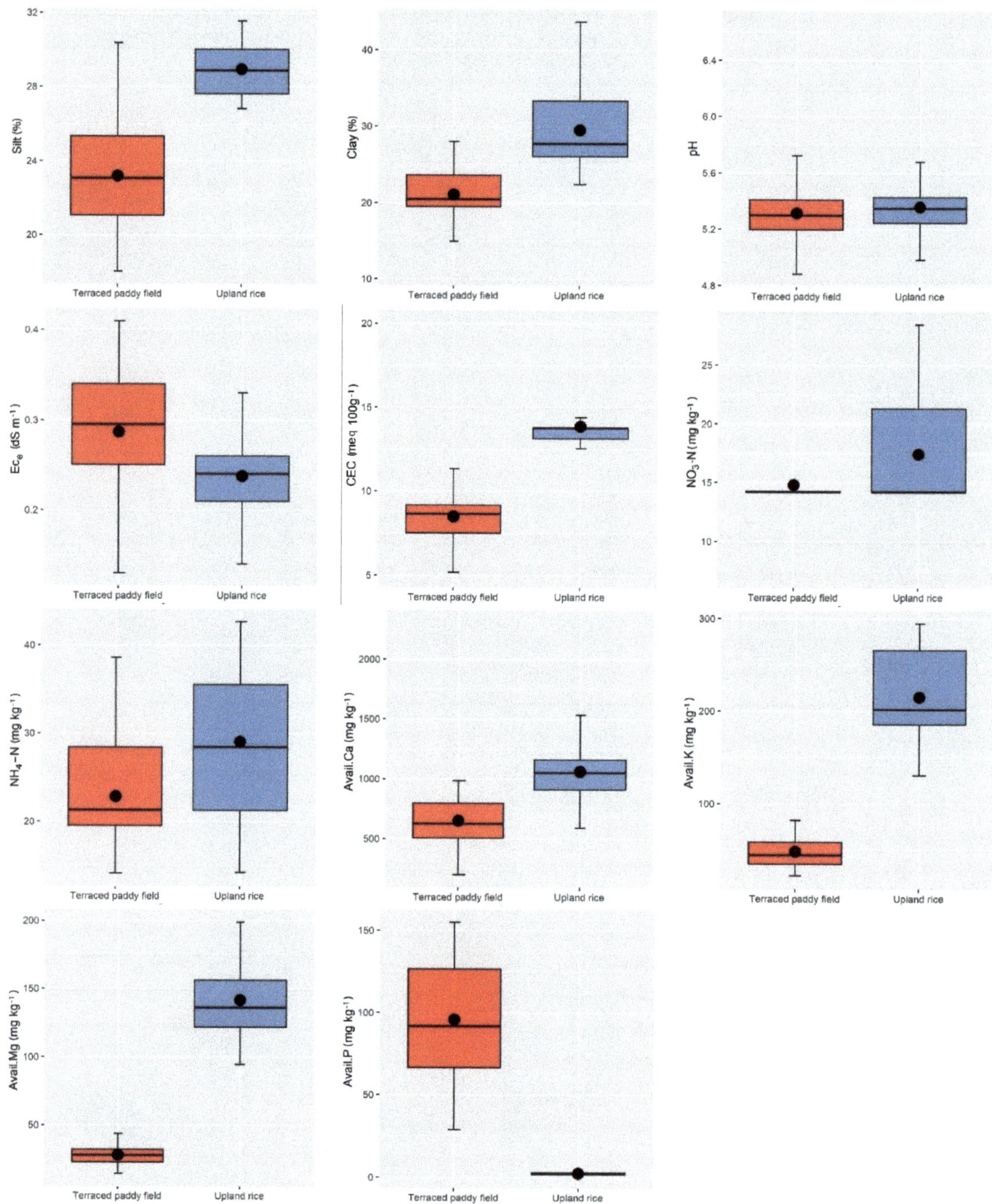

Figure 3. Comparison of soil physical and chemical properties between terraced paddy field and upland rice.

Based on trend analysis (Figure 4), there were no significant trends of soil physical and chemical properties with the topographic slopes. It is implied that the slope gradients did not influence to the variation of soil physical and chemical properties of both upland rice and terraced paddy fields. The trend line of the terraced paddy field showed that OM, silt content, pH, ECe, CEC, NH_4-N, available Ca, available K and available P decreased from the uppermost toposequence downwards the slope. Meanwhile, the trend line of upland rice field indicated that bulk density, OM, silt and clay contents, available K and available P decreased from the top slope to the lowermost toposequence point. This is mainly due to the transportation of nutrients and sediments from upstream to downstream locations under the plot-to-plot irrigation system, which detected the significant differences ($p < 0.01$) of silt and clay particles among terraced paddy soils (Table 1). This is consistent with Miyamoto et al. [61] and Mori et al. [62], who observed that fine particles in the surface were washed and transported downstream by surface flow. Terracing does not always increase soil nutrient status, as found in the present study, as most soil nutrients (NH_4-N, NO_3-N, available K, available Ca and available Mg) in the terraced paddy field were lower than the upland rice field (Table 1). Probably, the continuous water flows from plot-to-plot until downstream locations cause dissolved nutrients to be washed and removed from the flat surface (cultivation section), indicating a short period for accumulating nutrients into the soil that results in lower remaining nutrients in the soil. The draining of water at least 7 days before harvesting is another hypothesized cause for the reduction in soil nutrients. Schmitter et al. [63] suggested that soil fertility in rice paddy terraces can be maintained by balancing sediment inputs of different sources. However, continuous observation is needed to estimate any significant trends of these two cropping systems in our study area.

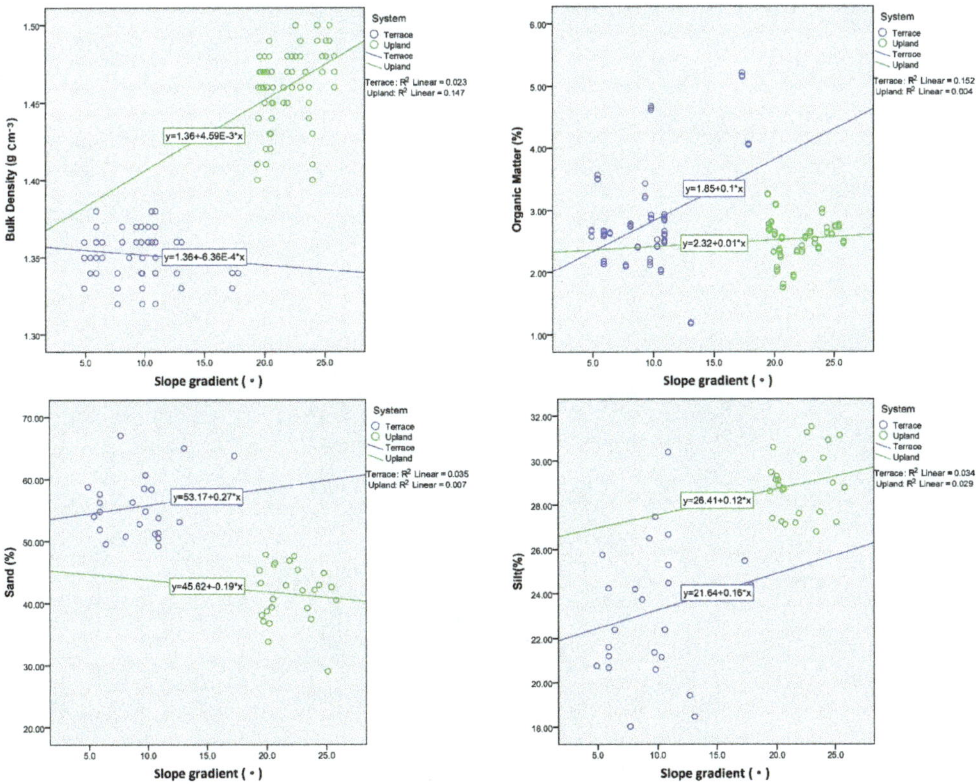

Figure 4. *Cont.*

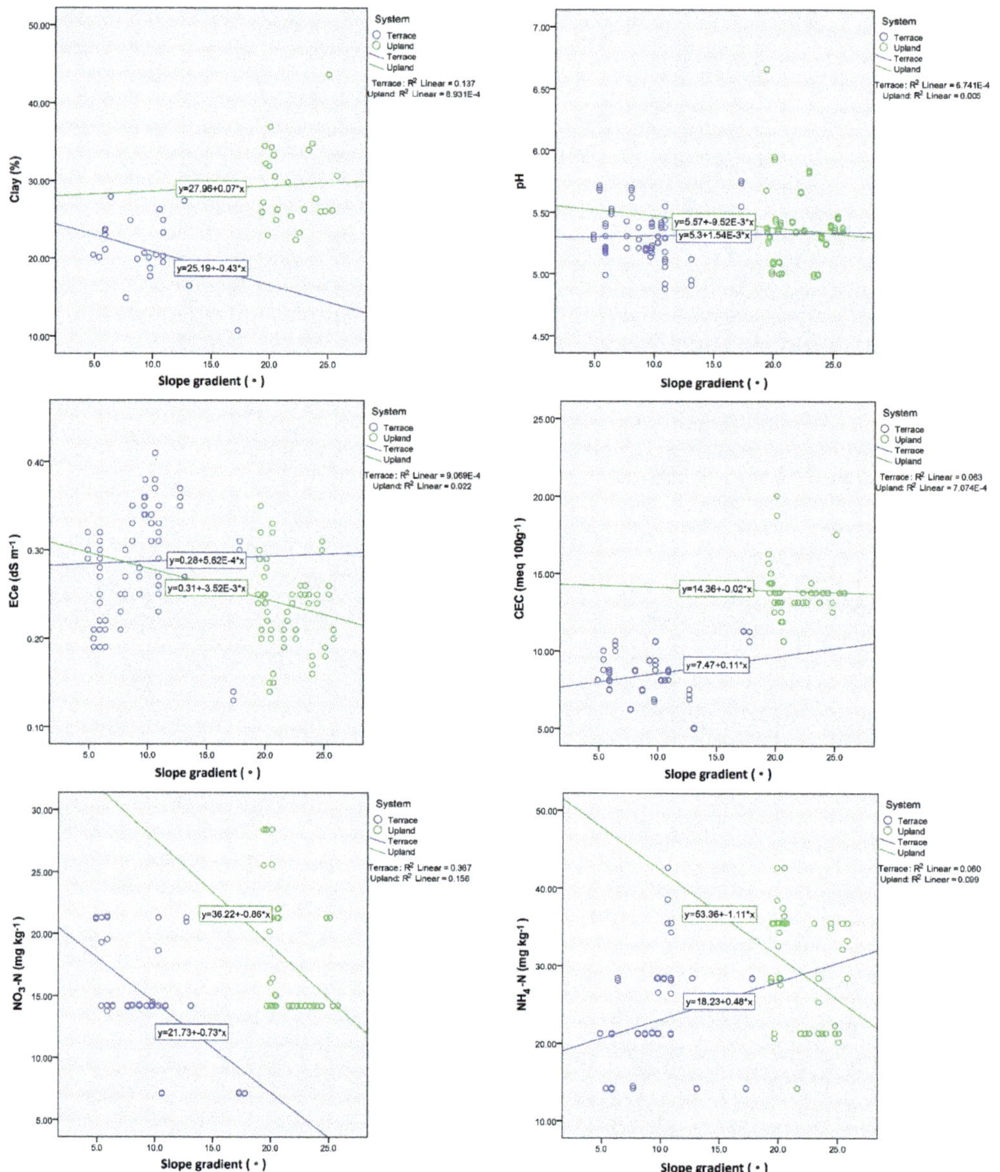

Figure 4. *Cont.*

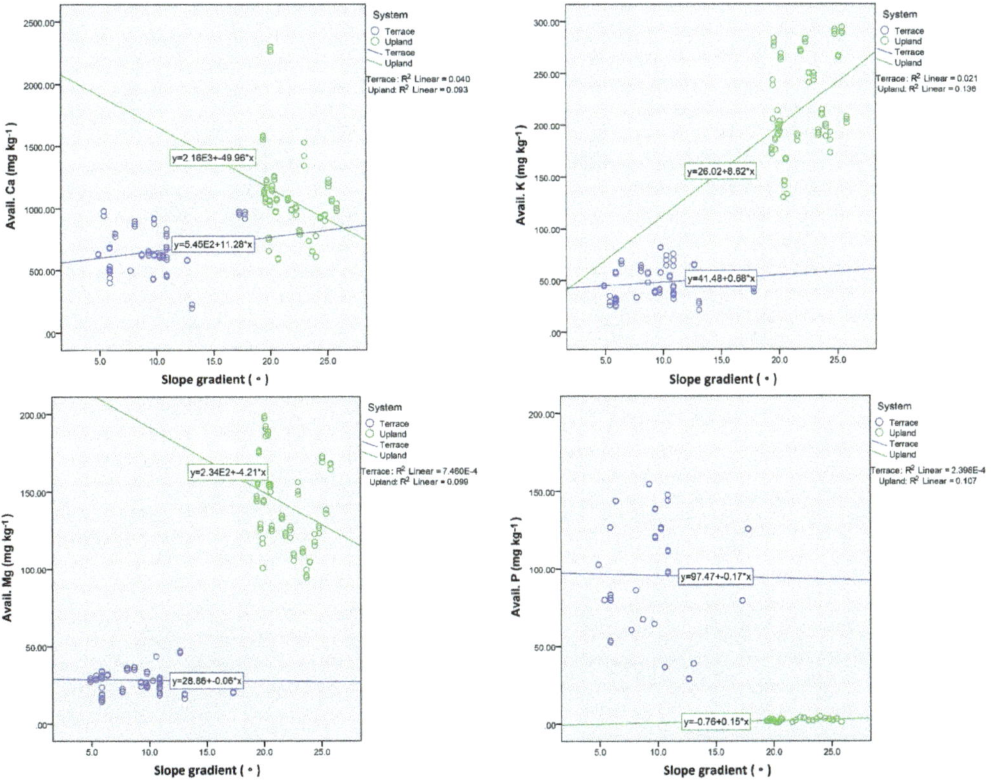

Figure 4. Trend analysis charts of soil physical and chemical properties between terraced paddy field and upland rice.

3.2. Soil Organic Carbon Stock under Terraced Paddy Field and Upland Rice

The OM fraction was significantly higher ($p < 0.01$) in terraced paddy soils ($2.8 \pm 0.9\%$) than in the upland rice soils ($2.6 \pm 0.3\%$) (Table 1). There was no significant difference in SOC stocks in both cropping systems, which was because the combination of OM and bulk density variations caused insignificant differences of SOC stocks between both systems. In the upland rice system, the lowest SOC stock (15.0 ± 0.2 Mg·C·ha^{-1}) was found in U1, which is the uppermost toposequence plot (Figure 4). Meanwhile, the sample soil at U12 had the greatest SOC content (26.7 ± 0.1 Mg·C·ha^{-1}) (Figure 5a). However, significant variations of SOC among different toposequences were not detected under upland rice soils. In the terraced paddy field, conversely, SOC stock shows significant differences among toposequence patches ($p < 0.01$, Table 1). It was found that T1, which is the uppermost toposequence plot, contained the largest SOC stock (40.2 ± 0.4 Mg·C·ha^{-1}) (Figure 5b). Meanwhile, the lowest SOC stock was found in T24 (9.3 ± 0.2 Mg·C·ha^{-1}), which was located at the lowest toposequence (Figure 5b). This can be explained by previous evidence of the effect of toposequence on soil properties. Baskan et al. [64] revealed that soil properties in different topographic positions were significantly different in terms of pedogenic processes shaped by the movement and transport of soil particles. In this study, a higher level of SOC content in upper terraces compared to the upland site indicates its potential on SOC sequestration from the above position. De Blécourt et al. [65] quantified SOC stock changes induced by terrace construction in the area of forest conversion to rubber plantation. They found that topsoil removal at the cut section caused a reduction in SOC stocks in the youngest plantation, and then the recovery of SOC stocks came from roots and litter and deposition of topsoil materials from the upper slope.

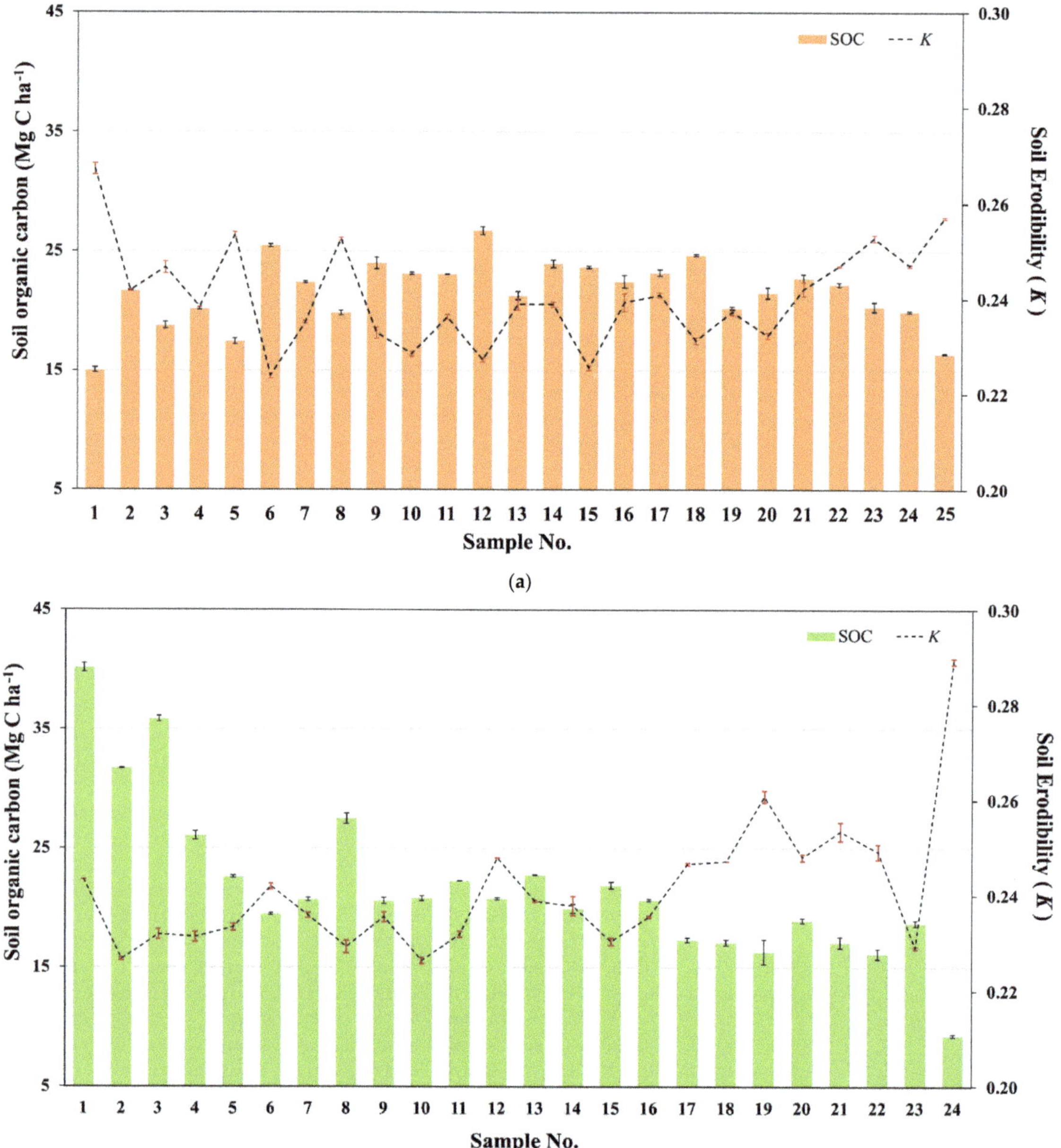

(a)

(b)

Figure 5. *Cont.*

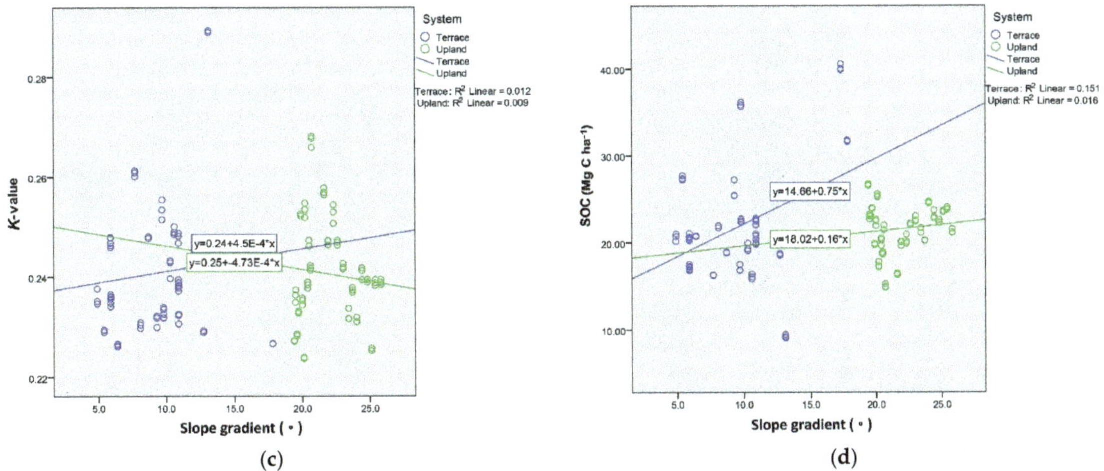

(c)

(d)

Figure 5. Soil organic carbon (Mg·C·ha^{-1}, bar area) and K-value (line) of under different toposequences, and trend analysis charts of soil organic carbon and K-value. (**a**) Upland rice area, (**b**) terraced paddy field, (**c**) trend analysis chart of K-value and (**d**) trend analysis chart of SOC stock.

As presented in Table 1, SOC stocks in the terraced paddy and upland rice fields were 21.84 and 21.61 Mg·C·ha^{-1}, respectively, but a significant difference was not detected. Previous studies [37,66,67] have reported that terracing can reduce SOC loss by modifying hillslopes to small flat fields, which was supported by the findings of the present study. During the dry period after harvest, the rice straw remaining in the terraced field is a great practice to improve SOC sequestration as well as restoring soil nutrients with fewer losses from the field, which is conserved by the structure of terracing, such as stair-rice paddy. Moreover, flooding in the terraced paddy field supplies suspended particles and soluble nutrients to the fields [68], while puddling facilitates the incorporation of organic inputs into the soil and creates low breakdowns of OM [69]. On the other hand, the loss in upland rice may be higher than terraced paddy fields due to having no riser or wall to slow down erosion, together with fewer weeds and plants to cover the soil after harvest. Chen et al. [70] reported higher SOC in the surface layer (0–20 cm) than the deeper soil layer (20–100 cm), indicating that protection of surface soil of terraced field is the key to enhancing SOC [33,71]. As the meta-analysis by Chen et al. [72] points out, terraces in China increased 32.4% of SOC sequestration compared with sloping areas. Tadesse et al. [73], Arunrat et al. [74] and Arunrat et al. [75] suggested that the application of manure, crop residues and soil conservation could increase SOC. However, SOC stock in the present study reflects a visual tendency, but no significant difference was found, highlighting that continuous investigation is necessary to conserve soil nutrients and SOC sequestration.

3.3. Soil Erodibility under Terraced Paddy Field and Upland Rice

Soil erodibility is used to calculate the K-value in the universal soil loss equation (USLE) and the revised universal soil loss equation (RUSLE), which is an important factor for soil erosion assessments as well as soil and water conservation planning [76,77]. In the present study, there was no significant difference in K-values, ranging from 0.2261–0.2893 t·h·MJ^{-1}·mm^{-1} and 0.2238–0.2681 t·h·MJ^{-1}·mm^{-1} between terraced paddy and upland rice soils ($p > 0.05$, Table 1), respectively. In upland rice soil, the difference in K-values across different toposequences was not significant ($p > 0.05$, Table 1). We found that the K-value was slightly higher when SOC content dropped, especially in U1 and U25 (Figure 5a). As noted previously, the finer particles tended to be dominantly exported by erosion [78]; thus, SOC was relatively eroded [79,80]. Conversely, soil erodibility was significantly variable under the terraced paddy soil ($p < 0.01$, Table 1). The highest K-value was, as expected, detected in T24 (lowermost location point), which contained the lowest SOC stock (Figure 5b). This can probably be explained by particle distribution differences being reduced due to the terracing technique. Moreover, SOC stock was possibly conserved under terracing cultivation by its contribution to plant root distribution [61], and its impact varied strongly among management practices [81]. This is because puddling in land preparation and transplanting method for rice planting caused a decrease in bulk density and enhanced root length density [82]. However, our findings merely reflect a tendency that was not a significant difference between upland rice and terrace paddy soils.

As a significant difference of soil erodibility under terraced paddy soil was detected ($p < 0.01$) (Table 1), it indicated the high fluctuation of soil erodibility among flat sections across the terraced paddy field. Moreover, the trend analysis chart of K-values along the topographical gradients between terraced paddy field and upland rice can be observed in Figure 5c. This is because each flat section of the terraced paddy field contains a different proportion of sand, silt, clay and OC contents, and vice versa under upland rice (Table 1). Thus, maintaining the fractions of sand, silt and clay contents as well as improving OC contents is the primary important factor to control the erodibility of both cropping systems. Among these factors, increasing OC content seems to be the most possible strategy and does not disturb current farmer's management practices by retaining rice straw and stubbles, applying animal manure and reducing tillage. Once SOC is increased, the benefits can result in the stability of soil aggregates and enhanced soil structure, resulting in resistance to erosion [83]. Moreover, terraces have the potential to reduce sediment yield and runoff [84] as well as increase water infiltration, soil moisture and soil water holding capacities in several areas [21,71]. It can be observed in Figure 5d that SOC stocks decreased from the uppermost toposequence to the downward slope.

It should be noted that the K-value of the EPIC model is dependent on soil particle size and organic carbon, which may not be sufficient to estimate soil erodibility with the different climate and cropping systems. For example, the studies of Zhang et al. [85], Chen et al. [20] and Zhang et al. [86] have developed a database of K factors for China's agricultural soils to reduce the biases of soil loss estimation. Therefore, the study of the feasibility of combining methods (e.g., K-value from nomographs method [49], K-value of EPIC model method [50] and the soil geometric mean diameter method [51]) to provide accurate estimations of K-values in Thailand is required for future studies. Moreover, it should be constructed as Thailand's database of K factor and K factor maps.

3.4. Correlation Coefficient Matrix and PCA Analysis

The Pearson's correlation matrix among soil physicochemical properties is presented in Table 2. In terraced paddy soils, SOC showed a high positive correlation with both CEC (0.88, $p < 0.05$) and available Ca (0.85, $p < 0.05$). Available Ca had significant positive correlation with pH ($r = 0.80$, $p < 0.05$). Meanwhile, it was found that the available Ca positively correlated with CEC ($r = 0.78$, $p < 0.05$) and pH ($r = 0.75$, $p < 0.05$) in upland rice soils. A significant negative correlation was also found between clay and sand contents in upland rice soils and terraced paddy soils ($r = -0.96$ and -0.78, respectively). Interestingly, the negative values of correlation coefficients of the relationship between SOC or CEC and clay content were found in the upland rice field, indicating that clay minerals might not be active (Table 2).

These results, together with principal component analysis, explained 62.6% of the total variance (Figure 6, PC1: 46.3% and PC2: 16.3%) and allowed a better understanding of the correlation between the physicochemical properties of the soils collected in different cropping systems. Additionally, factor loading analyses showed that the first 6 of 16 PCs can explain 90.5% of the total variance with eigenvalue greater than 0.7. The significant loading factors with 10% of the highest factor loading in each significant PCs are underlined in Table 3. Three variables, available Mg, CEC and sand content, were obviously weighted in PC1. Meanwhile, in PC2, SOC, soil erodibility and toposequence were considered significant. In contrast, PC3 was strongly related to EC, NH_4-N and NO_3-N. Both EC and NO_3-N were significantly included in PC4 and PC5. While pH and EC were considered in PC4 and PC5, respectively. In PC6, soil erodibility, clay content and bulk density were significantly weighted.

Table 2. Pearson's correlation matrix of soil properties in terraced paddy field samples ($n = 72$, white background) and upland rice ($n = 75$, grey background).

Variable	pH	BD	SOC	EC$_e$	CEC	NH$_4$-N	NO$_3$-N	P	K	Ca	Mg	%Sand	%Silt	%Clay
pH		−0.24 *	0.45 *	0.18	0.54 *	−0.08	0.24 *	−0.18	0.23 *	0.75 *	0.41 *	−0.01	0.31 *	−0.08
BD	−0.14		−0.13	0.06	−0.24 *	0.01	−0.26 *	0.14	0.44 *	−0.30 *	−0.10	0.00	0.17	−0.05
SOC	0.52 *	−0.33 *		0.36 *	0.66 *	0.02	0.28 *	0.17	0.21	0.47 *	0.09	−0.25 *	0.43 *	0.10
ECe	−0.42 *	0.16	−0.30 *		0.15	0.24 *	0.51 *	0.11	0.40 *	0.23 *	0.10	0.11	0.13	−0.14
CEC	0.49 *	−0.13	0.88 *	−0.25 *		−0.08	0.37 *	−0.29 *	0.18	0.78 *	0.48 *	−0.65 *	0.02 *	0.57 *
NH$_4$-N	−0.17	0.30 *	−0.09	0.66 *	0.12		0.27 *	−0.17	−0.32 *	0.15	0.39 *	0.09	0.12	−0.11
NO$_3$-N	−0.17	0.09	−0.35 *	0.05	−0.38 *	−0.23		−0.04	0.05	0.37 *	0.37 *	−0.07	−0.29	0.14
P	−0.13	−0.07	0.44 *	−0.08	0.61 *	0.15	−0.04		0.25 *	−0.53 *	−0.66 *	0.40 *	0.36 *	−0.46 *
K	0.31 *	−0.15	0.33 *	0.30 *	0.39 *	0.37	−0.11	0.25 *		0.02	−0.01	−0.04	0.25	−0.03
Ca	0.80 *	−0.19	0.82 *	−0.32 *	0.83 *	−0.03	−0.32 *	0.31 *	0.41 *		0.70 *	−0.39 *	0.12 *	0.32 *
Mg	0.40 *	0.13 *	−0.02	0.45 *	0.03	0.44 *	0.02	−0.34 *	0.54 *	0.28 *		−0.51 *	−0.19 *	0.51 *
%Sand	−0.09	−0.41	−0.04	−0.24 *	−0.34 *	−0.54 *	−0.01	−0.33 *	−0.28 *	−0.27 *	−0.45 *		0.23 *	−0.96 *
%Silt	0.16	0.02 *	0.59 *	−0.06	0.62 *	0.20	−0.32 *	0.57 *	0.15	0.51 *	−0.02	−0.58 *		−0.48 *
%Clay	−0.01	0.49	−0.41 *	0.34 *	−0.07	0.50 *	0.25 *	−0.03	0.23 *	−0.07	0.56 *	−0.78 *	−0.06	

* Correlation is significant at 0.05 probability level ($p < 0.05$); BD = bulk density; SOC = soil organic carbon; ECe = electrical conductivity; CEC = cation exchange capacity; P = available P; K = available K; Ca = available Ca; Mg = available Mg.

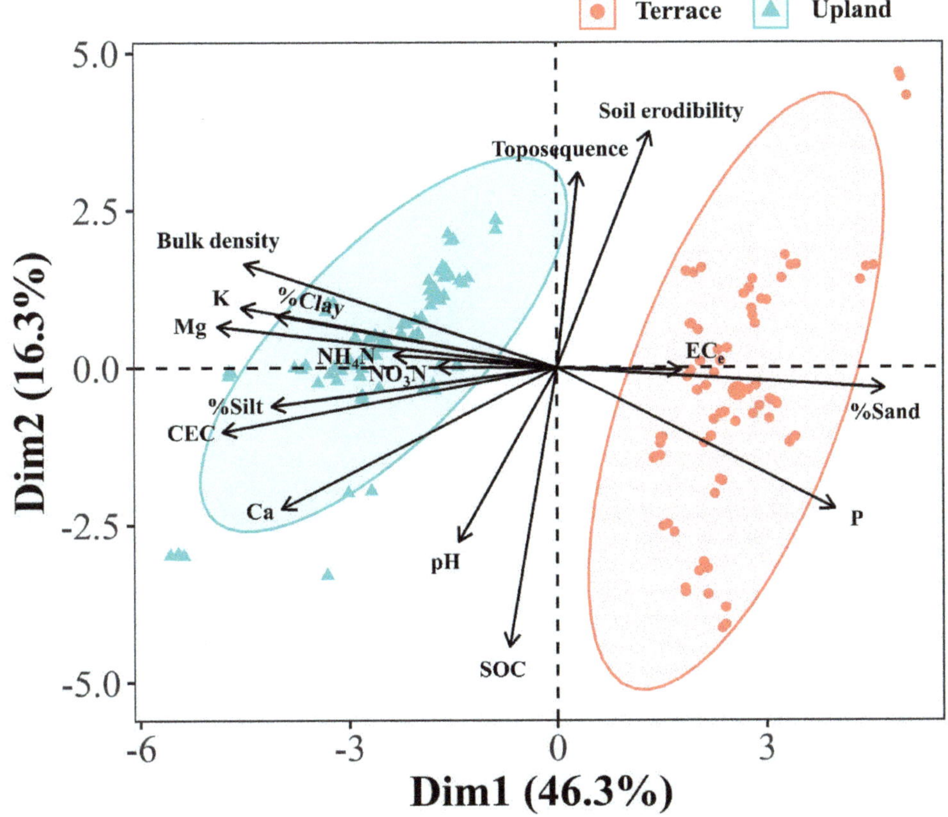

Figure 6. Principal component analysis (PCA) and the loading values of soil properties and toposequences for terraced paddy field samples (red area) and upland rice samples (blue area).

Table 3. Principal components (PCs), eigenvalues, percentage of variance explained by the PCs (% Var.) and cumulative percentage of variance explained by PCs of soil properties and toposequences for terraced paddy field and upland rice samples.

	PC1	PC2	PC3	PC4	PC5	PC6
Eigenvalue	7.408	2.607	1.502	1.196	0.909	0.866
Var. (%)	46.301	16.296	9.386	7.473	5.681	5.415
Cum. var. (%)	46.301	62.598	71.984	79.457	85.138	90.553
Factor loading/eigenvector						
Toposequence	0.022	**0.378**	0.266	**0.442**	**−0.409**	0.310
pH	−0.103	−0.337	0.276	**0.493**	−0.186	**−0.330**
Bulk density	−0.325	0.201	0.051	−0.124	−0.053	−0.009
SOC	−0.050	**−0.541**	0.158	−0.169	−0.141	0.044
ECe	0.129	−0.005	**−0.537**	0.263	**−0.552**	0.003
CEC	**−0.348**	−0.125	0.065	−0.007	0.011	0.049
NH_4-N	−0.170	0.024	**−0.506**	−0.286	−0.327	−0.291
NO_3-N	−0.125	0.001	**−0.396**	**0.445**	**0.494**	−0.239
P	0.289	−0.273	−0.120	−0.107	0.015	0.175
K	−0.329	0.116	0.087	0.032	−0.139	−0.003
Ca	−0.287	−0.273	0.055	0.198	−0.034	−0.205
Mg	**−0.353**	0.080	−0.033	−0.004	0.061	−0.157
%Sand	**0.342**	−0.040	0.087	0.052	−0.029	−0.298
%Silt	−0.297	−0.075	0.110	−0.285	−0.245	−0.027
%Clay	−0.292	0.101	−0.185	0.099	0.186	**0.426**
Soil erodibility	0.097	**0.457**	0.201	−0.149	−0.001	**−0.529**

BD = bulk density; SOC = soil organic carbon; ECe = electrical conductivity; CEC = cation exchange capacity; P = available P; K = available K; Ca = available Ca; Mg = available Mg.

As shown in Figure 6, the result of PC1 shows the direct correlation among some properties (bulk density, available Mg, CEC, available K, available Ca, percent clay, percent silt, NH_4-N and NO_3-N), which were inversely correlated with percent sand, available P and EC_e. A negative relationship between bulk density and sand content was detected. It indicated that upland rice soils were in higher bulk density, while the terraced paddy field had more sand content. It was clear that the upland rice and terraced paddy fields were different in the relationship between bulk density and sand content. Lower sand content and higher bulk density can make soil more fertile, while in the higher sand content area, more nutrients are expected to be gathered by irrigation. These results suggested that terraced paddies might be constructed to increase soil fertility rather than to reduce soil erosion in low upland crop production fields on natural hilly slopes. A strong correlation between the soil erodibility and toposequence was detected, indicating that higher soil erodibility occurred at lower toposequence points. As expected, by using the *K*-values of the EPIC model method, the correlation between soil erodibility and SOC was the opposite, meaning that a decrease in SOC stock was related with increasing soil erodibility (Figure 6). This is in line with the studies of Shabani et al. [87] and Ostovari et al. [88], who found a significant negative correlation between soil erodibility and OM.

3.5. Recommendations for Further Study

The findings of our study were investigated from a specific area, which may not be stated with high confidence about the differences between terraced paddy and upland rice fields, especially SOC stock and soil erodibility. Our study can be simply stated that SOC stocks and soil erodibility were not significantly different between the terraced paddy and upland rice fields in our study area. Therefore, more research should be conducted to validate the results in our study for providing appropriate management practices for these two systems. In future studies, experimental measurements should be conducted by measuring nutrient movement characteristics in terraced paddy and upland rice fields as well as soil erosion, water runoff and infiltration.

4. Conclusions

This study investigated hillslope cultivation fields that have been continuously managed as terrace paddy fields and upland rice cultivation, with the aim to explain how SOC sequestration, erodibility and physiochemical properties of topsoils are affected by terracing management. More variation in soil properties for each toposequence was found in terraced paddy soils rather than upland rice soils. Most soil nutrients (NH_4-N, NO_3-N, available K, available Ca and available Mg) in the terraced paddy field were lower than in the upland rice field. SOC stocks in the terraced paddy and upland rice fields were 21.84 and 21.61 $Mg \cdot C \cdot ha^{-1}$, respectively, but a significant difference was not detected. Similarly, there was no significantly difference in soil erodibility between terraced paddies (range 0.2261–0.2893 $t \cdot h \cdot MJ^{-1} \cdot mm^{-1}$) and upland rice (range 0.2238–0.2681 $t \cdot h \cdot MJ^{-1} \cdot mm^{-1}$). Higher soil erodibility and lower SOC stock were found at the lower toposequence points of both cropping systems.

Author Contributions: Conceptualization, N.A., S.S. and R.H.; methodology, N.A., S.S. and R.H.; investigation, N.A., S.S. and P.K.; writing—original draft preparation, N.A. and P.K.; writing—review and editing, N.A.; supervision, R.H. All authors have read and agreed to the published version of the manuscript.

Funding: This research project is supported by Thailand Science Research and Innovation (TSRI) and Office of the Higher Education Commission (OHEC) (MRG 6280146).

Institutional Review Board Statement: The study was conducted according to the guidelines of the Declaration of Helsinki and approved by the Institutional Review Board of Institute for Population and Social Research, Mahidol University (IPSR-IRB) (COA. No. 2019/05–154).

Informed Consent Statement: Not applicable.

Data Availability Statement: Not applicable.

Acknowledgments: The authors extend their appreciation to Thailand Science Research and Innovation (TSRI) and Office of the Higher Education Commission (OHEC) (MRG 6280146) for supporting this research project. Furthermore, the authors would like to thank the reviewers for their helpful comments to improve the manuscript.

Conflicts of Interest: The authors declare no conflict of interest.

References

1. Jobbágy, E.G.; Jackson, R.B. The vertical distribution of soil organic carbon and its relation to climate and vegetation. *Ecol. Appl.* **2000**, *10*, 423–436. [CrossRef]
2. Todd-Brown, K.E.O.; Randerson, J.T.; Hopkins, F.; Arora, V.; Hajima, T.; Jones, C.; Shevliakova, E.; Tjiputra, J.; Volodin, E.; Wu, T.; et al. Changes in soil organic carbon storage predicted by Earth system models during the 21st century. *Biogeosciences* **2014**, *11*, 2341–2356. [CrossRef]
3. Amelung, W.; Bossio, D.; de Vries, W.; Kögel-Knabner, I.; Lehmann, J.; Amundson, R.; Bol, R.; Collins, C.; Lal, R.; Leifeld, J.; et al. Towards a global-scale soil climate mitigation strategy. *Nat. Commun.* **2020**, *11*, 5427. [CrossRef]
4. Sanderman, J.; Hengl, T.; Fiske, G.J. Soil carbon debt of 12,000 years of human land use. *Proc. Natl. Acad. Sci. USA* **2017**, *114*, 9575–9580. [CrossRef]
5. Gibbs, H.K.; Salmon, J.M. Mapping the world's degraded lands. *Appl. Geogr.* **2015**, *57*, 12–21. [CrossRef]
6. Liu, X.; Herbert, S.; Hashemi, A.; Zhang, X.; Ding, G. Effects of agricultural management on soil organic matter and carbon transformation—A review. *Plant Soil Environ.* **2011**, *52*, 531–543. [CrossRef]
7. Freibauer, A.; Rounsevell, M.D.A.; Smith, P.; Verhagen, J. Carbon sequestration in the agricultural soils of Europe. *Geoderma* **2004**, *122*, 1–23. [CrossRef]
8. Arunrat, N.; Kongsurakan, P.; Sereenonchai, S.; Hatano, R. Soil Organic Carbon in Sandy Paddy Fields of Northeast Thailand: A Review. *Agronomy* **2020**, *10*, 1061. [CrossRef]
9. Mueller, L.; Schindler, U.; Mirschel, W.; Shepherd, T.G.; Ball, B.C.; Helming, K.; Rogasik, J.; Eulenstein, F.; Wiggering, H. Assessing the productivity function of soils. A review. *Agron Sustain. Dev.* **2010**, *30*, 601–614. [CrossRef]
10. Lal, R.; Delgado, J.A.; Groffman, P.M.; Millar, N.; Dell, C.; Rotz, A. Management to mitigate and adapt to climate change. *J. Soil Water Conserv.* **2011**, *66*, 276–285. [CrossRef]
11. Haddaway, N.R.; Hedlund, K.; Jackson, L.E.; Kätterer, T.; Lugato, E.; Thomsen, I.K.; Jørgensen, H.B.; Söderström, B. What are the effects of agricultural management on soil organic carbon in boreo temperate systems? *Environ. Evid.* **2015**, *4*, 23. [CrossRef]

12. Sattler, D.; Murray, L.T.; Kirchner, A.; Lindner, A. Influence of soil and topography on aboveground biomass accumulation and carbon stocks of afforested pastures in South East Brazil. *Ecol. Eng.* **2014**, *73*, 126–131. [CrossRef]

13. Xin, Z.; Qin, Y.; Yu, X. Spatial variability in soil organic carbon and its influencing factors in a hilly watershed of the Loess Plateau, China. *CATENA* **2016**, *137*, 660–669. [CrossRef]

14. Arunrat, N.; Pumijumnong, N.; Sereenonchai, S.; Chareonwong, U. Factors Controlling Soil Organic Carbon Sequestration of Highland Agricultural Areas in the Mae Chaem Basin, Northern Thailand. *Agronomy* **2020**, *10*, 305. [CrossRef]

15. Cao, L.; Liang, Y.; Wang, Y.; Lu, H. Runoff and soil loss from Pinus massoniana forest in southern China after simulated rainfall. *CATENA* **2015**, *129*, 1–8. [CrossRef]

16. Budry, B.; Curtis, J. Environmental perceptions and behavioral change of hillside farmers: The case of Haiti. *J. Caribb. Agro.-Econ. Soc. (CAES)* **2007**, *7*, 122–138.

17. Tadele, A.; Terefe, A.; Selassie, Y.G.; Yitaferu, B.; Wolfgramm, B.; Hurni, H. Soil properties and crop yields along the terraces and topo-sequece of Anjeni watershed, Central Highlands of Ethiopia. *J. Agric. Sci.* **2013**, *5*, 134.

18. Morgan, J.M.; Condon, A.G. Water Use, Grain Yield, and Osmoregulation in Wheat. *Funct. Plant Biol.* **1986**, *13*, 523–532. [CrossRef]

19. Li, Y.M.; Wang, K.Q.; Liu, Z.Q.; Wang, J.Y.; Zhou, X. Effect of measure of engineering preparation to soil water in Yunnan dry-hot river valley. (English Abstract). *J. Soil Water Conserv.* **2006**, *1*, 15–19.

20. Chen, S.-K.; Liu, C.-W.; Chen, Y.-R. Assessing soil erosion in a terraced paddy field using experimental measurements and universal soil loss equation. *CATENA* **2012**, *95*, 131–141. [CrossRef]

21. Shao, H.; Baffaut, C.; Gao, J.E.; Nelson, N.O.; Janssen, K.A.; Pierzynski, G.M.; Barnes, P.L. Development and application of algorithms for simulating terraces within swat. *Trans. Asabe* **2013**, *56*, 1715–1730.

22. Petanidou, T.; Kizos, T.; Soulakellis, N. Socioeconomic Dimensions of Changes in the Agricultural Landscape of the Mediterranean Basin: A Case Study of the Abandonment of Cultivation Terraces on Nisyros Island, Greece. *Environ. Manag.* **2008**, *41*, 250–266. [CrossRef]

23. Tarolli, P.; Cao, W.; Sofia, G.; Evans, D.; Ellis, E.C. From features to fingerprints: A general diagnostic framework for anthropogenic geomorphology. *Prog. Phys. Geogr. Earth Environ.* **2019**, *43*, 95–128. [CrossRef]

24. Chang, T.T. The rice cultures. *Philos. Trans. R. Soc. Lond. B* **1976**, *275*, 143–157.

25. Price, S.; Nixon, L. Ancient Greek Agricultural Terraces: Evidence from Texts and Archaeological Survey. *Am. J. Archaeol.* **2005**, *109*, 665–694. [CrossRef]

26. Chen, S.-K.; Chen, Y.-R.; Peng, Y.-H. Experimental study on soil erosion characteristics in flooded terraced paddy fields. *Paddy Water Environ.* **2012**, *11*, 433–444. [CrossRef]

27. Yuan, Z.; Lun, F.; He, L.; Cao, Z.; Min, Q.; Bai, Y.; Liu, M.; Cheng, S.; Li, W.; Fuller, A.M. Exploring the State of Retention of Traditional Ecological Knowledge (TEK) in a Hani Rice Terrace Village, Southwest China. *Sustainability* **2014**, *6*, 4497–4513. [CrossRef]

28. Deng, C.; Zhang, G.; Liu, Y.; Nie, X.; Li, Z.; Liu, J.; Zhu, D. Advantages and disadvantages of terracing: A comprehensive review. *Int. Soil Water Conserv. Res.* **2021**, *9*, 344–359. [CrossRef]

29. Tian, H.; Lu, C.; Yang, J.; Banger, K.; Huntzinger, D.N.; Schwalm, C.R.; Michalak, A.M.; Cook, R.; Ciais, P.; Hayes, D.; et al. Global patterns and controls of soil organic carbon dynamics as simulated by multiple terrestrial biosphere models: Current status and future directions. *Glob. Biogeochem. Cycles* **2015**, *29*, 775–792. [CrossRef]

30. Haas, H.J.; Willis, W.O.; Boatwright, G.O. Moisture Storage and Spring Wheat Yields on Level-Bench Terraces as Influenced by Contributing Area Cover and Evaporation Control. *Agron. J.* **1966**, *58*, 297–299. [CrossRef]

31. Al Ali, Y.; Touma, J.; Zante, P.; Nasri, S.; Albergel, J. Water and sediment balances of a contour bench terracing system in a semi-arid cultivated zone (El Gouazine, central Tunisia). *Hydrol. Sci. J.* **2008**, *53*, 883–892. [CrossRef]

32. Homburg, J.A.; Sandor, J.A. Anthropogenic effects on soil quality of ancient agricultural systems of the American Southwest. *CATENA* **2011**, *85*, 144–154. [CrossRef]

33. Li, X.H.; Yang, J.; Zhao, C.Y.; Wang, B. Runoff and sediment from orchard terraces in southeastern China. *Land Degrad. Dev.* **2012**, *25*, 184–192. [CrossRef]

34. Liu, X.; He, B.; Li, Z.; Zhang, J.; Wang, L.; Wang, Z. Influence of land terracing on agricultural and ecological environment in the loess plateau regions of China. *Environ. Earth Sci.* **2011**, *62*, 797–807. [CrossRef]

35. Fukamachi, K. Sustainability of terraced paddy fields in traditional satoyama landscapes of Japan. *J. Environ. Manag.* **2017**, *202*, 543–549. [CrossRef]

36. Garcia-Franco, N.; Wiesmeier, M.; Goberna, M.; Martínez-Mena, M.; Albaladejo, J. Carbon dynamics after afforestation of semiarid shrublands: Implications of site preparation techniques. *For. Ecol. Manag.* **2014**, *319*, 107–115. [CrossRef]

37. Xu, G.; Lu, K.; Li, Z.; Li, P.; Wang, T.; Yang, Y. Impact of soil and water conservation on soil organic carbon content in a catchment of the middle Han River, China. *Environ. Earth Sci.* **2015**, *74*, 6503–6510. [CrossRef]

38. Zhang, J.H.; Wang, Y.; Li, F.C. Soil organic carbon and nitrogen losses due to soil erosion and cropping in a sloping terrace landscape. *Soil Res.* **2015**, *53*, 87–96. [CrossRef]

39. McConchie, J.A.; Huan-Cheng, M. A discussion of the risks and benefits of using rock terracing to limit soil erosion in Guizhou Province. *J. For. Res.* **2002**, *13*, 41–47. [CrossRef]

40. Ramos, M.C.; Cots-Folch, R.; Martínez-Casasnovas, J.A. Effects of land terracing on soil properties in the Priorat region in Northeastern Spain: A multivariate analysis. *Geoderma* **2007**, *142*, 251–261. [CrossRef]

41. Sahahirun, A.; Baconguis, R.D. A Model in Promoting Highland Terrace Paddy Cultivation Technology in Northern Thailand. *IJERD-Int. J. Environ. Rural Dev.* **2018**, *9*, 9–10.
42. Damrongkiet, C. Upland Rice and Food Security Development in Highland Area (in Thailand). 2012. Available online: http://www.ricethailand.go.th/rkb3/Eb_024.pdf (accessed on 8 August 2021).
43. Sang-Arun, J.; Mihara, M.; Horaguchi, Y.; Yamaji, E. Soil erosion and participatory remediation strategy for bench terraces in northern Thailand. *CATENA* **2006**, *65*, 258–264. [CrossRef]
44. United States Department of Agriculture (USDA). *Diagnosis and Improvement of Saline and Alkali Soils, Agriculture. Handbook No. 60, U.S. Salinity Laboratory*; Government Printing Office: Washington, DC, USA, 1954.
45. National Soil Survey Center. *Soil Survey Laboratory Methods Manual. Soil Survey Investigations Report No. 42, Version 3.0, Natural Conservation Service*; USDA: Washington, DC, USA, 1996.
46. Walkley, A.; Black, J.A. An examination of the dichromate method for determining soil organic matter and a proposed modification of the chromic acid titration method. *Soil Sci.* **1934**, *37*, 29–38. [CrossRef]
47. Bray, R.H.; Kurtz, L.T. Determination of total organic and available form of phosphorus in soil. *Soil Sci.* **1945**, *59*, 39–46. [CrossRef]
48. Thomas, G.W. *Soil pH and Soil Acidity. Method of Soil Analysis, Part 3: Chemical Methods*; Sparks, D.L., Page, A.L., Helmke, P.A., Loeppert, R.H., Soltanpour, P.N., Tabatabai, M.A., Johnston, C.T., Sumner, M.E., Eds.; SSSA Inc.: Madison, WI, USA; ASA Inc.: Madison, WI, USA, 1996; pp. 475–490.
49. Wischmeier, W.H.; Johnson, C.B.; Cross, B.V. A soil erodibility nomograph for farmland and construction sites. *J. Soil Water Conserv.* **1971**, *26*, 189–193.
50. Williams, J.; Renard, K.; Dyke, P. EPIC: A new method for assessing erosion's effect on soil productivity. *J. Soil Water Conserv.* **1983**, *38*, 381–383.
51. Shirazi, M.A.; Boersma, L. A Unifying Quantitative Analysis of Soil Texture. *Soil Sci. Soc. Am. J.* **1984**, *48*, 142–147. [CrossRef]
52. Du, J.; Shi, C.; Fan, X.; Zhou, Y. Impacts of socio-economic factors on sediment yield in the Upper Yangtze River. *J. Geogr. Sci.* **2011**, *21*, 359–371. [CrossRef]
53. Wu, L.; Long, T.-Y.; Liu, X.; Mmereki, D. Simulation of soil loss processes based on rainfall runoff and the time factor of governance in the Jialing River Watershed, China. *Environ. Monit. Assess.* **2012**, *184*, 3731–3748. [CrossRef]
54. Chen, J.; He, B.; Wang, X.; Ma, Y.; Xi, W. The effects of Herba Andrographitis hedgerows on soil erodibility and fractal features on sloping cropland in the Three Gorges Reservoir Area. *Environ. Sci. Pollut. Res.* **2013**, *20*, 7063–7070. [CrossRef]
55. Liu, X.; Zhang, Y.; Li, P. Spatial Variation Characteristics of Soil Erodibility in the Yingwugou Watershed of the Middle Dan River, China. *Int. J. Environ. Res. Public Health* **2020**, *17*, 3568. [CrossRef] [PubMed]
56. Sharpley, A.N.; Wiliams, J.R. EPIC The Erosion Productivity Impact Calculator. 1. Model Documentation. U.S. Department of Agriculture (Technical Buletin No. 1768). USA Government Printing Office: Washington, DC, USA, 1990.
57. De Mendiburu, F. Package "agricolae" Title Statistical Procedures for Agricultural Research. 2020. Available online: http://tarwi.lamolina.edu.pe/~{}fmendiburu (accessed on 15 September 2021).
58. Wickham, H. *ggplot2. ggplot2*; Springer: New York, NY, USA, 2009.
59. Kassambara, A.; Mundt, F. Package "factoextra" Type Package Title Extract and Visualize the Results of Multivariate Data Analyses. 2020. Available online: https://github.com/kassambara/factoextra/issues (accessed on 15 September 2021).
60. Cui, B.; Zhao, H.; Li, X.; Zhang, K.; Ren, H.; Bai, J. Temporal and spatial distributions of soil nutrients in Hani terraced paddy fields, Southwestern China. *Procedia Environ. Sci.* **2010**, *2*, 1032–1042. [CrossRef]
61. Miyamoto, T.; Kawahara, M.; Mori, Y.; Somura, H.; Ide, J.; Takahashi, E.; Yone, M.; Suetsugu, A. Evaluation of management practices in forest soil environments using a multi-frequency electromagnetic sounding system. *J. Jpn. Soc. Soil Phys.* **2013**, *124*, 17–24.
62. Mori, Y.; Sasaki, M.; Morioka, E.; Tsujimoto, K. When do rice terraces become rice terraces? *Paddy Water Environ.* **2019**, *17*, 323–330. [CrossRef]
63. Schmitter, P.; Dercon, G.; Hilger, T.; Le Ha, T.T.; Thanh, N.H.; Lam, N.; Vien, T.D.; Cadisch, G. Sediment induced soil spatial variation in paddy fields of Northwest Vietnam. *Geoderma* **2010**, *155*, 298–307. [CrossRef]
64. Baskan, O.; Dengiz, O.; Gunturk, A. Effects of toposequence and land use-land cover on the spatial distribution of soil properties. *Environ. Earth Sci.* **2016**, *75*, 448. [CrossRef]
65. De Blécourt, M.; Hänsel, V.M.; Brumme, R.; Corre, M.D.; Veldkamp, E. Soil redistribution by terracing alleviates soil organic carbon losses caused by forest conversion to rubber plantation. *For. Ecol. Manag.* **2014**, *313*, 26–33. [CrossRef]
66. Qiu, Y.J.; Xu, M.X.; Shi, C.D.; Zhang, Z.X.; Zhang, S. Dynamic accumulation of soil organic carbon of terrace changed from slope cropland in the hilly loess plateau of eastern Gansu Province. *J. Plant Nutr. Fertil.* **2014**, *20*, 87–98.
67. Pabst, H.; Gerschlauer, F.; Kiese, R.; Kuzyakov, Y. Land Use and Precipitation Affect Organic and Microbial Carbon Stocks and the Specific Metabolic Quotient in Soils of Eleven Ecosystems of Mt. Kilimanjaro, Tanzania. *Land Degrad. Dev.* **2015**, *27*, 592–602. [CrossRef]
68. Schmitter, P.; Dercon, G.; Hilger, T.; Hertel, M.; Treffner, J.; Lam, N.; Duc Vien, T.; Cadisch, G. Linking spatio-temporal variation of crop response with sediment deposition along paddy rice terraces. *Agric. Ecosyst. Environ.* **2010**, *140*, 34–45. [CrossRef]
69. Huang, L.-M.; Thompson, A.; Zhang, G.-L.; Chen, L.-M.; Han, G.-Z.; Gong, Z.-T. The use of chronosequences in studies of paddy soil evolution: A review. *Geoderma* **2015**, *237–238*, 199–210. [CrossRef]

70. Chen, D.; Wei, W.; Daryanto, S.; Tarolli, P. Does terracing enhance soil organic carbon sequestration? A national-scale data analysis in China. *Sci. Total Environ.* **2020**, *721*, 137751. [CrossRef] [PubMed]
71. Wei, W.; Chen, D.; Wang, L.; Daryanto, S.; Chen, L.; Yu, Y.; Lu, Y.; Sun, G.; Feng, T. Global synthesis of the classifications, distributions, benefits and issues of terracing. *Earth-Sci. Rev.* **2016**, *159*, 388–403. [CrossRef]
72. Chen, D.; Wei, W.; Chen, L. How can terracing impact on soil moisture variation in China? A meta-analysis. *Agric. Water Manag.* **2019**, *227*, 105849. [CrossRef]
73. Tadesse, B.; Mesfin, S.; Tesfay, G.; Abay, F. Effect of integrated soil bunds on key soil properties and soil carbon stock in semi-arid areas of northern Ethiopia. *South Afr. J. Plant Soil* **2016**, *33*, 297–302. [CrossRef]
74. Arunrat, N.; Pumijumnong, N.; Hatano, R. Practices sustaining soil organic matter and rice yield in tropical monsoon area. *Soil Sci. Plant Nutr.* **2017**, *63*, 274–287.
75. Arunrat, N.; Sereenonchai, S.; Hatano, R. Impact of burning on soil organic carbon of maize-upland rice system in Mae Chaem Basin of Northern Thailand. *Geoderma* **2021**, *392*, 115002. [CrossRef]
76. Bryan, R.; Govers, G.; Poesen, J. The concept of soil erodibility and some problems of assessment and application. *CATENA* **1989**, *16*, 393–412. [CrossRef]
77. Tejada, M.; Gonzalez, J. The relationships between erodibility and erosion in a soil treated with two organic amendments. *Soil Tillage Res.* **2006**, *91*, 186–198. [CrossRef]
78. Römkens, M.J.M.; Wells, R.R.; Wang, B.; Zheng, F.; Hickey, C.J. Soil Erosion on Upland Areas by Rainfall and Overland Flow. In *Advances in Water Resources Engineering. Handbook of Environmental Engineering*; Yang, C., Wang, L., Eds.; Springer: Berlin/Heidelberg, Germany, 2015; Volume 14.
79. Dlugoß, V.; Fiener, P.; Van Oost, K.; Schneider, K. Model based analysis of lateral and vertical soil carbon fluxes induced by soil redistribution processes in a small agricultural catchment. *Earth Surf. Process. Landf.* **2011**, *37*, 193–208. [CrossRef]
80. Wiaux, F.; Cornelis, J.-T.; Cao, W.; Vanclooster, M.; Van Oost, K. Combined effect of geomorphic and pedogenic processes on the distribution of soil organic carbon quality along an eroding hillslope on loess soil. *Geoderma* **2014**, *216*, 36–47. [CrossRef]
81. Qi, Y.; Wei, W.; Li, J.; Chen, C.; Huang, Y. Effects of terracing on root distribution of Pinus tabulaeformis Carr. forest and soil properties in the Loess Plateau of China. *Sci. Total Environ.* **2020**, *721*, 137506. [CrossRef] [PubMed]
82. Bajpai, R.K.; Tripathi, R.P. Evaluation of non-pudding under shallow water tables and alternative tillage methods on soil and crop parameters in a rice—Wheat system in Uttar Pradesh. *Soil Tillage Res.* **2000**, *55*, 96–106. [CrossRef]
83. Deng, L.; Kim, D.-G.; Peng, C.; Shangguan, Z. Controls of soil and aggregate-associated organic carbon variations following natural vegetation restoration on the Loess Plateau in China. *Land Degrad. Dev.* **2018**, *29*, 3974–3984. [CrossRef]
84. Shi, Z.H.; Cai, C.F.; Ding, S.W.; Wang, T.W.; Chow, T.L. Soil conservation planning at the small watershed level using RUSLE with GIS: A case study in the Three Gorge Area of China. *CATENA* **2004**, *55*, 33–48. [CrossRef]
85. Zhang, K.L.; Shu, A.P.; Xu, X.L.; Yang, Q.K.; Yu, B. Soil erodibility and its estimation for agricultural soils in China. *J. Arid Environ.* **2008**, *72*, 1002–1011. [CrossRef]
86. Zhang, K.; Yu, Y.; Dong, J.; Yang, Q.; Xu, X. Adapting & testing use of USLE K factor for agricultural soils in China. *Agric. Ecosyst. Environ.* **2019**, *269*, 148–155. [CrossRef]
87. Shabani, F.; Kumar, L.; Esmaeili, A. Improvement to the prediction of the USLE K factor. *Geomorphology* **2014**, *204*, 229–234. [CrossRef]
88. Ostovari, Y.; Ghorbani-Dashtaki, S.; Bahrami, H.-A.; Naderi, M.; Dematte, J.A.M.; Kerry, R. Modification of the USLE K factor for soil erodibility assessment on calcareous soils in Iran. *Geomorphology* **2016**, *273*, 385–395. [CrossRef]

Article

Considerations on Field Methodology for Macrofungi Studies in Fragmented Forests of Mediterranean Agricultural Landscapes

Abel Fernández Ruiz [1], David Rodríguez de la Cruz [1,2,*], José Luis Vicente Villardón [3], Sergio Sánchez Durán [4,5], Prudencio García Jiménez [1] and José Sánchez Sánchez [1,2]

[1] Institute for Agrobiotechnology Research (CIALE), Universidad de Salamanca, Río Duero 12, 37185 Villamayor, Spain; abel@usal.es (A.F.R.); prudenciogarjim@gmail.com (P.G.J.); jss@usal.es (J.S.S.)

[2] Department of Botany and Plant Physiology, Universidad de Salamanca, Licenciado Méndez Nieto s/n, 37007 Salamanca, Spain

[3] Department of Statistics, Campus Miguel de Unamuno, Universidad de Salamanca, Alfonso X El Sabio s/n, 37007 Salamanca, Spain; villardon@usal.es

[4] Unidad de Recursos Forestales, Centro de Investigación y Tecnología Agroalimentaria de Aragón (CITA), Avda de Montañana 930, 50059 Zaragoza, Spain; ssanchezd@cita-aragon.es

[5] Instituto Agroalimentario de Aragón IA2 (CITA-Universidad de Zaragoza), 50013 Zaragoza, Spain

* Correspondence: droc@usal.es; Tel.: +34-677-584-172

Abstract: The methodology used for the determination of macrofungal diversity in Mediterranean areas differs in the time of sampling and the number of years displayed, making it difficult to compare results. Furthermore, the results could be refuted because the studies are being conducted over an insufficient number of years or without considering the variation of the meteorological conditions from one year to the next and its effects on fruiting time, which might not fit the sampling. In order to optimize field work on fungal fruiting in Mediterranean environments dominated by holm oak (*Quercus ilex* L.), a weekly field analysis of macrofungal diversity from February 2009 to June 2013 was carried out in a Mediterranean holm oak forest in the middle-west of the Iberian Peninsula. The results revealed that fruiting bodies appeared throughout the year and that there was a delay in autumn fruiting, overlapping with spring. All this seems to indicate that weekly collection throughout the year and for a period of two years could be sufficient to estimate the macrofungal biodiversity of this ecosystem.

Keywords: macrofungi; field sampling; fungal diversity; Mediterranean forests; *Quercus ilex*

Citation: Fernández Ruiz, A.; Rodríguez de la Cruz, D.; Vicente Villardón, J.L.; Sánchez Durán, S.; García Jiménez, P.; Sánchez Sánchez, J. Considerations on Field Methodology for Macrofungi Studies in Fragmented Forests of Mediterranean Agricultural Landscapes. *Agronomy* **2022**, *12*, 528. https://doi.org/10.3390/agronomy12020528

Academic Editor: Adamo Domenico Rombolà

Received: 12 January 2022
Accepted: 18 February 2022
Published: 20 February 2022

Publisher's Note: MDPI stays neutral with regard to jurisdictional claims in published maps and institutional affiliations.

1. Introduction

It is estimated that the number of fungal species on earth is between 5.1 and 6.0 M [1,2]. Fungal species play an essential role in forest ecosystems, whether they act as decomposers, parasites, pathogens or mutualists [3]. Studies on ecology and fungi have traditionally focused on fungi that form macroscopic spore-producing structures, thus called macrofungi, as opposed to microfungi and their microscopic reproductive structures [4]. There are a large number of papers dealing with fungal diversity and the ecological aspects derived from it in temperate ecosystems [5–7]. However, little is known about the composition and structure of the fungal communities in holm oak-dominated forests (*Quercus ilex* subsp. *ballota* (Desf.) Samp.) that characterize much of the western landscape of the Mediterranean basin [8]. The agricultural landscape of the Mediterranean Sea basin is made up of numerous plots of continuous crop formations, both arable and woody, interrupted by forest formations [9]. In many cases these forest communities are relict forest ecosystems that used to shape the landscape. The intensification of cultivation in recent centuries has contributed to a greater homogeneity of the landscape, although some traditional patches dominated by tree species have been maintained due to various forestry or livestock uses [10]. The existence of these small woodlots is of great interest for a possible natural succession and conversion of disturbed ecosystems [11]. In addition, some of these residual isolated forests in the landscape could contribute to the design of future ecological corridors [12]. The performance of these fragmented forest formations is affected by various ecological processes linked to adaptation and acclimation, including, among others, plant–fungus interactions [13]. The diversity of soil fungi promotes numerous functionalities in forest ecosystems, not only the decomposition or recycling of nutrients, but also increasing their productivity and providing a better response to environmental stress [14].

One of the reasons for the little knowledge of macro-fungal communities in Mediterranean holm oak-dominated forests could be that this type of study requires several years of work due to the influence of meteorological variations on fungal fruiting [15]. In addition, the complexity is increased by the heterogeneity of the methods used, particularly the sampling frequency and different time periods.

Regarding the sampling period, early work revealed that the main period of fungal fruiting was concentrated in autumn [16], noting that environmental factors could explain the results of diversity obtained in several consecutive study years [17]. Other studies advised concentrating the estimation effort in the season of maximum production, since with weekly frequency it was possible to register 75% of the total production [18]. Most of the studies on fungal diversity in Mediterranean ecosystems are based on these results obtained for temperate ecosystems. Other researchers worked on personal observations not based on previous working hypotheses, which restrict collections at certain times of the year [19–21].

The conclusions obtained in a specific ecosystem could not be premises for others, and the quantitative methods based on fungal production may not be assumptions for qualitative studies of absence/presence. On the other hand, previous studies [17,18], when carried out in autumn or during the peak production period, took into account (i) the significant changes that the meteorological conditions cause not only in the fungal life cycle, but also in the composition of the community, and (ii) the seasonal fluctuations in the appearance of the fruiting bodies throughout the year, which could even go so far as to not occur [8,22]. Seasonal weather conditions varied significantly and successively on the planet during recent decades and could be related to these observations. These changes appear to be due to a rapid anthropogenic climate change [23] with a delay in autumn fruiting, a spring advance and a shorter duration of fruiting [24,25]. As noted above, most studies of macrofungal diversity for *Quercus ilex* L. in the Mediterranean area focused on a time of the year considered to be of maximum fungal fructification, the autumn. However, these studies do not consider if the new meteorological conditions have consequences in terms of the period of fructification and pay little attention to the specimens collected during summer and winter. The methodology used could be refuted, partially or totally, to the detriment of the results obtained.

Another problem when comparing studies is the different number of years invested in their development and the level of intensity. The patterns of variation in the latitudinal fungal distribution, its abundance, and the possible drivers of these patterns are insufficiently known [1,26]. However, several studies indicate that fungal communities have a fairly variable taxonomic composition over small distances. This high level of spatial variation has been attributed to the irregular development of fungi, to the natural heterogeneity of the resources in the soil matrix [27,28], and to fine-scale spatial variations [29]. Estimating fungal diversity in an ecosystem is an expensive and slow process which requires several years to record most of the species present [30–32] until the species accumulation curve reaches an equilibrium [33]. Another study [34] analyzed collections made during 30 years in one ecosystem, finding that the species accumulation curve began to level after six years, then grew more slowly and stabilized, reaching a plateau after 19 years. Therefore, and according to this work, it would take about 20 years to know the fungal biodiversity of an ecosystem, which is too high of an economic cost to carry out any study. The number of years used in the works reviewed for Mediterranean ecosystems differs, varying from a single year [35], through two [36], three [37], four [8,19,38] to twelve years [21].

The main aim of this work is to estimate the minimum number of years needed to know the macrofungal diversity in a Mediterranean ecosystem dominated by the holm oak. To this end, weekly transects were carried out over four years, eliminating, a priori, seasonal estimation errors and paying attention to the appearance of fruiting bodies in the months theoretically more adverse to fungal fructification. Other objectives considered were to know if it is possible to compare the data obtained in studies that have used a different number of years, and to check if restricted samples to the autumn are able to characterize the fungal biodiversity.

2. Materials and Methods

2.1. Study Area

The Forest of La Orbada is located in the middle-west part of the Iberian Peninsula (41°8'0.48" N, 5°29'0.01" W) at an altitude of 820-856 m.a.s.l. It consists of a set of plots of mature holm oaks with a shrub stratum formed by *Cistus ladanifer* L., *Rosa canina* L., *Lavandula stoechas* Lam., *Daphne gnidium* L., *Thymus mastichina* L. and *Thymus zygis* Loefl. ex. L. Geologically it is composed of sandstones and conglomerates—tertiary sediments (Priabonian age), which rest on a Paleozoic plinth formed by granites and rocks of the Cambrian and Siluric, thin detritic packages, strongly reddish, where thick sandstones are very abundant [39]. This area has a flat topography and Luvisol calcium soil type (reddish-brown soil, with silica gravel), with poor regular drainage [40]. The pH of the soil ranges between 6.6 and 7.2 [41]. The climate is temperate continental cold with a dry season of two months [42], induced by the combined effect of altitude, length and orographic insulation. This climate is characterized by cold winters, relatively cool summers, a superior annual thermal oscillation at 17 °C and an annual rainfall of 372 mm. The average annual temperature is 12.2 °C, January is the coldest month with average temperatures around 4 °C, and July the warmest with an average temperature of about 21.5 °C [43].

2.2. Data Collection

A plot of forest land with an area of 7.5 Ha was selected from the analysis of aerial photographs and previous visits. The experimental design consisted of weekly sampling from February 2009 to June 2013. A linear transect in the center of the forest land of 1030 m was performed, with a bandwidth of 5 m, covering an area recommended for this kind of survey [44]. It was a qualitative study, assessing the presence or absence of epigeous Ascomycota and Basidiomycota species with medium or large fruit bodies. The collected specimens were identified at the Institute for Agrobiotechnology Research (CIALE, University of Salamanca) using classical techniques in mycology, analyzing both macroscopic and microscopic characters and using different chemical reactions [21]. The different taxa and groups were identified and classified using general and specific literature, while CABI Index Fungorum [45] and Kirk et al. [46] were referenced mainly for the nomenclature.

Species collected in other parts of the La Orbada forest by the University of Salamanca were also used to accomplish biodiversity tests.

2.3. Diversity Analysis

2.3.1. Alpha and Beta Diversity

The most direct measure of biodiversity is richness [47,48], so we assessed the fungal biodiversity of La Orbada forest based on the species richness recorded. Alpha diversity was the number of species obtained in transects. In spite of the lack of botanic homogeneity characterizing Mediterranean ecosystems [49], five Mediterranean communities dominated by *Quercus* spp. with a known macrofungal diversity (Figure 1) were selected to obtain Beta diversity:

Campanarios de Azaba Biological Reserve: Located in central-western Spain, it presents the traditional agrosilvopastoral system known as "dehesa" (in Spanish) or "montado" (in Portuguese). In this area, there is a predominance of holm oaks mixed with Pyrenean oaks (*Quercus pyrenaica* Willd.) and meadows of different types and compositions. The study of macrofungal diversity was developed between 2009 and 2012 [37].

Foros: A "montado" dominated by cork oak (*Quercus suber* L.) with holm oaks, located in central Portugal (western Iberian Peninsula). Fungal fruiting bodies were collected between 2003–2006 [33].

Peloponnese: A study made in the southern part of the Peloponnese (southern Greece), in deciduous forests of *Quercus* spp., mainly Italian oak (*Quercus frainetto* Ten.) and downy oak (*Quercus humilis* Mill.), in the 1996–2001 period [19].

Fango: Located in Corsica (France). It is a mature holm oak forest with many large trees and a large shrub cover of 7 m high. Although the vegetation appears to be similar to La Orbada forest, there are other woody species capable of associating with ectomycorrhizal fungi such as strawberry tree (*Arbutus unedo* L.). This work was developed between 1999 and 2002 [20].

Collestrada: A study carried out during 2011 in the Collestrada forest (Central Italy). The vegetation is heterogeneous, with several species of *Quercus* spp., European hornbeam (*Carpinus betulus* L.) and plantations of stone pine (*Pinus pinea* L.) and maritime pine (*Pinus pinaster* Aiton) [35].

We compared ecosystems using the Sørensen Similarity Index (Is), which allowed us to describe the spatial differentiation and differences in species richness between these communities [28]. For this analysis, collections determined at species level or synonymized after publication were not taken into account.

Figure 1. Location of Mediterranean habitats used in the analysis of Alpha and Beta Diversity. 1: La Orbada Forest (CW, Spain), 2: Campanarios de Azaba Biological Reserve (CW, Spain), 3: Foros (C, Portugal), 4: Peloponnese (S, Greece), 5: Fango forest (Corsica, France), 6: Collestrada forest (C, Italy).

2.3.2. Non-Parametric Estimators

Non-parametric estimators are based on the study of rare species and allow us to estimate new species numbers from the ratios of species already detected in the samples. There are two types of non-parametric estimators: based on abundance and incidence models [50]. We used estimators based on incidence models because it was a presence-absence study. The incidence model considers the number of samples in which each species is present. Four non-parametric estimators were applied to the dataset: Jacknife 1, Jacknife 2, Chao 2 and Incidence-based Coverage Estimator (ICE) [51]. Estimations were made using EstimateS vers. 9.0 [52], and the results of the random rearrangements were exported to a Microsoft Office Excel file, where the bias was calculated and the graphs were developed. Therefore, we performed a rarefaction curve fitted, by extrapolation, to the asymptotic species accumulation functions. This resulted in smoothed curves calculated with non-parametric estimators that allowed us to estimate the real richness of the species. The smooth richness accumulation curve, S(est), is a sampling-based rarefaction curve, which estimates the statistical expectation of the number of species observed when resampling over the total number of sampling units at each site with increasing efforts [53]. The upper and lower limits of the 95% confidence intervals for S(est) were also plotted. Finally, we determined the real richness by fitting the cumulative richness curve to one of the species accumulation models and extrapolating for an infinite sampling effort [54]. Chao 2 estimates the presence-absence data of a species in a given sample, i.e., whether the species is present and how many times it is present in a given sample [55].

2.4. Statistical Analysis

Principal Coordinate Analysis (PCA), a multivariate ordination analysis, was performed using a similarity coefficient that took into account the binary character of the data (Sørensen coefficient). The PCA identifies the points in a weekly scatter plot with the combination of both variables. January 2009 is not represented because there is no data, nor other combinations in which no species were found in the corresponding weeks.

The analyses were performed with R (R Core Team 2016) using the vegan packages [56] and MultBiplotR [57]. In the graphical representation of PCA analyses, the dimensions obtained can be interpreted as hypothetical (environmental) gradients that adequately capture the structure of the data [58]. Two weeks of sampling will appear next when their species composition is similar and, indirectly, when the environmental conditions necessary for the development of fungal fruiting bodies are the same.

3. Results

In the four years studied, 173 different taxa were identified. If the number of species collected weekly is grouped by months and accumulated by years (Figure 2), greater fungal fruiting in late autumn and early winter (November, December and January) was observed, followed by smaller and shorter fruiting in early spring (April). Species were collected during all months of the year.

The most abundant species (collected in more than 40 weeks) were *Leccinellum lepidum* H. Bouchet ex Essette) Bresinsky & Manfr. Binder, *Infundibulicybe gibba* (Pers.) Harmaja (=*Clitocybe gibba* (Pers.) P. Kumm.), *Laccaria laccata* (Scop.) Cooke, *Lepista nuda* (Bull.) Cooke and *Russula fragilis* Fr. Figure 3 shows the species collected in at least 18 samplings, with annual fruiting during autumn-winter and extensive collection over a period of several weeks.

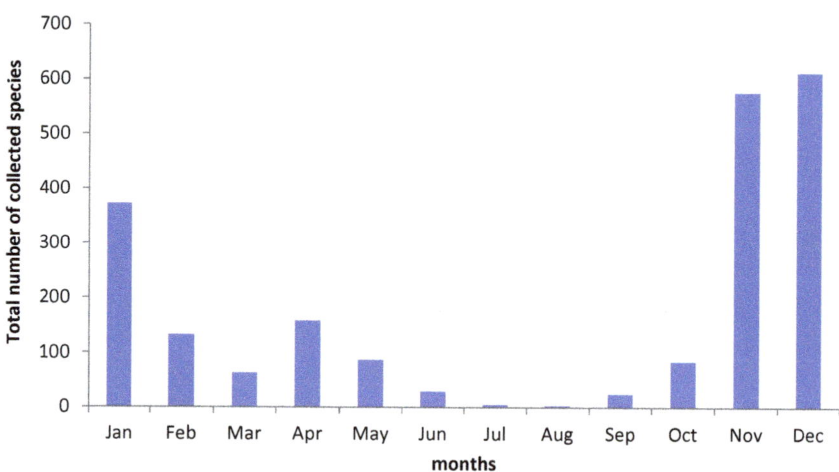

Figure 2. Total number of collected species accumulated by months.

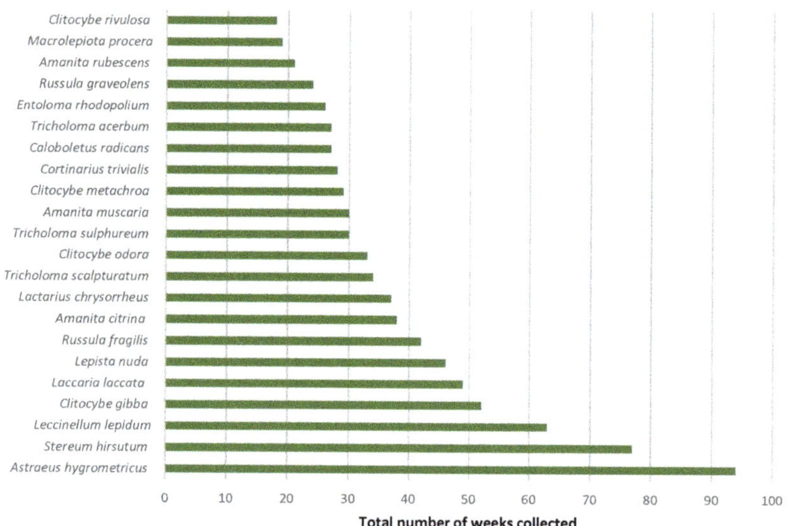

Figure 3. Number of weeks of collection for the most abundant species.

The species collected in the months of July and August were all mycorrhizal: *Boletus aereus* Bull., *Suillellus queletii* (Schulzer) Vizzini, Simonini & Gelardi, *Caloboletus radicans* (Pers.) Vizzini, *Leccinellum lepidum, Amanita rubescens* Pers., *Russula pectinatoides* Peck. and *Russula heterophylla* (Fr.) Fr., with *Caloboletus radicans* being the most abundant, collected for 27 weeks.

Macrofungal species richness (S), total number of species collected, and Sørensen similarity indexes (Is), obtained by comparing the La Orbada forest (189 species) with other Mediterranean ecosystems dominated by *Quercus* spp., are shown in Table 1. The number of species identified in the study area was higher than those recorded in Foros, Campanarios de Azaba Biological Reserve and Collestrada (133 species), similar to southern Peloponnese and lower than those collected in Fango forest. Similarity rates were low.

Table 1. Diversity parameters between total collections in La Orbada forest (189 species) and other Mediterranean ecosystems dominated by *Quercus* species.

Sites	S	SC	Is
Campanarios de Azaba BR (Spain)	148	55	0.32
Foros "montado" (Portugal)	172	60	0.33
Fango forest (Corcega)	234	44	0.24
Collestrada (Italy)	133	41	0.26
Peloponnese (Greece)	197	42	0.22

S: Species richness; SC: Species in common; Is: Sørensen similarity index.

Figure 4 shows the total number of species collected per season, displaying also an annual fluctuation of the fungal fruiting with great interannual variation. It should be noted that for the year 2013 data only appear until spring because the study ended in June 2013. The macrofungal fruiting bodies were collected during the four seasons, with a maximum production period that begins in autumn and does not end until spring. The season with the highest number of species was autumn, except in the year 2011, where it took place in winter.

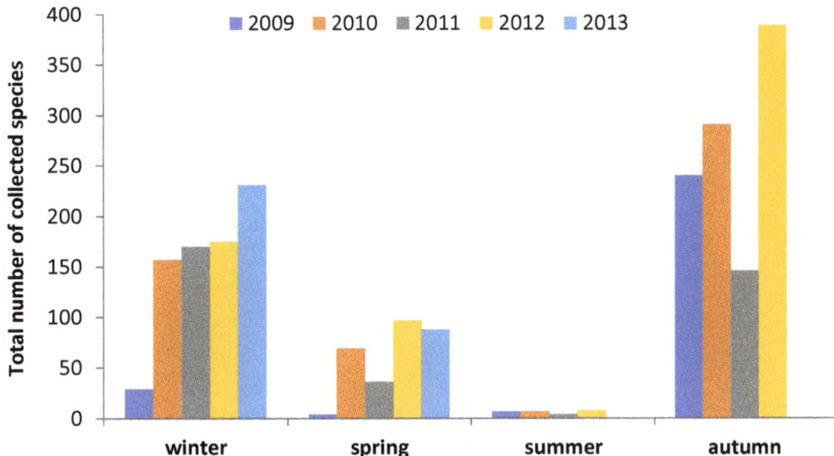

Figure 4. Total number of species collected by seasons.

The greatest diversity of species collected during the study period occurred in the months of December and November, with an average of 153 and 144 species identified, respectively. January also recorded a high macrofungal diversity, with 93 species on average, although the remaining winter months showed a lower number of species, with 27 and 13 species on average in February and March, respectively (Figure 5). This decline continued during the following months, except in the month of April with a slight average increase of 32 species for all the years analyzed. In some months of July and August in the 2009–2012 period, no species were found. Likewise, a small increase in fungal diversity was observed in late summer and early autumn, with 21 species on average collected in October of the 2009–2012 period, a similar average number for the month of May (17 species).

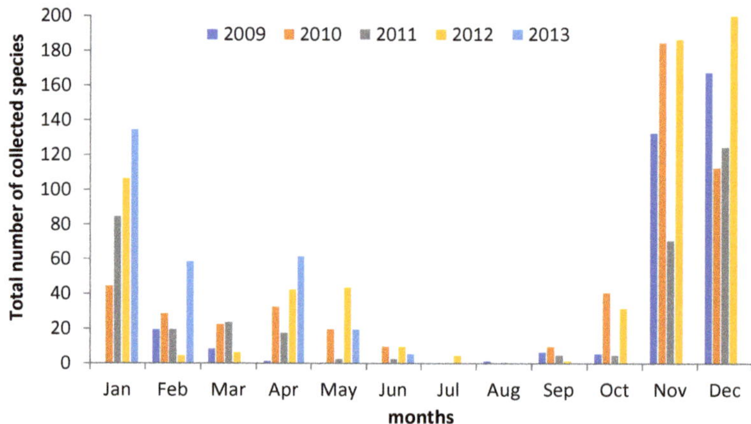

Figure 5. Total number of species collected per years and months.

In any case, the interannual variation indicated that fructification took place throughout the year with higher production in December, a decrease throughout the winter, a slight increase in April and very low average diversity values in the summer months (six in June and September, one in July and August, respectively).

A weekly analysis of fungal diversity in the transition months between autumn and winter (December and January) indicates that the last weeks of autumn (second and third week of December) had a greater variety of fruiting bodies, except in 2009 where it took place in the first week of December (Figure 6). Subsequently, the number of species collected decreased in early winter (last week of December and first two weeks of January) but remained more or less stable for one month.

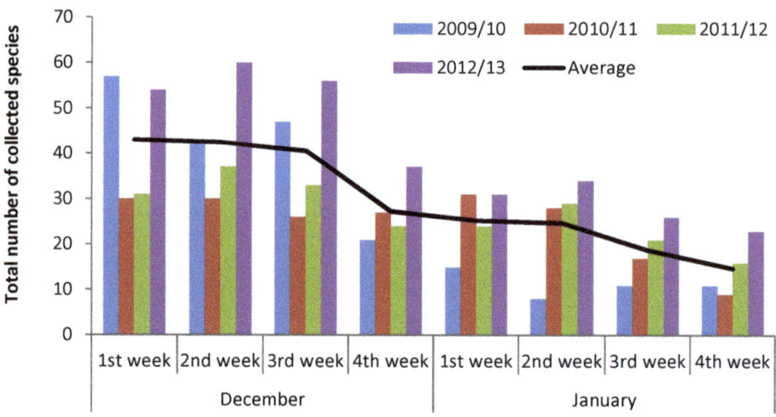

Figure 6. Total number of collections per week in December and January for the four years of survey.

An analysis of the number of taxa identified weekly in the La Orbada forest (Figure 7) indicated a first spring with low diversity, which increased exponentially in autumn 2009. In subsequent years, there was the same seasonal behavior on a smaller scale, increasing in the spring and autumn periods and remaining stable the rest of the year. This stability did not imply that there were no slight increases in diversity as the amount of sampling progressed, although at the end of the study (June 2013), there did not seem to be a plateau in the number of taxa collected. The number of new species collected during the first

12 months of the study was 110, 33 new species were collected in the 2010/2011 period (143), 13 in the 2011/2012 period (156) and 17 in the last period.

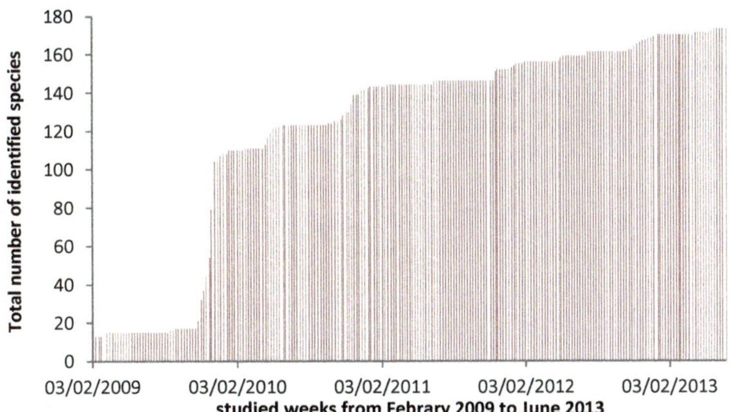

Figure 7. New species (accumulated) identified during the studied period in La Orbada forest by weeks of sampling.

The smooth richness accumulation curve gradually lost slope as the number of sample units increased, clearly tending towards an asymptote (Figure 8a). The statistical expectation of the number of species observed at the end of the work was close to the asymptotic richness, exceeding the logarithmic phase of the curve, for 229 transects it was 172, with a confidence interval between 162 and 182 species.

Non-parametric estimators smoothly approximate the asymptotic richness with increasing sampling efforts. Some of them were above the observed richness values, underestimating the asymptotic richness and producing a positive bias (Figure 8b). The most efficient species accumulation models were ICE, 87.60%, and Chao 2, 85.14%.

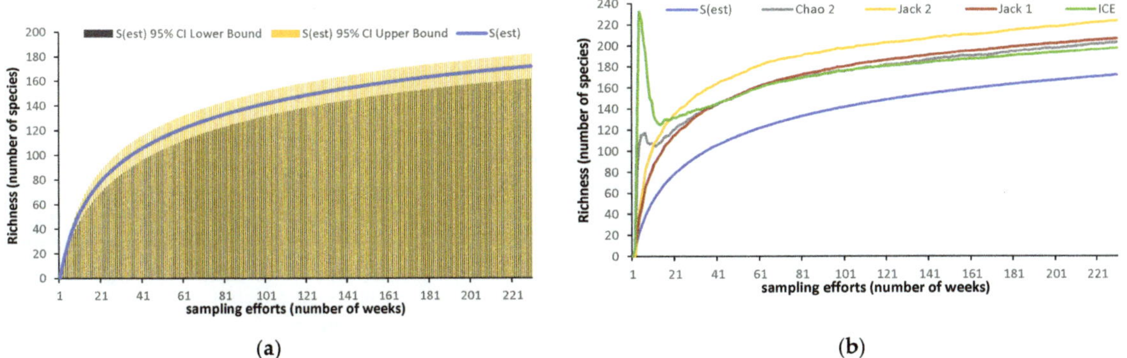

(a)

(b)

Figure 8. (a) Species rarefaction curve S(est), Upper and lower bounds of the 95% confidence interval. (b) Smooth richness accumulation curves for the non-parametric estimators ICE, Chao 2, Jack 1, Jack 2.

In the graphic representation of the weeks studied by means of a Principal Coordinate Analysis (PCA), four patterns of behavior were observed, four groups of weeks isolated and with very little similarity among them and among the species collected in each of those weeks (Figure 9). Group A, located in the lower left corner, was formed by a large number of weeks with great similarity (short distances); two subgroups can be distinguished, one located at the upper part, around week 154 that would be the centroid, and another in the lower part, around centroid week 204. Group B, located in the upper region of the figure, was less numerous than the previous one, and it was composed of a smaller number of weeks and was more distant from each other than the previous one, displaying also less homogeneity in the species that integrate them. This group appeared to be divided into two subgroups organized around different centroids, a left subgroup (week 174) and a right subgroup (week 167). The third group, group C, appears in the central right, and was formed by a small number of weeks quite dispersed and consequently different in their composition. Finally, group D, located at the bottom right, was made up of very few weeks, and it was very separated from the rest of the groups.

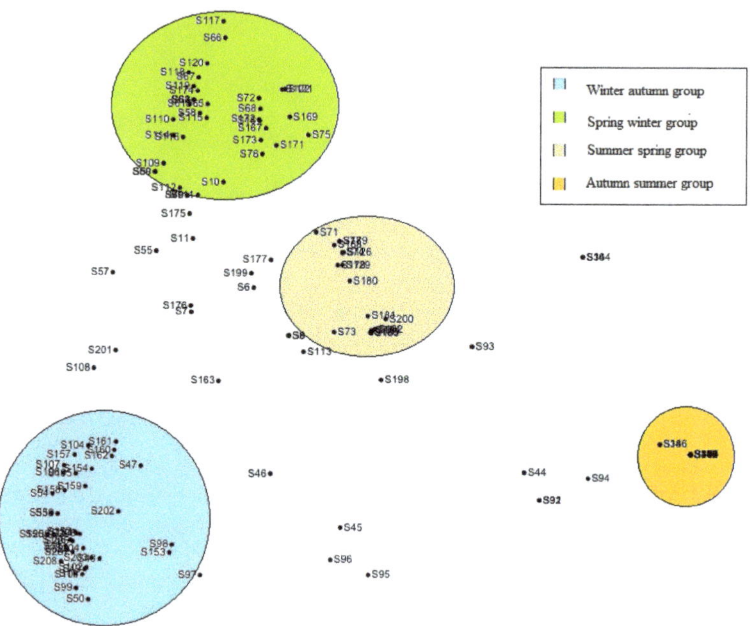

Figure 9. Groups identified in the PCA by week. The points (S) indicate the weeks of sampling.

Figure 10 presents the results of applying the PCA to macrofungal diversity by month in the period 2009–2012. January 2009 was not included in the analysis because no data were available; we also excluded February 2013 for better comparison among full years. It was observed that the four groups described in Figure 9 mainly belonged to pairs of months that were not dispersed and were constant over the years studied. The months of November and December had the least number of differences between them and in the set of years studied. Both months covered the weeks of group A, while the month of January presented a significant correlation with the upper subgroup of group A. March and April corresponded to group B; May and June to group C; and August and September to group D. February and October were the months that differed the most over the years, meaning that the species identified in these months varied from year to year. For example, February 2009 behaved similarly to May and June of that year, while the same month in 2010 and 2011 displayed more similarities with March and April of both years.

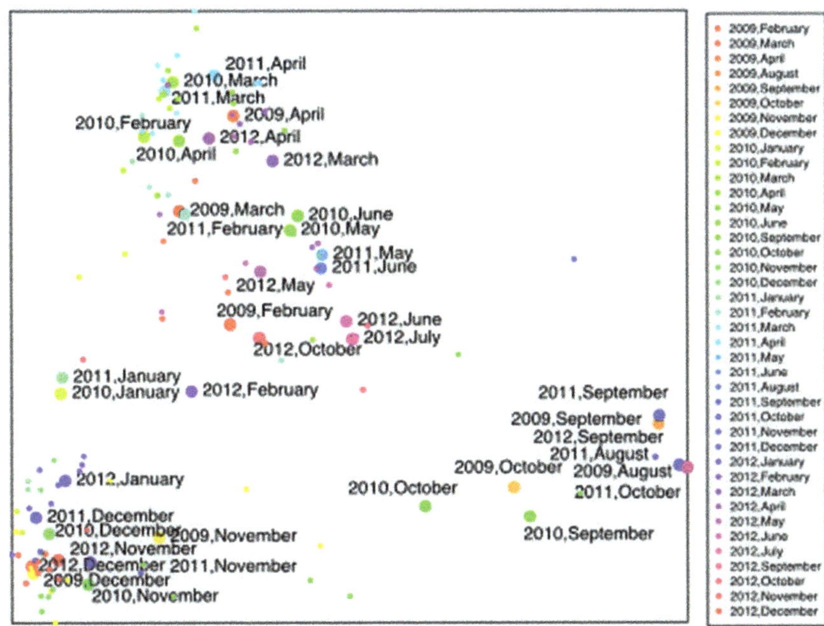

Figure 10. PCA for the months in the 2009–2012 period. The points correspond to the year and month.

4. Discussion

4.1. Macrofungal Fructification

The monthly distribution indicated that fungal fructifications did not disappear throughout the year, although the number of species collected in the different months was very different. In addition, a high macrofungal diversity was observed during the months of January and February, and the monthly fruiting for years was very variable. This difference was observed in the spring of 2009, when the number of species collected was significantly lower than in the rest of the years during the same season.

The similarity indices (Is) obtained were low, with none greater than 0.33, which could indicate a distancing of populations among the Mediterranean ecosystems compared, as result of latitudinal variation in fungal diversity [59] or the effect of rainfall [8,60,61]. However, two other factors could be taken into account to explain these differences. On the one hand, this could be explained by the plant variability that characterizes Mediterranean areas, housing almost 20% of the vascular plants when they occupy just under 5% of the land surface [49], and that in the case of the ecosystems compared, there were several vegetation units as opposed to the almost exclusive vegetation unit in the Orbada forest. On the other hand, the different prospecting methodologies used should not be ignored. In the forest of La Orbada and Collestrada, specimens were collected throughout the year; in Foros, it was carried out from September to December; in Campanarios de Azaba Biological Reserve, during the autumn and spring seasons; in the southern part of the Peloponnese, there were no collections during summer, and Fango forest was prospected from 15th September to 15th March. The number of years employed also varied: one in Collestrada forest, three in Campanarios de Azaba Biological Reserve and in Fango forest, four in Foros and five in southern Peloponnese and in La Orbada forest.

Many factors could influence fungal fruiting in a given ecosystem that cannot be controlled, such as plant biodiversity [49], climatic conditions [62], latitude [59] or pollution [63], among others. All this could make it difficult to compare results among different studies on macrofungal fruiting in Mediterranean climates. However, we consider it essential to try to minimize the biases due to heterogeneous sampling methodologies in intensity and duration. At this point, studies that are more or less exhaustive in time and frequency, with different numbers of qualified personnel or more or less precise identifications, could make it impossible to compare richness indexes in similar ecosystems [41].

4.2. Climatic and Meteorological Factors

The temporal evolution of the main meteorological variables in the study area displayed a clearly positive trend of annual temperatures (mean, maximum and minimum), with a negative trend, not statistically significant, in rainfall [43]. In addition, climate models predict an increase in average temperature between 2.2 °C and 5.1 °C, with a potential decrease of 4%–27% in annual rainfall during the 21st century, as well as dramatic changes in the distribution of the rains in the Mediterranean basin [23]. The period of highest fungal production during the study comprised the months of September to May and it was not limited to autumn, resulting in a delay to the winter months that joined the traditional periods of autumn and spring. This fact could be explained by climate change that, as well as delaying the appearance of ectomycorrhizal fruiting bodies in forest areas dominated by deciduous trees [24,64], could delay the fructification towards the first weeks of winter in forests with holm oak dominance [25,65]. On the other hand, the PCA results for the analyzed weeks showed four separated groups with a greater similarity of species, which seemed to indicate that the formation of carpophores was not restricted to a certain time of the year, but neither could it be included in a single fruiting period. If the sampling season in our study had been restricted to the autumn season, 16% of the species collected would not have been recorded, a bias that could influence the results of surveys trying to describe macrofungal diversity in Mediterranean ecosystems.

Global climate change is generating debate about its impact on the ecosystem biodiversity in the near future [66]. In particular, the Mediterranean region could be especially vulnerable, with the extinction of more than 2000 plant species [67] and negative consequences on the structure, dynamics and functioning of forests [68,69]. Therefore, populations will have to adapt to and compete in new environmental conditions, and the response of the fungal communities will be part of that ecological competition [70,71], producing changes in the composition of the species in the ectomycorrhizal communities [8,72]. For this reason, taxa collected in the summer season, although not important from a quantitative point of view, should be of great importance as they are able to fructify in the most adverse weather conditions, and potentially, they are the best adapted to some conditions that could be extended throughout the year. A methodology that does not include the most adverse months could carry a qualitatively significant bias by not paying attention to these taxa and overlooking the possible response of a dynamic system to variable conditions and disturbances, a fundamental aspect for its analysis. The presence of a late-fruiting pattern could indicate that a method that attempted to estimate 75% of the biodiversity by concentrating the sampling effort, would have to include winter and autumn (Figure 11) but it could still omit the species that fruit in spring and summer. All this could lead to the loss of information on the response of fungal communities to climate change in Mediterranean forest ecosystems.

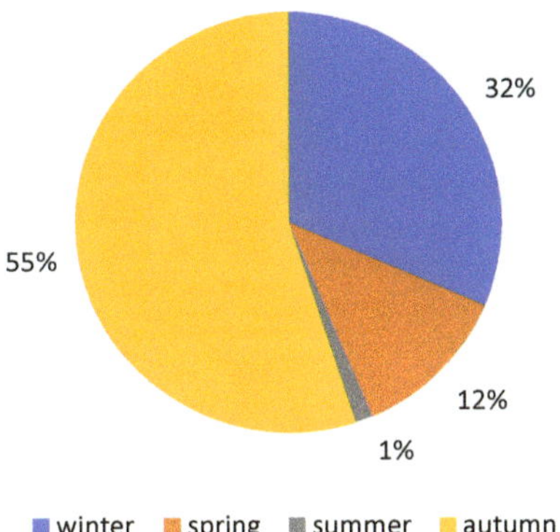

Figure 11. Average species collected by seasons. Percentages (%) over the studied period.

4.3. Macrofungal Lifestyles

The grouping of identified taxa by their form of nutrition (Figure 12) can contribute to our understanding of how symbiotic fungi, saprophytes and parasites develop their life cycle, and to a better understanding of their ecological functions and their impact on plant communities [73]. The saprophytic species were more represented in winter and less in summer, disappearing in July and August (all the species collected were mycorrhizal), which seemed to indicate that macrofungal saprophytes were more sensitive to adverse summer conditions and perhaps to climate change. The mycelium of saprophytic taxa could be more affected by atmospheric changes, since they are more superficial (in humus and litter) than symbionts [36,74]. The exclusive appearance of symbiotic species could be related to large-scale disturbances in ecosystems, which could lead to collapse or increase of the overall density of the vegetation [75,76] and the associated fungal community. Similarly, the increase of CO_2 as a result of global warming leads to an increase in the growth of plants [77,78] and subsequently the mycorrhizal symbionts of trees and shrubs [79,80]. The changes in the mycorrhizal community are the basis for the carbon demands of fungi and the variation of carbon costs for the host in a changing environment [70,81]. On the other hand, variations in nitrogen levels, important in the competitive relationships between plants and fungi, could affect the fungal composition [82,83].

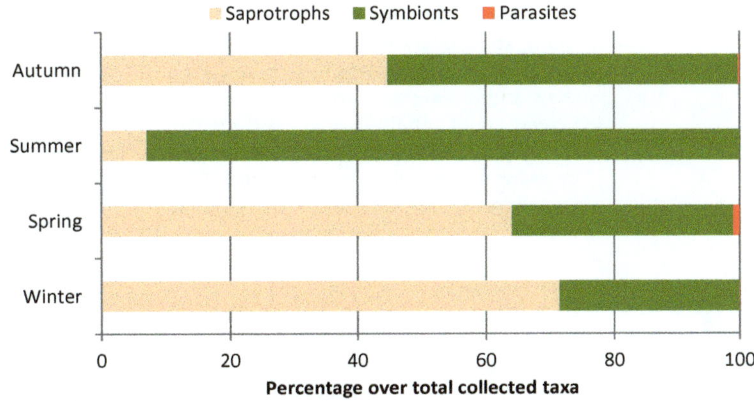

Figure 12. Percentage of total fungal collections sorted by season and form of nutrition.

4.4. Viable Methodology in Field Studies

Field methodology usually only includes macrofungi [84], as they form more easily observable spore-bearing structures than microfungi [85]. Furthermore, microfungi are rarely considered in relation to conservation [86]. However, some studies suggest that most fungi could be termed "microfungi" and many are probably rare and threatened [87]. At this point, the valuable contribution of molecular techniques should be highlighted, not only for the assessment of fungal diversity in different substrates of a given ecosystem [88,89], but also in the evaluation of fungal fruiting [90]. In any case, the importance of field studies to assess macrofungal diversity and evaluate their fruiting remains evident [91], not only because of their lower methodological and economic cost but also because of possible limitations in delimitations at the specific level with DNA sequences [92].

The estimation of the macrofungal diversity in holm oak-dominated Mediterranean ecosystems may seem simple due to the low plant diversity of these formations with respect to other Mediterranean forests [93,94] and the low specific fungal diversity in fruiting bodies, in which three fungal species may account for more than 50% of the carpophores found [95,96]. However, the type of vegetation has a low influence on the mechanisms that regulate fungal fruiting [36,97] and the fungal diversity does not follow the same latitudinal distribution pattern as other organisms. In fact, it seems that fungal diversity depends on the type of vegetation and the composition of the forest, correlating inversely with plant diversity due to a high level of fungal speciation [59]. For this reason, to the great fungal diversity expected in Mediterranean forest systems, we must add other problems when establishing a viable study methodology: the large number of known fungal taxa, the intraspecific variation, the inconstant appearance of fruiting bodies over the years, and the need for specialized personnel in taxonomic determination [41]. This could include another problem that this study highlighted: the need to prospect throughout the year and over a period of several years. Viable fungal diversity studies cannot avoid having biases, but they should be minimized if reliable results are to be obtained, with constant collection throughout the year being one of the most important factors to achieve this goal. This type of sampling would make it possible to include a number of prevalent species, very abundant species and species of annual appearance that have been successful in competing for available habitats, and with greater ecological plasticity than a random distribution would indicate [17]. In the same way, it would also allow for the inclusion of saprophytic leaf and soil indicator species for each plant type, with high fidelity but relatively low abundance [98], a series of early-stage mycorrhizal species [99] and a last series that would include the species that could fructify in the most adverse weather conditions.

There is no doubt that the more years a study lasts, the more complete it will be, but doing such long studies is too expensive. The rarefaction curve obtained did not reach equilibrium and the number of new species collected during the last year is high, but the exponential phase has been overcome and it approaches the plateau phase, with a percentage of variability that has decreased over the years. Some studies have shown that the collection of specimens depends on the intensity of sampling in the same season or year, or on the number of samples in different seasons and years [100,101]. In our study, we chose weekly sampling in order to better characterize the diversity of some groups that seem to be favored by continuous sampling, such as agarics [86]. In the form of a summary by year of sampling, in the first two years 83% of the total species were collected in La Orbada forest reaching 90% during the third year (Figure 13). Alpha and beta diversity obtained for La Orbada forest and other Mediterranean ecosystems dominated by *Quercus* sp. gave an approximate insight into fungal richness. On the other hand, the exhaustive samples of 229 weeks allowed the performance of the non-parametric estimators to be assessed even if they underestimated real richness [48,102]. The Chao 2 estimator appeared to be the best adapted to the transects of our study, possibly because it studies rare species and allows species to be estimated from incidence ratios [53]. The estimate for our field work was 203 species, compared to 173 collected. The differences in the asymptotic estimates were small, 85.15% of efficiency, so we believe that the obtained inventory could optimally reflect fungal diversity.

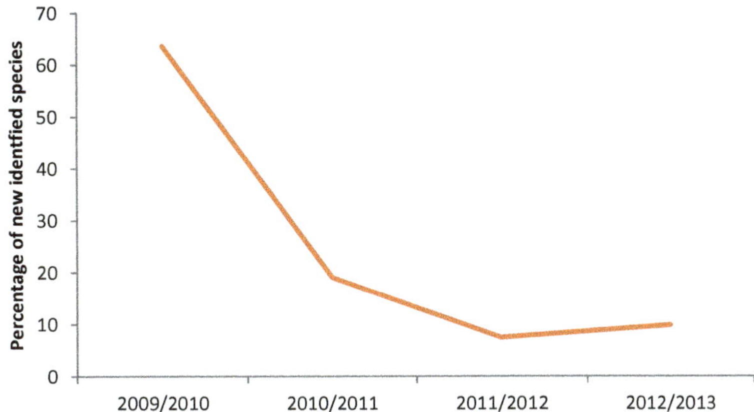

Figure 13. Total percentage of new identified species per collection years from February 2009 to June 2013.

Taking into account that fruiting species do not reflect the real fungal diversity of the mycorrhizal community [62], the larger inter-annual variation of species, and that is not possible to collect fruiting bodies of all fungal species present in an environment [17], we considered Chao 2 efficiency values. We estimated the minimum collection time to obtain 75% of the species was 133 weeks—two and a half years. With two years smooth richness accumulation curve reaching 144 species, we would obtain an efficiency of 70.60%, with a confidence interval between 65.95% and 75.79%. Data obtained over two years could allow us to assess the fungal diversity of a Mediterranean ecosystem with holm oak dominance and to compare results by collecting a high percentage of species. However, many of the questions concerning the estimation of diversity may remain unanswered, so we think it is necessary to carry out a greater number of studies in these environments that could contribute to solving them.

In any case, we agree with Cannon [103] and Hyde [104] that rapid initial studies developed by unqualified personnel are also important to allow an approach to diversity and a comparison among different ecosystems. Prior knowledge of the prevalent species in similar ecosystems, the identification at the genus level of the species collected, the consideration of their form of nutrition and the qualitative proportion among them could be a good starting point.

A suitable sampling method for macrofungi in Mediterranean ecosystems should pay particular attention to summer fruiting and the displacement of fungal fruiting towards winter. The monitoring of organisms in their own environment and their changes are essential to achieving a true renewal of science [105]. We believe that sampling carried out during a minimum period of two years, periodically and regularly throughout the year, could take into account the possible variations in fructification as a consequence of the meteorological factors, and could make it possible to know the fungal diversity of the studied forest ecosystem.

Author Contributions: Conceptualization, A.F.R., P.G.J., D.R.d.l.C. and J.S.S.; methodology, A.F.R., D.R.d.l.C. and J.L.V.V.; software, J.L.V.V. and A.F.R.; validation, J.L.V.V., A.F.R., S.S.D. and J.S.S.; formal analysis, A.F.R., D.R.d.l.C. and J.S.S.; investigation, A.F.R., D.R.d.l.C., P.G.J. and J.S.S.; resources, J.S.S., D.R.d.l.C. and A.F.R.; data curation, A.F.R. and P.G.J.; writing—original draft preparation, A.F.R., D.R.d.l.C. and J.S.S.; writing—review and editing, A.F.R., S.S.D., D.R.d.l.C., P.G.J. and J.S.S.; visualization, A.F.R., D.R.d.l.C. and J.L.V.V.; supervision, A.F.R., P.G.J. and J.S.S.; project administration, A.F.R.,

P.G.J., D.R.d.l.C. and J.S.S.; funding acquisition, A.F.R., D.R.d.l.C. and J.S.S. All authors have read and agreed to the published version of the manuscript.

Funding: This research received no external funding.

Data Availability Statement: Not applicable.

Acknowledgments: The authors would like to thank Sergio Pérez Gorjón for his advice on several identifications.

Conflicts of Interest: The authors declare no conflict of interest.

References

1. Taylor, D.L.; Hollingsworth, T.N.; McFarland, J.W.; Lennon, N.J.; Nusbaum, C.; Ruess, R.W. A first comprehensive census of fungi in soil reveals both hyperdiversity and fine-scale niche partitioning. *Ecol. Monogr.* **2014**, *84*, 3–20. [CrossRef]
2. Wijayawardene, N.N.; Hyde, K.D.; Al-Ani, L.K.; Tedersoo, L.; Haelewaters, D.; Rajeshkumar, K.C.; Zhao, R.L.; Aptroot, A.; Leontyev, D.V.; Saxena, R.K.; et al. Outline of Fungi and fungus-like taxa. *Mycosphere* **2020**, *11*, 1060–1456. [CrossRef]
3. Perotto, S.; Angelini, P.; Bianciotto, V.; Bonfante, P.; Girlanda, M.; Kull, T. Interaction of fungi with other organisms. *Plant Biosyst.* **2013**, *147*, 208–218. [CrossRef]
4. Cannon, P.F.; Sutton, B.C. Microfungi on wood and plant debris. In *Biodiversity of Fungi*; Mueller, G.M., Bills, G.F., Foster, M.S., Eds.; Elsevier Academic Press: Cambridge, MA, USA, 2004; pp. 217–239.
5. Rubino, D.L.; McCarthy, B.C. Composition and ecology of macrofungal and myxomycete communities on oak woody debris in a mixed-oak forest of Ohio. *Can. J. For. Res.* **2003**, *33*, 2151–2163. [CrossRef]
6. Landi, M.; Salerni, E.; Ambrosio, E.; D'Aguanno, M.; Nucci, A.; Saveri, C.; Perini, C.; Angiolini, C. Concordance between vascular plant and macrofungal community composition in broadleaf deciduous forest in central Italy. *Iforest* **2015**, *8*, 279–286. [CrossRef]
7. Kutszegi, G.; Siller, I.; Dima, B.; Merényi, Z.; Varga, T.; Takács, K.; Turcsányi, G.; Bidló, A.; Ódor, P. Revealing hidden drivers of macrofungal species richness by analyzing fungal guilds in temperate forests, West Hungary. *Community Ecol.* **2021**, *22*, 13–28. [CrossRef]
8. Richard, F.; Roy, M.; Shahin, O.; Sthultz, C.; Duchemin, M.; Joffre, R.; Selosse, M.A. Ectomycorrhizal communities in a Mediterranean forest ecosystem dominated by *Quercus ilex*: Seasonal dynamics and response to drought in the surface organic horizon. *Ann. For. Sci.* **2011**, *68*, 57–68. [CrossRef]
9. Vogiatzakis, I.N.; Mannion, A.M.; Griffiths, G.H. Mediterranean ecosystems: Problems and tools for conservation. *Prog. Phys. Geogr.* **2006**, *30*, 175–200. [CrossRef]
10. Biasi, R.; Brunori, E.; Ferrara, C.; Salvati, L. Towards sustainable rural landscapes? a multivariate analysis of the structure of traditional tree cropping systems along a human pressure gradient in a mediterranean region. *Agrofor. Syst.* **2017**, *91*, 1199–1217. [CrossRef]
11. Bazzato, E.; Lallai, E.; Serra, E.; Mellis, M.T.; Marignani, M. Key role of small woodlots outside forest in a Mediterranean fragmented landscape. *For. Ecol. Manag.* **2021**, *496*, 119389. [CrossRef]
12. Hidalgo, P.J.; Hernández, H.; Sánchez-Almendro, A.J.; López-Tirado, J.; Vessella, F.; Porras, R. Fragmentation and Connectitivy of Island Forests in Agricultural Mediterranean Environments: A Comparative Study between the Guadalquivir Valley (Spain) and the Apulia Region. *Forests* **2021**, *12*, 1201. [CrossRef]
13. Scarascia-Mugnozza, G.; Oswald, H.; Piussi, P.; Radoglou, K. Forests of the Mediterranean region: Gaps in knowledge and research needs. *For. Ecol. Manag.* **2000**, *132*, 97–109. [CrossRef]
14. Rincón, A.M.; Pérez-Izquierdo, L.; de Miguel, S.; Parladé, J. Mycorrhizae in Mediterranean Pine and Mixed Forests. In *Pines and Their Mixed Forest Ecosystems in the Mediterranean Basin*; Ne'eman, G., Osem, Y., Eds.; Springer: Cham, Switzerland, 2021; Volume 38.
15. Büntgen, U.; Kauserud, H.; Egli, S. Linking climate variability to mushroom productivity and phenology. *Front. Ecol. Environ.* **2012**, *10*, 14–19. [CrossRef]
16. Grainger, J. Ecology of the larger fungi. *Trans. Br. Mycol. Soc.* **1946**, *29*, 52–63. [CrossRef]
17. Parker-Rhodes, A.F. The Basidiomycetes of Skokholm Island VII. Some floristic and ecological calculations. *New Phytol.* **1951**, *50*, 227–243. [CrossRef]
18. Richardson, M. Studies of *Russula emetica* and other agarics in a Scots pine plantation. *Trans. Br. Mycol. Soc.* **1970**, *55*, 217–229. [CrossRef]
19. Zervakis, G.I.; Polemis, E.; Dimou, D.M. Mycodiversity studies in selected ecosystems of Greece: III. Macrofungi recorded in *Quercus* forests from southern Peloponnese. *Mycotaxon* **2002**, *84*, 141–162.
20. Richard, F.; Moreau, P.A.; Selosse, M.A.; Gardes, M. Diversity and fruiting patterns of ectomycorrhizal and saprobic fungi in an old-growth Mediterranean forest dominated by *Quercus ilex* L. *Can. J. Bot.* **2004**, *82*, 1711–1729. [CrossRef]
21. Dimou, D.M.; Polemis, E.; Konstantinidis, G.; Kaounas, V.; Zervakis, G.I. Diversity of macrofungi in the Greek islands of Lesvos and Agios Efstratios, NE Aegean Sea. *Nova Hedwig.* **2016**, *102*, 439–475. [CrossRef]
22. Mehus, H. Fruit Body Production of Macrofungi in Some North Norwegian Forest Types. *Nord. J. Bot.* **1986**, *6*, 679–702. [CrossRef]

23. IPCC. *Climate Change 2013: The Physical Science Basis. Contribution of Working Group I to the Fifth Assessment Report of the Intergovernmental Panel on Climate Change*; Stocker, T.F., Qin, D., Plattner, G.-K., Tignor, M., Allen, S.K., Boschung, J., Nauels, A., Xia, Y., Bex, V., Midgley, P.M., et al., Eds.; Cambridge University Press: Cambridge, UK; New York, NY, USA, 2013; 1535p.

24. Gange, A.C.; Gange, E.G.; Sparks, T.H.; Boddy, L. Rapid and recent changes in fungal fruiting patterns. *Science* **2007**, *316*, 71. [CrossRef] [PubMed]

25. Kauserud, H.; Heegaard, E.; Büntgen, U.; Halvorsen, R.; Egli, S.; Senn-Irlet, B.; Høiland, K. Warming-induced shift in European mushroom fruiting phenology. *Proc. Natl. Acad. Sci. USA* **2012**, *109*, 14488–14493. [CrossRef]

26. Martiny, J.B.; Bohannan, B.J.; Brown, J.H.; Colwell, R.K.; Fuhrman, J.A.; Green, J.L.; Horner-Devine, M.C.; Kane, M.; Krumins, J.A.; Kuske, C.R.; et al. Microbial biogeography: Putting microorganisms on the map. *Nat. Rev. Microbiol.* **2006**, *4*, 102–112. [CrossRef] [PubMed]

27. Buée, M.; De Boer, W.; Martin, F.; Van Overbeek, L.; Jurkevitch, E. The rhizosphere zoo: An overview of plant-associated communities of microorganisms, including phages, bacteria, archaea, and fungi, and of some of their structuring factors. *Plant Soil* **2009**, *321*, 189–212. [CrossRef]

28. Aponte, C.; García, L.V.; Marañón, T.; Gardes, M. Indirect host effect on ectomycorrhizal fungi: Leaf fall and litter quality explain changes in fungal communities on the roots of co-occurring Mediterranean oaks. *Soil Biol. Biochem.* **2010**, *42*, 788–796. [CrossRef]

29. Robertson, G.P.; Crum, J.R.; Ellis, B.G. The spatial variability of soil resources following long-term disturbance. *Oecologia* **1993**, *96*, 451–456. [CrossRef]

30. Hering, T.F. The terricolous higher fungi of four lake district woodlands. *Trans. Br. Mycol. Soc.* **1966**, *49*, 369–383. [CrossRef]

31. Durall, D.M.; Gamiet, S.; Simard, S.W.; Kudrna, L.; Sakakibara, S.M. Effects of clearcut logging and tree species composition on the diversity and community composition of epigeous fruit bodies formed by ectomycorrhizal fungi. *Can. J. Bot.* **2006**, *84*, 966–980. [CrossRef]

32. Halme, P.; Heilmann-Clausen, J.; Rämä, T.; Kosonen, T.; Kunttu, P. Monitoring fungal biodiversity—Towards an integrated approach. *Fungal Ecol.* **2012**, *5*, 750–758. [CrossRef]

33. Langer, E.; Langer, G.; Striegel, M.; Riebesehl, J.; Ordynets, A. Fungal diversity of the Kellerwald-Edersee National Park–indicator species of nature value and conservation. *Nova Hedwig.* **2014**, *99*, 129–144. [CrossRef]

34. Watling, R. Assessment of fungal diversity: Macromycetes, the problems. *Can. J. Bot.* **1995**, *73* (Suppl. S1), S515–S524. [CrossRef]

35. Angelini, P.; Compagno, R.; Arcangeli, A.; Bistocchi, G.; Gargano, M.L.; Venanzoni, R.; Venturella, G. Macrofungal diversity and ecology in two Mediterranean forest ecosystems. *Plant Biosyst.* **2016**, *150*, 540–549. [CrossRef]

36. Salerni, E.; Laganà, A.; Perini, C.; Loppi, S.; de Dominicis, V. Effects of temperature and rainfall on fruiting of macrofungi in oak forests of the Mediterranean area. *Isr.J. Plant Sci.* **2002**, *50*, 189–198. [CrossRef]

37. García Jiménez, P.; Fernández Ruiz, A.; Sánchez Sánchez, J.; Rodríguez de la Cruz, D. Mycological Indicators in Evaluating Conservation Status: The Case of *Quercus* spp. Dehesas in the Middle-West of the Iberian Peninsula (Spain). *Sustainability* **2020**, *12*, 10442. [CrossRef]

38. Azul, A.M.; Castro, P.; Sousa, J.P.; Freitas, H. Diversity and fruiting patterns of ectomycorrhizal and saprobic fungi as indicators of land-use severity in managed woodlands dominated by *Quercus suber*—A case study from southern Portugal. *Can. J. For. Res.* **2009**, *39*, 2404–2417. [CrossRef]

39. Crusafont, M.; Truyols, J. Algunas precisiones sobre la edad y extensión del Paleógeno de las provincias de Salamanca y Zamora. *Cursos Conf. Inst. Lucas Mallada* **1958**, *4*, 83–85.

40. García-Rodríguez, A.; Forteza, J.; Prat-Pérez, L.; Gallardo, J.F.; Lorenzo-Martín, L.F. Suelos. Estudio integrado y multidisciplinario de la dehesa salmantina 1. Estudio fisiográfico descriptivo. *Cons. Super. Investig. Científicas Salamanca-Jaca.* **1979**, *3*, 65–100.

41. Fernández, A.; Sánchez, S.; García, P.; Sánchez, J. Macrofungal diversity in an isolated and fragmented Mediterranean Forest ecosystem. *Plant Biosyst.* **2020**, *154*, 139–148. [CrossRef]

42. Capel Molina, J.J. *Los climas de España*; Oikos-tau S.A.: Barcelona, Spain, 1981.

43. Barbancho, A.C.; Tejeda, E.M.; Hernández, M.Q. Evolución de las temperaturas y precipitaciones en las capitales de Castilla y León en el período 1961–2006. *Polígonos. Rev. Geogr.* **2012**, *17*, 59–81. [CrossRef]

44. Feest, A. Establishing baseline indices for the quality of the biodiversity of restored habitats using a standardized sampling process. *Restor. Ecol.* **2006**, *14*, 112–122. [CrossRef]

45. CABI Index Fungorum. Available online: http://www.indexfungorum.org/names/names.asp (accessed on 8 December 2021).

46. Kirk, P.M.; Cannon, P.; Minter, D.; Stalpers, J. *Dictionary of the Fungi*, 10th ed.; CAB International: Wallingford, UK, 2008.

47. Sarkar, S. Defining "biodiversity": Assessing biodiversity. *Monist* **2002**, *85*, 131–155. [CrossRef]

48. Magurran, A.E. *Measuring Biological Diversity*; Blackwell: Malden, MA, USA, 2004.

49. Cowling, R.M.; Rundel, P.W.; Lamont, B.B.; Arroyo, M.K.; Arianoutsou, M. Plant diversity in Mediterranean-climate regions. *Trends Ecol. Evol.* **1996**, *11*, 362–366. [CrossRef]

50. Oreja, J.G.; de la Fuente-Díaz-Ordaz, A.A.; Hernández-Santín, L.; Buzo-Franco, D.; Bonache-Regidor, C. Evaluación de estimadores no paramétricos de la riqueza de especies. Un ejemplo con aves en áreas verdes de la ciudad de Puebla, México. *Anim. Biodivers. Conserv.* **2010**, *33*, 31–45.

51. Basualdo, C.V. Choosing the best non-parametric richness estimator for benthic macroinvertebrates databases. *Rev. Soc. Entomol. Argent.* **2011**, *70*, 27–38.

52. Colwell, R.K. EstimateS: Statistical Estimation of Species Richness and Shared Species from Samples, Version 8. User's Guide and Application. 2006. Available online: http://purl.oclc.org/estimates (accessed on 1 September 2021).

53. Gotelli, N.J.; Colwell, R.K. Quantifying bio-diversity: Procedures and pitfalls in the measurement and comparison of species richness. *Ecol. Lett.* **2001**, *4*, 379–391. [CrossRef]

54. Díaz–Frances, E.; Soberón, J. Statistical estimation and model selection of species–accumulation functions. *Conserv. Biol.* **2005**, *19*, 569–573. [CrossRef]

55. Chao, A. Estimating the population size for capture-recapture data with unequal catchability. *Biometrics* **1987**, *43*, 783–791. [CrossRef]

56. Oksanen, J.; Blanchet, F.G.; Friendly, M.; Kindt, R.; Legendre, P.; McGlinn, D.; Minchin, P.R.; O'Hara, R.B.; Simpson, G.L.; Solymos, P.; et al. *R Package Version 2.4-0. Vegan: Community Ecology Package.* 2016. Available online: https://cran.r-project.org/web/packages/vegan/index.html (accessed on 1 September 2021).

57. Vicente-Villardón, J.L. MultBiplot(R): Multivariate Analysis using Biplots. R Package Version 0.3.3. 2016. Available online: https://biplot.usal.es/classicalbiplot/multbiplot-in-r/ (accessed on 1 September 2021).

58. Gallego-Álvarez, I.; Vicente-Villardón, J.L. Analysis of environmental indicators in international companies by applying the logistic biplot. *Ecol. Indic.* **2012**, *23*, 250–261. [CrossRef]

59. Shi, L.L.; Mortimer, P.E.; Slik, J.F.; Zou, X.M.; Xu, J.; Feng, W.T.; Qiao, L. Variation in forest soil fungal diversity along a latitudinal gradient. *Fungal Divers.* **2014**, *64*, 305–315. [CrossRef]

60. Ágreda, T.; Águeda, B.; Olano, J.M.; Vicente-Serrano, S.M.; Fernández-Toirán, M. Increase evapotranspiration demand in a Mediterranean climate might cause a decline in fungal yields under global warming. *Glob. Chang. Biol.* **2015**, *21*, 3499–3510. [CrossRef] [PubMed]

61. Barnes, C.J.; Van der Gast, C.J.; McNamara, N.P.; Rowe, R.; Bending, G.D. Extreme rainfall affects assembly of the root-associated fungal community. *New Phytol.* **2018**, *220*, 1172–1184. [CrossRef]

62. Jang, S.K.; Kim, S.W. Relationship between Ectomycorrhizal Fruiting Bodies and Climatic and Environmental Factors in Naejangsan National Park. *Mycobiology* **2015**, *43*, 122–134. [CrossRef]

63. Sun, Q.; Liu, Y.; Yuan, H.; Lian, B. The effect of environmental contamination on the community structure and fructification of ectomycorrhizal fungi. *MicrobiologyOpen* **2017**, *6*, e00396. [CrossRef] [PubMed]

64. Baptista, P.; Martins, A.; Tavares, R.M.; Lino-Nieto, T. Diversity and fruiting pattern of macrofungi associated with chestnut (*Castanea sativa*) in the Trás-os-Montes region (Northeast Portugal). *Fungal Ecol.* **2010**, *3*, 9–19. [CrossRef]

65. Kauserud, H.; Stige, L.C.; Vik, J.O.; Økland, R.H.; Høiland, K.; Stenseth, N.C. Mushroom fruiting and climate change. *Proc. Natl. Acad. Sci. USA* **2008**, *105*, 3811–3814. [CrossRef] [PubMed]

66. Martin, L.M.; Polley, H.W.; Daneshgar, P.P.; Harris, M.A.; Wilsey, B.J. Biodiversity, photosynthetic mode, and ecosystem services differ between native and novel ecosystems. *Oecologia* **2014**, *175*, 687–697. [CrossRef] [PubMed]

67. Malcolm, J.R.; Liu, C.; Neilson, R.P.; Hansen, L.; Hannah, L. Global warming and extinctions of endemic species from biodiversity hotspots. *Conserv. Biol.* **2006**, *20*, 538–548. [CrossRef] [PubMed]

68. Allard, V.; Ourcival, J.M.; Rambal, S.; Joffre, R.; Rocheteau, A. Seasonal and annual variation of carbon exchange in an evergreen Mediterranean forest in southern France. *Glob. Chang. Biol.* **2008**, *14*, 1–12. [CrossRef]

69. Mair, L.; Jönsson, M.; Räty, M.; Bärring, L.; Strandberg, G.; Lämås, T.; Snäll, T. Land use changes could modify future negative effects of climate change on old-growth forest indicator species. *Divers. Distrib.* **2018**, *24*, 1416–1425. [CrossRef]

70. Markkola, A.M.; Saravesi, K.; Aikio, S.; Taulavuori, E.; Taulavuori, K. Light-driven host-symbiont interactions under hosts' range shifts caused by global warming: A review. *Environ. Exp. Bot.* **2016**, *121*, 48–55. [CrossRef]

71. Thakur, M.P.; Quast, V.; Van Dam, N.M.; Eisenhauer, N.; Roscher, C.; Biere, A.; Martinez-Medina, A. Interactions between functionally diverse fungal mutualists inconsistently affect plant performance and competition. *Oikos* **2019**, *128*, 1136–1146. [CrossRef]

72. Martinová, V.; Van Geel, M.; Lievens, B.; Honnay, O. Strong differences in *Quercus robur*-associated ectomycorrhizal fungal communities along a forest-city soil sealing gradient. *Fungal Ecol.* **2016**, *20*, 88–96. [CrossRef]

73. Martin, F.; Cullen, D.; Hibbett, D.; Pisabarro, A.; Spatafora, J.W.; Baker, S.E.; Grigoriev, I.V. Sequencing the fungal tree of life. *New Phytol.* **2011**, *190*, 818–821. [CrossRef] [PubMed]

74. Hawkes, C.V.; Kivlin, S.N.; Rocca, J.D.; Huguet, V.; Thomsen, M.A.; Suttle, K.B. Fungal community responses to precipitation. *Glob. Chang. Biol.* **2011**, *17*, 1637–1645. [CrossRef]

75. Zelnik, Y.R.; Meron, E.; Bel, G. Localized states qualitatively change the response of ecosystems to varying conditions and local disturbances. *Ecol. Complex.* **2016**, *25*, 26–34. [CrossRef]

76. Hui, N.; Liu, X.; Jumpponen, A.; Setälä, H.; Kotze, D.J.; Biktasheva, L.; Romantschuk, M. Over twenty years farmland reforestation decreases fungal diversity of soils, but stimulates the return of ectomycorrhizal fungal communities. *Plant Soil.* **2018**, *427*, 231–244. [CrossRef]

77. Bazzaz, F.A. The response of natural ecosystems to the rising global CO_2 levels. *Annu. Rev. Ecol. Evol. Syst.* **1990**, *21*, 167–196. [CrossRef]

78. Kern, M.; Caldararu, S.; Engel, J.; Rammig, A.; Vicca, S.; Zaehle, S. A novel plant-mycorrhiza interaction model improves representation of plant responses to elevated CO_2. In Proceedings of the 20th EGU General Assembly, EGU2018, Vienna, Austria, 4–13 April 2018; p. 7526.

79. Pickles, B.J.; Egger, K.N.; Massicotte, H.B.; Green, D.S. Ectomycorrhizas and climate change. *Fungal Ecol.* **2012**, *5*, 73–84. [CrossRef]
80. Morgado, L.N.; Semenova, T.A.; Welker, J.M.; Walker, M.D.; Smets, E.; Geml, J. Summer temperature increase has distinct effects on the ectomycorrhizal fungal communities of moist tussock and dry tundra in Arctic Alaska. *Glob. Chang. Biol.* **2015**, *21*, 959–972. [CrossRef]
81. Akroume, E.; Maillard, F.; Bach, C.; Hossann, C.; Brechet, C.; Angeli, N.; Zeller, B.; Saint-André, L.; Buée, M. First evidences that the ectomycorrhizal fungus *Paxillus involutus* mobilizes nitrogen and carbon from saprotrophic fungus necromass. *Environ. Microbiol.* **2018**, *21*, 197–208. [CrossRef]
82. Egerton-Warburton, L.M.; Johnson, N.C.; Allen, E.B. Mycorrhizal community dynamics following nitrogen fertilization: A cross-site test in five grasslands. *Ecol. Monogr.* **2007**, *77*, 527–544. [CrossRef]
83. Terrer, C.; Vicca, S.; Stocker, B.D.; Hungate, B.A.; Phillips, R.P.; Reich, P.B.; Finzi, A.C.; Prentice, C.P. Ecosystem responses to elevated CO_2 governed by plant-soil interactions and the cost of nitrogen acquisition. *New Phytol.* **2017**, *217*, 507–522. [CrossRef] [PubMed]
84. Gonçalves, S.C.; Haelewaters, D.; Furci, G.; Mueller, G.M. Include all fungi in biodiversity goals. *Science* **2021**, *373*, 403. [CrossRef] [PubMed]
85. Mueller, G.M.; Schmit, J.P.; Leacock, P.R.; Buyck, B.; Cifuentes, J.; Desjardin, D.E.; Halling, R.E.; Hjortstam, K.; Iturriaga, T.; Larsson, K.H.; et al. Global diversity and distribution of macrofungi. *Biodivers. Conserv.* **2007**, *16*, 37–48. [CrossRef]
86. Arnolds, E. The future of fungi in Europe: Threats, conservation and management. *Br. Mycol. Symp. Ser.* **2001**, *22*, 64–80.
87. Cannon, P.F.; Mibey, R.K.; Siboe, G.M. Microfungus diversity and the conservation agenda in Kenya. *Br. Mycol. Symp. Ser.* **2001**, *22*, 197–208.
88. Piazza, G.; Ercoli, L.; Nuti, M.; Pellegrino, E. Interaction between conservation tillage and nitrogen fertilization shapes prokaryotic and fungal diversity at different soil depths: Evidence from a 23-year field experiment in the Mediterranean Area. *Front. Microbiol.* **2019**, *10*, 2047. [CrossRef]
89. Girometta, C.E.; Bernicchia, A.; Baiguera, R.M.; Bracco, F.; Buratti, S.; Cartabia, M.; Picco, A.M.; Savino, E. An Italian research culture collection of wood decay fungi. *Diversity* **2020**, *12*, 58. [CrossRef]
90. De la Varga, H.; Águeda, B.; Ágreda, T.; Martínez-Peña, F.; Parladé, J.; Pera, J. Seasonal dynamics of Boletus edulis and Lactarius deliciosus extraradical mycelium in pine forests of central Spain. *Mycorrhiza* **2013**, *23*, 391–402. [CrossRef]
91. Halme, P.; Kotiaho, J.S. The importance of timing and number of surveys in fungal biodiversity research. *Biodivers. Conserv.* **2012**, *21*, 205–209. [CrossRef]
92. Zamora, J.C.; Svensson, M.; Kirschner, R.; Olariaga, I.; Ryman, S.; Parra, L.A.; Geml, J.; Rosling, A.; Adamčík, S.; Ahti, T.; et al. Considerations and consequences of allowing DNA sequence data as types of fungal taxa. *IMA Fungus* **2018**, *9*, 167–175. [CrossRef]
93. Rey-Benayas, J.M.; Scheiner, S.M. Plant diversity, biogeography and environment in Iberia: Patterns and possible causal factors. *J. Veg. Sci.* **2002**, *13*, 245–258. [CrossRef]
94. Teobaldelli, M.; Cona, F.; Saulino, L.; Migliozzi, A.; D'Urso, G.; Langella, G.; Manna, P.; Saracino, A. Detection of diversity and stand parameters in Mediterranean forests using leaf-off discrete return LiDAR data. *Remote Sens. Environ.* **2017**, *192*, 126–138. [CrossRef]
95. Clavería, V.; de Miguel, A.M. Diversidad ectomicorrícica en una formación natural de carrasca (*Quercus ilex* L. subsp. *ballota* (Desf.) Samp.). In *Actas del 4° Congreso Forestal Español*; Sociedad Española de Ciencias Forestales: Zaragoza, Spain, 2005.
96. Sarrionandia, E.; Salcedo, I. Macrofungal diversity of holm-oak forests at the northern limit of their distribution range in the Iberian Peninsula. *Scand. J. For. Res.* **2018**, *33*, 23–31. [CrossRef]
97. Carrie, A.; Heegaard, E.; Høiland, K.; Senn-Irlet, B.; Kuyper, T.W.; Krisai-Greilhuber, I.; Kirk, P.M.; Heilmann-Clausen, J.; Gange, A.C.; Egli, S.; et al. Explaining European fungal fruiting phenology with climate variability. *Ecology* **2018**, *99*, 1306–1315.
98. Bahnmann, B.; Mašínová, T.; Halvorsen, R.; Davey, M.L.; Sedlák, P.; Tomšovský, M.; Baldrian, P. Effects of oak, beech and spruce on the distribution and community structure of fungi in litter and soils across a temperate forest. *Soil Biol. Biochem.* **2018**, *119*, 162–173. [CrossRef]
99. Moreno, G.; Manjón, J.L.; Álvarez-Jiménez, J. Hypogeous desert fungi. In *Desert Truffles*; Kagan-Zur, V., Sitrit, Y., Roth-Bejerano, N.A., Morte, A., Eds.; Springer: Berlin, Germany, 2013; pp. 129–135.
100. Abrego, N.; Halme, P.; Purhonen, J.; Ovaskainen, O. Fruit body based inventories in wood-inhabiting fungi: Should we replicate in space or time? *Fungal Ecol.* **2016**, *20*, 225–232. [CrossRef]
101. Ambrosio, E.; Mariotti, M.G.; Zotti, M.; Cecchi, G.; Di Piazza, S.; Feest, A. Measuring macrofungal biodiversity quality using two different survey approaches: A case study in broadleaf Mediterranean forests. *Ecol. Indic.* **2018**, *85*, 1210–1230. [CrossRef]
102. Walther, B.A.; Moore, J.L. The concepts of bias, precision, and accuracy, and their use in testing the performance of species richness estimators, with a literature review of estimator performance. *Ecography* **2005**, *28*, 815–829. [CrossRef]
103. Cannon, P. Options and constraints in rapid diversity analysis of fungi in natural ecosystems. *Fungal Divers.* **1999**, *2*, 1–15.
104. Hyde, K.D. Can we rapidly measure fungal diversity? *Mycologist* **1997**, *11*, 176–178. [CrossRef]
105. Manzo, S. Experimentación, instrumentos científicos y cuantificación en el método de Francis Bacon. *Manuscrito* **2001**, *24*, 49–84.

Article

Multitemporal Analysis of Land Use Changes and Their Effect on the Landscape of the Jerte Valley (Spain) by Remote Sensing

Yolanda Sánchez Sánchez [1,*], Antonio Martínez Graña [1], Fernando Santos-Francés [2], Joan Leandro Reyes Ramos [1] and Marco Criado [2]

[1] Department of Geology, Faculty of Sciences, University of Salamanca, Plaza de la Merced s/n, 37008 Salamanca, Spain; amgranna@usal.es (A.M.G.); reyesramos.joanleandro@usal.es (J.L.R.R.)

[2] Department of Soil Sciences, Faculty of Environmental Sciences, Avenue Filiberto Villalobos, 119, University of Salamanca, 37007 Salamanca, Spain; fsantos@usal.es (F.S.-F.); marcocn@usal.es (M.C.)

* Correspondence: yolanda.ss@usal.es

Abstract: In recent years, the interest of institutions in land use has increased, creating the need to determine the changes in use through spatial-temporal and statistical analysis. This study analyzes the changes over the last 40 years, based on a cartography of landscape units obtained from the study of geo-environmental parameters in the Jerte Valley (Spain) with satellite images, Landsat 5 and 7. Subsequently, through the analysis of spatial patterns and diversity and fragmentation indices, and with the Fragstat software, the landscape was characterized from 1994 to the present. The results show that wooded areas decreased slightly, crops increased in altitude and major environmental disturbances (mainly forest fires) negatively affected the environmental mosaic. Land uses affect the landscape by developing larger tesserae (+5 ha), which are less fragmented (−0.15), but more isolated (0.12). This study demonstrates that landscape metrics can be used to understand changes in spatial pattern, help in decision making to implement appropriate management measures in the conservation of traditional land uses, and allow the maintenance of connecting areas between fragments to avoid the loss of natural corridors to increase landscape quality.

Keywords: land use; landscape fragmentation; remote sensing; climate change

Citation: Sánchez Sánchez, Y.; Martínez Graña, A.; Santos-Francés, F.; Reyes Ramos, J.L.; Criado, M. Multitemporal Analysis of Land Use Changes and Their Effect on the Landscape of the Jerte Valley (Spain) by Remote Sensing. *Agronomy* **2021**, *11*, 1470. https://doi.org/10.3390/agronomy11081470

Academic Editor: Francisco Manzano Agugliaro

Received: 9 June 2021
Accepted: 22 July 2021
Published: 24 July 2021

Publisher's Note: MDPI stays neutral with regard to jurisdictional claims in published maps and institutional affiliations.

1. Introduction

Landscape can be defined as a spatial configuration of patches of dimensions relevant to the phenomenon under consideration or to the selected organism, which exists only at the moment in which it is perceived by the senses. The landscape can also be considered as a portion of the real world within which we are interested in describing and interpreting processes and patterns; this context can lead to different conclusions, depending on whether we use abiotic and/or biotic factors [1]. These factors make it intrinsically dynamic, both at a temporal [2] and spatial scale [3], since it is conditioned by environmental conditions [4], the ecological processes that take place in it, changes in land use, and anthropic disturbances [5].

Landscapes with marked heterogeneity have a complex structure of habitats, which translates into a high index of diversity [6]. This heterogeneity, as a factor of organization of ecological systems, presents a permanent character of landscapes and determines the generation of differentiated environmental mosaics [7]; which makes their study at a large scale extremely difficult. To the natural dynamics perpetuated by geomorphological [8–10], hydrological [11], and biotic [12] processes, to cite some significant examples, we must add human activities, which are currently the main factor in landscape evolution worldwide [13–16].

Remote sensing imagery is widely used for land cover classification, target identification, and thematic mapping from local to global scales due to its technical advantages such as as multiresolution, wide coverage, repeatable observation, and multispectral/hyperspectral–spectral records [17,18]. Several classification methods based on satellite images can be classified as supervised or unsupervised classification methods giving higher priority to supervised classification [19] because of its learning method [20,21].

Each land use class is to an image what a patch is to a landscape mosaic, so with this premise in mind, landscape metrics can be applied to measure the effect of land use changes on landscape shape [22]. Different studies calculated multiple levels of landscape metrics to measure landscape patterns in order to analyze land use, mainly focused on urban land use [23–25], and on the classification of satellite images rather than on the analysis of spatial patterns [26]. The research studies where a landscape analysis was performed used metrics based on average patch characteristics [27], ignoring the distribution, size, and changes occurring in the last decades, therefore, this study focuses on diversity indices and indices of the entire landscape.

The implementation of geographic information systems and remote sensing [28] have helped to carry out large-scale landscape studies, and with the support of different spatial pattern analysis programs [29], they have allowed a more precise study of the temporal dynamics of the landscape, which has provided a significant leap in the quality of the studies. For the analysis of spatial patterns, multiple indices have been developed that allow the study of landscape configurations at different temporal moments, evaluating their composition and configuration, the proportion of each class, or the shape of the elements [30].

The objective of this article was to know the effect of changes in land use on the landscape configured as a fluvial-structural valley during the last three decades, comparing the potential landscape with the landscape in 1994, and analyzing the spatial patterns in the subsequent years 2000, 2010, and 2019. Using Landsat 5 and Landsat 8 satellite images, and by analyzing spatial patterns and indices of diversity, dominance, shape, and fragmentation with Fragstat software, four scenarios were characterized and studying their tendency in the configuration of the environmental mosaic allowed us to understand the natural dynamics and the influences of human activities on the landscape.

Landscape metrics and the indices obtained have shown an important role in the analysis of changes in land use, so this study will enable land managers to implement appropriate measures for the maintenance of physical and functional connectivity in an anthropized environment, in order to achieve the objectives, set by the strategies of conservation and improvement of the landscape, for the sake of future sustainability.

2. Materials and Methods

2.1. Study Area

The Jerte valley is located in the north of the province of Cáceres (Spain) (Figure 1). It presents a geomorphological structure of a valley sandwiched between two mountain complexes (Sierra de Béjar, Sierra de Tormantos).

The geology of the study area is mainly formed by granitoids with some quartz outcrops, in the dykes of the Alentejo-Plasencia fault. On the southwestern edge of the study area there are outcrops of metamorphic rocks.

The geomorphological component was obtained based on the mapping of geomorphological units [31], synthesized in a mapping of geomorphological domains. The main geomorphological domains are as follows: summit surfaces and fluvial divides, embedded fluvial-glacial valleys, slopes and colluvial slopes, polygenic surfaces, hills and hillocks, glaciers, and fluvial terraces, alluvial and valley bottoms.

As for the vegetation that develops in the Jerte Valley, it is conditioned by the slopes of the hillsides, the differences in altitude, and the climate. In the areas located to the west, areas of low slope and glacial geomorphology, open formations such as pastures have developed. As one ascends in latitude and altitude there are large extensions of tree crops (cherry trees) very typical of this area. Halfway up the slope, both to the south and north of the river, there are large areas of wooded formations (pine forest, chestnut grove, oak forest, ...). On reaching the timberline, there are different scrublands, and later, ascending in altitude, the summit pastures. Some summits have a steep slope which hinders the development of vegetation while the rock remains on the surface.

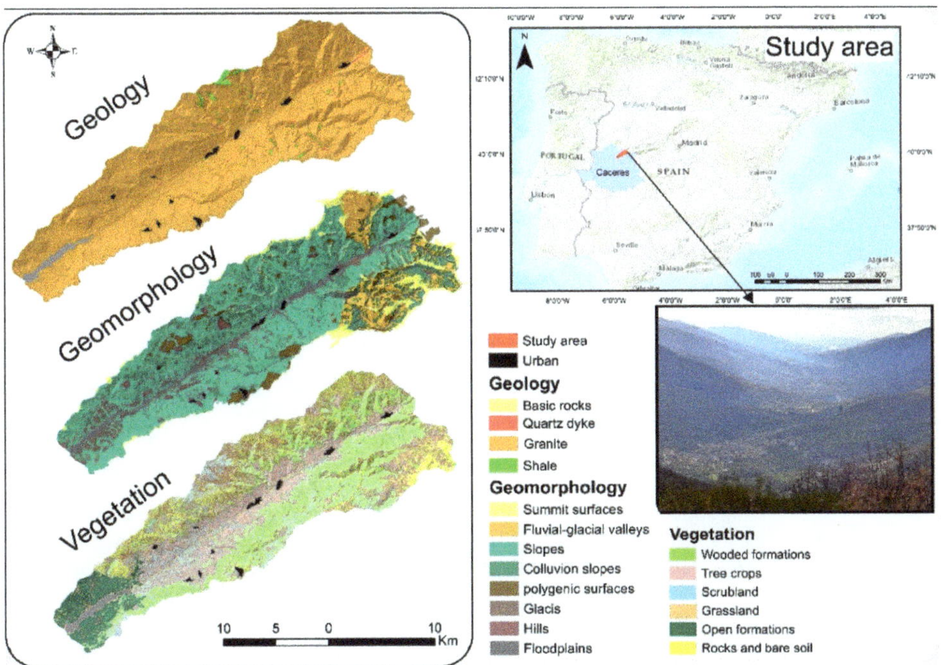

Figure 1. Study area.

2.2. Methods

The methodology followed (Figure 2) for landscape mapping was that described by Martínez-Graña [32], using a supervised classification of Landsat images as vegetation mapping for the years 1994, 2000, 2010, 2019.

With ArcGIS 10.8 software, the mapping of homogeneous units was carried out, from the union between lithology and geomorphological domain mapping. Once the union of these mappings was carried out, 29 homogeneous units were obtained. To simplify these mappings, all units smaller than 2 hectares were filtered out, as they were not representative. The units with similar landscape development behavior, such as quartz slopes and colluviated slopes, basic rock slopes and colluviated slopes, and schist slopes and colluviated slopes, were unified. Finally, the 18 homogeneous units, most representative of the study area, were mapped.

Landsat images were downloaded from the U.S. Geological Survey (https://earthexplorer. usgs.gov/, accessed on 21 July 2021). These images were taken between the months of April, May, or June of the years 1994, 2000, 2010, and 2019. It was decided to use images from Landsat satellites since they were the free satellites with the highest resolution available during the entire study period. The use of images from other satellites, such as Sentinel 2, was considered, but only images from 2015 were available, so we would only have had the latest image to study and with a different resolution, so the surfaces to be compared could have varied. In the end, this option was discarded to homogenize the data as much as possible.

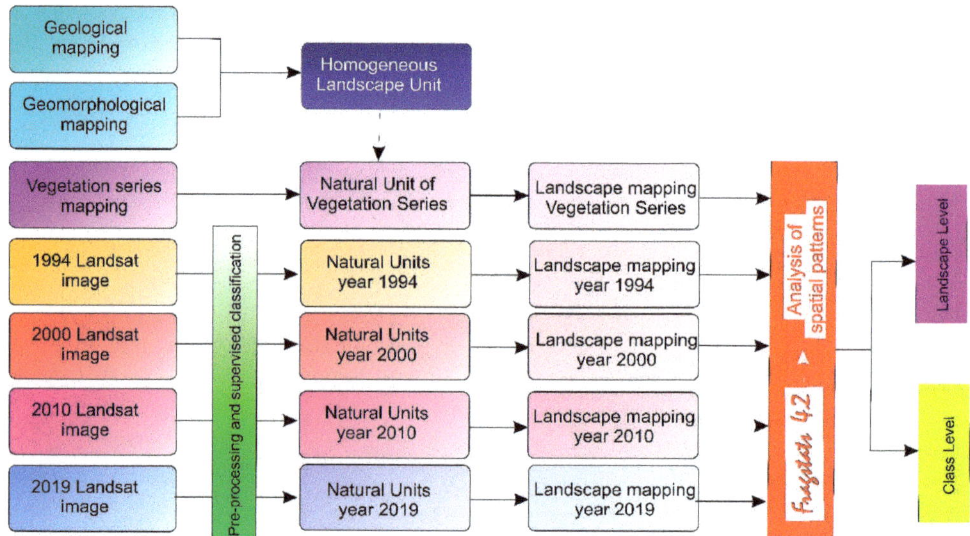

Figure 2. General methodology.

Each image was subjected to radiometric calibration preprocessing, atmospheric corrections, and topographic corrections. Once the images were obtained, training areas were selected after a field reconnaissance and with the help of orthophotos from previous years. Supervised classifications were performed on each image to obtain the land cover. It was decided to carry out the land cover mapping based on supervised classification since high-resolution orthophotos of the study years and field sampling of the last year were available and thus training areas could be delimited with great precision and a highly accurate mapping could be obtained.

Once the study areas were delimited, we proceeded to calculate the spectral signatures of each defined class, in order to extrapolate each pixel value of each class of the training area from the rest of the image. The supervised classification was performed by the maximum likelihood process, this method makes a statistical study (mean and standard deviation) of the pixel values of the training areas and calculates the probability of the values of the external indices to the training areas of belonging to one class or another, while the class with the highest probability is the one assigned to it.

After the supervised classification, a Majority filter was performed, thus filtering the neighboring contiguous cells of the larger size classes and avoiding the salt and pepper effect of supervised classification.

To validate the historical land use mapping, the Kappa index of each supervised classification was calculated. For the mapping to be accepted for further study, the Kappa index had to have been greater than 0.75 [33,34].

Land uses were classified into 10 main classes (water, forest, forest, cherry crops, treeless, scrub, snow, rocky, urban, burned area, clouds) to make the data homogeneous and to be able to analyze their parameters in the landscape as a whole.

The vegetation series mapping used was obtained from the Vegetation Series Map developed by Rivas Martínez in 1981 and revised in 1987 [35]. In the study area there are 5 vegetation series of the 37 existing in Spain. The vegetation series present in the study area have been reclassified (Table 1) in the denomination of the vegetation classes used in the present work in order to make a comparison of potential vegetation and real vegetation.

Table 1. Reclassification of the vegetation series to the legend of the vegetation mapping used for the study.

Vegetation Series	Vegetation Unit
Serie bejarano- gradense silicícola from *Festuca indigesta*	Grassland
Serie bejarano-gredense occidental and salmantina silicícola from *Cytisus purgans*	Scrubland
Serie carpetano-ibérico-alcarrena sub-humid silicícola from *Q. pirenaica*	Wooded formations
Serie luso-extremadurense humid from *Q.pyrenaica*	Wooded formations
Serie luso- extremadurense silicícola from *Q.rotundifolia*	Wooded formations
Geoserie riparia mediterráneo	Wooded formations

Then, the corresponding unions of the homogeneous unit mappings with the different vegetation mappings obtained from the vegetation series map and supervised classifications from Landsat images of the years 1994, 2000, 2010, 2020 were carried out, giving rise to the natural unit mappings.

In the mapping of natural units, 53 units were obtained and simplified in the following way:

1. Very small units, smaller than 2 hectares, were eliminated.
2. Units adjacent and similar to the most representative neighboring units were unified, as long as the extension of the different units was not significant.
3. The field study showed that the simplified landscape units were not representative of the landscape to be studied (Figure 3).
4. The result was 8 landscape units (Table 2):

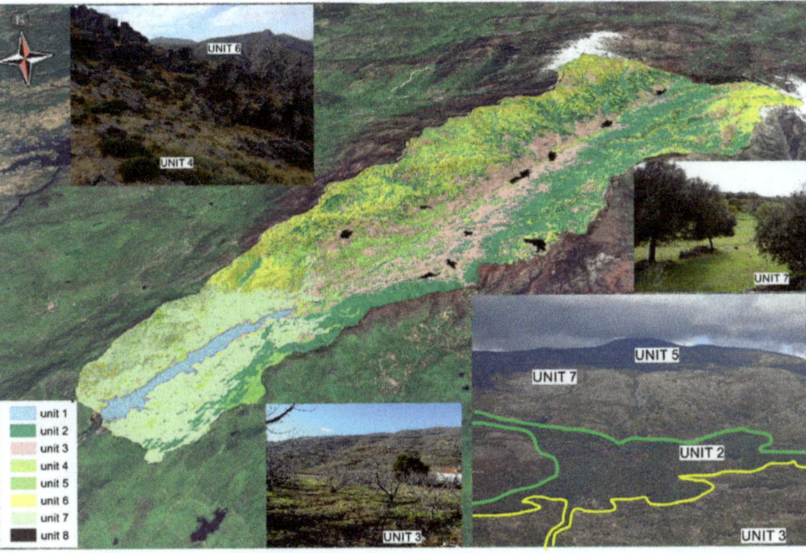

Figure 3. 3D modeling of landscape units with reconnaissance field images of the most representative landscape units.

Table 2. Ending landscape unit.

Landscape Units	Characteristics of the Unit
Unit 8	Urban landscape.
Unit 7	Glacis and slope in granites with open formations.
Unit 6	Summits and slopes in rocky outcrop granitoids and bare soil.
Unit 5	Polygenic surfaces in scrubland granitoids.
Unit 4	Summits and fluvial-glacial valleys in grassland granite.
Unit 3	Slope y colluvion slope in granites of tree crops.
Unit 2	Slopes and colluvial slopes in granites of wooded formations.
Unit 1	Water landscape

Finally, with the data collected, with the software Fragstat v4.2.1 [36] the analyses of the spatial and ecological patterns of the Jerte Valley were carried out. Fragstat is a software that allows the quantification of landscape patterns, which is a prerequisite for the study of landscape relationships with the processes that degrade it. Fragstat calculates a set of indices and variables that quantitatively describe the level of fragmentation and spatial distribution of land use and land cover.

These analyses were carried out with the help of indices that describe the different classes or patches of the units or allow a description of the landscape as a whole. The indices used are those described in Table 3.

Table 3. Index Fragstat.

Index	Observations
Total Area (TA)	TA equals the total area (ha) of the landscape
Number of Patches (NP)	NP is the number of patches of the corresponding patch type.
Mean Patch Area (ÁREA_MN	It is the average of the area of the patches (ha)

Index	Equation	Observations
Radius of Gyration (GYRATE)	$$\sum_{r=1}^{z} \frac{Hijr}{Z}$$ $Hijr$ = distance (m) between cell ijr and the centroid of patch jr, based on cell center-to-cell center distance; Z = number of cells in patch ij.	It is equal to the distance in m, between each pixel of the fragment and its corresponding centroid.
Patch Density (PD)	$$PD = \frac{Ni}{A} \times (10.000) \times (100)$$ Ni = number of patches in the landscape of patch type (class) i; A = total landscape area (m^2).	Number of fragments in this category for the total area.
Interspersion and Juxtaposition Index (IJI)	$$IJI = \frac{\sum_{i=1}^{m} \sum_{k=i+1}^{m} \left[\left(\frac{e_{ik}}{E} \times \ln\left(\frac{e_{ik}}{E}\right)\right)\right]}{\ln(0.5[m(m-1)])}(100)$$ e_{ik} = total length (m) of edge in landscape between patch types i and k; E = total length (m) of edge in landscape, excluding background; m = number of patch types present in the landscape, including the landscape border, if present.	This is the interspersion observed over the maximum probable dispersion according to the number of categories.
Shannon's Diversity Index (SHDI)	$$SHDI = -\sum_{i=1}^{m}(P_i \times \ln P_1)$$ P_i = proportion of the landscape occupied by patch type (class) i.	Relative abundance of different types of coverage in the landscape.
Shannon's Evenness Index (SHEI)	$$SHEI = \frac{\sum_{i=1}^{m}(P_i \times \ln P_i)}{\ln m}$$ P_i = proportion of the landscape occupied by patch type I; m = number of patch types present in the landscape, excluding the landscape border if present.	It is the Shannon diversity index divided by the maximum diversity index expected for the classes.
Simpson's Diversity Index (SIDI)	$$SIDI = 1 - \sum_{i=l}^{m} P_i^2$$ P_i = proportion of the landscape occupied by patch type (class) i.	Probability that two patches belong to the same class.

Table 3. *Cont.*

Index	Observations
Patch Cohesión Index (COHESION)	$$\text{COHESION} = \left[1 - \frac{\sum_{j=i}^{n} P_{ij}^{*}}{\sum_{j=1}^{n} P_{ij}^{*}\sqrt{a_{ij}^{*}}}\right] \times \left[1 - \frac{1}{\sqrt{Z}}\right]^{-1} \times (100)$$ P_{ij}^{*} = perimeter of patch ij in terms of number of cell surfaces; a_{ij}^{*} = area of patch ij in terms of number of cells. Z = total number of cells in the landscape.	It measures the physical connectivity of the analysed category.
Landscape Shape Index (LSI)	$$\text{LSI} = \frac{25E^{*}}{\sqrt{A}}$$ E^{*} = total length (m) of edge in landscape; includes the entire landscape boundary and some or all background edge segments; A = total landscape area (m^2).	Provides a standardised measure of total edge or edge density to suit the size of the landscape.
Contagion Index (CONTAG)	$$\text{CONTAG} = \left[1 + \frac{\sum_{i=1}^{m}\sum_{k=1}^{m}\left[P_{i}\times\frac{g_{ik}}{\sum_{k=1}^{m}g_{ik}}\right]\times\left[\ln\left(P_{i}\times\frac{g_{ik}}{\sum_{k=1}^{m}g_{ik}}\right)\right]}{2\ln(m)}\right] \times (100)$$ P_{i} = proportion of the landscape occupied by patch type; g_{ik} = number of adjacencies between pixels of patch types i and k based on the double-count method; m = number of patch types present in the landscape, including the landscape border if present.	It measures the percentage of adjacency between classes, in relation to the maximum possible considering the frequency of these.
Landscape Division Index (DIVISION)	$$\text{DIVISION} = \left[1 - \sum_{i=1}^{m}\sum_{j=1}^{n}\left(\frac{a_{ij}}{A}\right)^{2}\right]$$ a_{ij} = area (m^2) of patch ij. A = total landscape area (m^2).	Probability that two areas of the landscape are not located in the same habitat fragment.
Connectance Index (CONNECT)	$$\text{CONNECT} = \left[\frac{\sum_{i=1}^{m}\sum_{j=k}^{n} c_{ijk}}{\sum_{i=1}^{m}\left(\frac{n_{i}(n_{i}-1)}{2}\right)}\right](100)$$ c_{ijk} = joining between patch j and k of the same patch type, based on a user-specified threshold distance; n_{i} = number of patches in the landscape of each patch type (i).	It is the percentage of the total tesserae or of a class connected according to a threshold distance.
Proximity Index (PROX)	$$\text{PROX} = \sum_{g=1}^{n}\frac{a_{ijg}}{h_{ijg}^{2}}$$ a_{ijs} = area (m^2) of patch ijs within specified neighborhood (m) of patch ij; h_{ijs} = distance (m) between patch ijs and patch ijs, based on patch edge-to-edge distance, computed from cell center to cell center.	Sum of the areas of tesserae of the same class whose edges are at a specific radius.
Fragmentation (F)	$$F = \frac{TA}{NP\times 2\times ENN_{MN}\times\left(\frac{PD}{\pi}\right)}$$	Spatial disaggregation of patches or habitat types in a given area.

3. Results

The supervised rankings performed for the years 1994, 2000, 2010, and 2019 obtained a Kappa index higher than 0.75 (Table 4), so the mapping was accepted for study:

Table 4. Kappa index and overall accuracy for supervised classification (material complementary).

Year	Parámetros de Imágenes	Kappa Index	Overall Accuracy
1994	LT05_L1TP_202032_19940314_20180217_01_T1	0.92	0.93
2010	LT05_L1TP_202032_20100310_20161016_01_T1	0.76	0.80
2000	LT05_L1TP_202032_20000314_20180312_01_T1	0.88	0.91
2019	LC08_L1TP_202032_20190303_20190309_01_T1	0.86	0.88

The changes in land use in the Jerte Valley can be seen in Figure 4. The land use changes obtained in the Jerte Valley can be seen in Figure 3. There is a clear increase in the cultivation of cherry trees, a minimal loss in the extent of forests, with a decrease in scrubland in the years 2000 and 2020 where wildfires were notable.

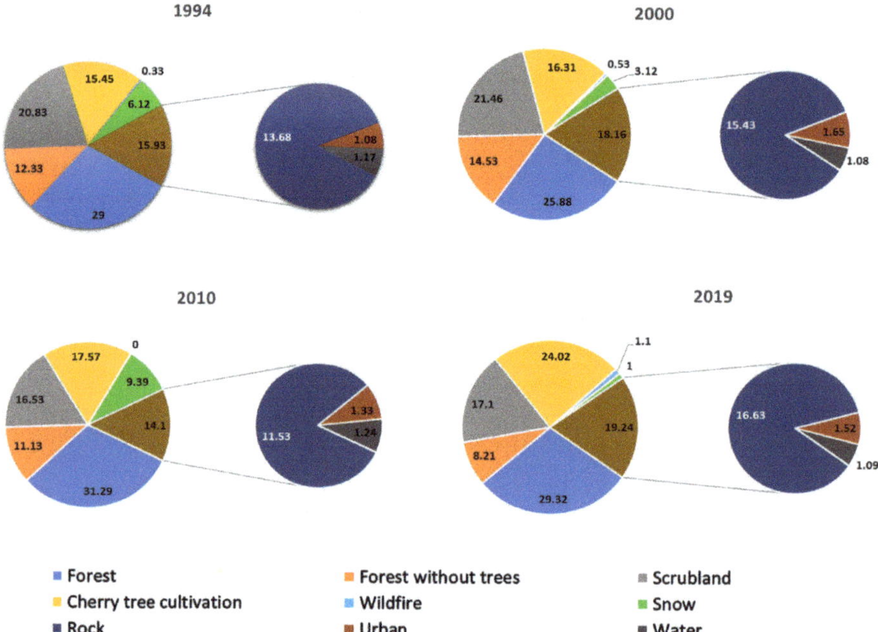

Figure 4. Evolution of land use in the last 4 decades in the Jerte Valley.

Based on the maps of landscape units for the study years: 1994, 2000, 2010, and 2019, the data obtained were analyzed (Figure 5). The statistical study began in 1994. Taking as a reference the vegetation series of Rivas Martínez, as potential vegetation, it can be observed that the proportion of land occupied by tree formations is much lower than would be expected (Table 5).

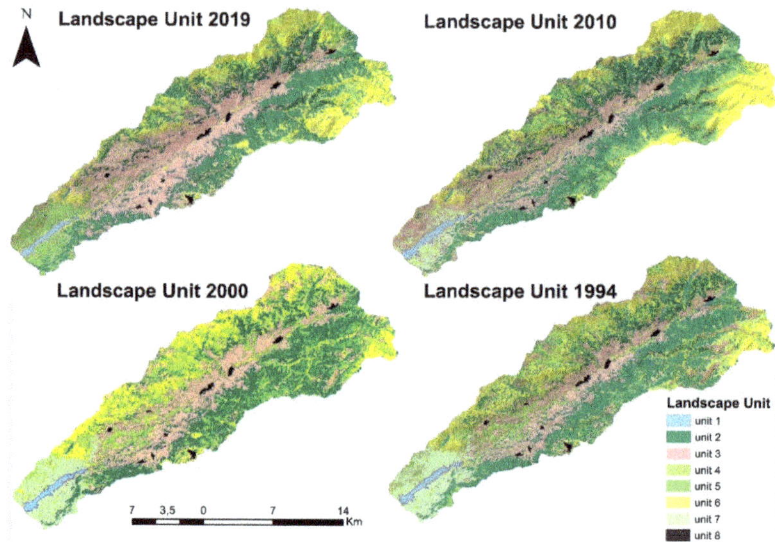

Figure 5. Multi-temporal spatial distribution of the natural units in the study area.

Table 5. Land cover indices in the Jerte Valley, in the years 1994, 2000, 2010, 2019 and in the vegetation series.

TA	SV		1994		2000		2010		2019	
	Total	%	Total	%	Total	%	Total	%	Total	%
Unit 1			433.7	1.10	433	1.10	380	0.98	358	0.92
Unit 2	30,957	82.34	11,103	28.06	9657	24.43	9836	25.29	10,951	28.13
Unit 3			7178	18.14	7486	18.93	7564	19.45	11,271	28.95
Unit 4	76.5	0.20	5257	13.28	5400	13.66	3333	8.57	2176	5.59
Unit 5	4087	10.87	3776	9.54	4913	12.43	3365	8.65	2728	7.01
Unit 6			6368	16.09	6609	16.72	8002	20.57	7187	18.46
Unit 7	2477	6.59	2594	6.55	2334	5.90	2280	5.86	2235	5.74
Unit 8			874	2.21	936	2.47	996	2.56	1386	3.56
Total	37,598		37,598		37,598		37,598		37,598	

Table 5 also shows how, in 1994, 28% of the surface area of the Jerte Valley was occupied by wooded formations (unit 2), such as oak (*Quercus pyrenaica*), chestnut (*Castanea sativa*), and pine (*Pinus sylvestris*), as opposed to 82% according to the optimum development of the vegetation series. It can also be observed that the presence of grassland areas (unit 3) is higher (13%) due to the traditional cultural uses of the land and its use for grazing, initial stages of recolonization of areas affected by forest fires, deforestation, etc. This stage of landscape development was taken as the initial situation, year 1994, and from there, the changes in the landscape over the last decades were studied.

The development of human activities in the environment has greatly conditioned the landscape in recent decades, occupying the optimal potential areas for the development of tree formations with cherry plantations (28.95%). On the other hand, there is an increase of 57% in the cultivation of cherry trees (unit 2) to the detriment of pasture and scrubland areas, 9% (unit 4 and 5).

The analysis of landscape structure was based on territorial changes in vegetation cover, geomorphology, and land use in recent decades. It was necessary to analyze land cover together with the models and management systems adopted in order to correctly assess the significance of the changes detected.

According to the values obtained in the diversity (SHDI and SIDI) and evenness (SHEI) indices, in the last three decades (Figure 6), heterogeneity and complexity are being lost in the landscape pattern. The number of patches is decreasing, becoming larger, more compact, and with simpler shapes. The grouping of tesserae causes a decrease in ecotones and has direct repercussions on the richness of species present in the environment. This evolution is largely due to the increase in agroforestry practices, forest management, and disturbances such as forest fires.

Once the spatial patterns and metrics of area, density, and landscape aggregation were analyzed (Table 6), it could be said that there is an increase in homogeneity at the landscape level in the study area, which corroborates the results obtained in reference to diversity.

The number of patches has almost halved from 16,023 in 1994 to 8183 in 2019. This has caused the patch density to decrease by 50%. Their average area of occupancy has doubled, indicating that we have fewer and larger patches. Likewise, the cohesion index, Landscape Shape Index, and the Interspersion and Juxtaposition Index have increased, which means that the masses maintain a moderately high degree of intermixing despite their tendency to homogenization.

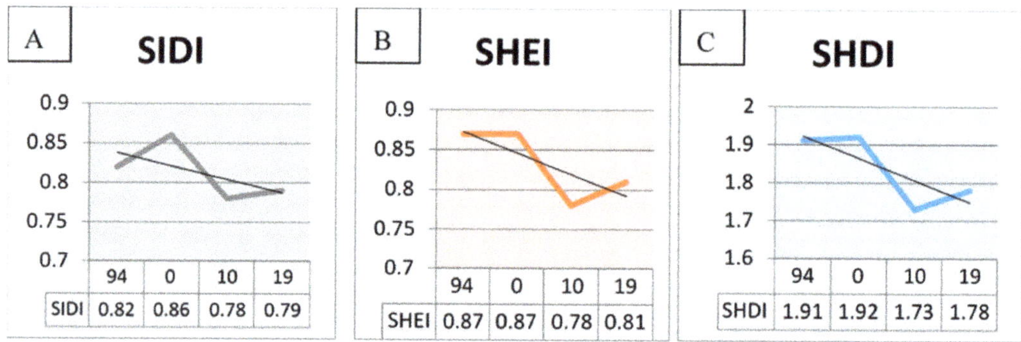

Figure 6. Diversity indices, the thick lines are the values of the indices in the different years, the thinner lines are the trend of each index. (**A**) Graph of Simpson's Diversity Index, (**B**) Graph of Shannon's Evenness Index, (**C**) Graph of Shannon's Diversity Index.

Table 6. Table of area, density and landscape aggregation metrics obtained by the FRAGSTAT software.

Area, Density and Aggregation Metrics	1994	2000	2010	2019
Total area (TA)	37,598	37,598	37,598	37,598
Number of Patches (NP)	16,023	15,939	9820	8183
Patch Density (PD)	40.48	40.31	25.48	20.69
Radius of Gyration (GYRATE_MN)	47.21	48.90	49.29	57.87
Landscape Shape Index (LSI)	83.32	86.71	50.54	55.25
Mean Patch Area (AREA_MN)	2.47	2.4805	3.92	4.83
Largest Patch Index (LPI)	16.72	10.53	27.25	23.33
Euclidean Nearest-Neighbor Distance Mean (EMM_MN)	141.70	142.01	149.61	166.43
Contagion Index (CONTAG)	30.20	29.10	39.3	40.85
Interspersion and Juxtaposition Index (IJI)	73.55	73.05	70.18	73.00
Connectance Index (CONNECT)	0.36	0.35	0.48	0.48
Landscape Division Index (DIVISION)	0.95	0.97	0.89	0.91
Proximity Index (PROX)	169	124	551	295
Patch Cohesion Index (COHESION)	96.94	96.20	98.38	98.02
Fragmentation Index (F)	0.07	0.07	0.16	0.22

The isolation of the spots increases, with a 74% increase of the proximity index and a 17% increase of the Euclidean Nearest-Neighbor Distance Mean. While the degree of fragmentation decreases by 200% (0.15, remember that it is inverse).

The study of the structure of the landscape at the class level makes it possible to discern the role of each tesserae in the environmental mosaic, assigning to each group a diversifying or homogenizing function in the whole.

The most outstanding results of the study in this area are the increase in the area under cultivation (+7000 ha), dedicated almost exclusively to cherry (Table 7). The tesserae that make up this class increase in number, but maintain the same patterns of isolation, shape, and area. Wooded areas decrease slightly but show changes in their arrangement indicating that the wooded tesserae are larger, less fragmented, but more isolated. On the other hand, units 4 and 5 decrease (3000 ha and 1000 ha respectively) and present very small, isolated, and fragmented tesserae. The open formations (unit 7), due to their singular character, being pasture areas subjected to a strong anthropic influence, require a detailed analysis and study. Large variations in fragmentation are observed, which can be explained by the agrosilvopastoral treatments to which this habitat is subjected.

Table 7. Table of area, density, and aggregation metrics of the patches obtained by the FRAGSTAT software. Where: TA: Total area; NP: Number of Patches; PD: Patch Density; AREA_MN: Mean Patch Area; IJI: Interspersion and Juxtaposition Index; Frag: Fragmentation Index.

		Unit 2	Unit 3	Unit 4	Unit 5	Unit 6	Unit 7	Unit 8
1994	TA	11103.25	7178.5	5257	3776	6368.5	874	2594
	NP	2517	2433	3555	3547	2728	825	19
	PD	6.36	6.148	8.983	8.962	6.893	2.085	0.048
	AREA_MN	4.41	2.95	1.47	1.06	2.33	1.05	136.52
	LSI	62.159	63.268	73.399	72.484	62.853	31.445	16.441
	ENN_MN	132.46	139.07	134.83	132.38	138.87	229.17	104.34
	IJI	74.018	78.363	76.914	64.811	72.291	72.84	66.494
	F	0.82	0.54	0.19	0.14	0.38	0.35	4281
2000	TA	9657	7486	5400	4913	6609	713	2334
	NP	2483	2407	3423	3740	2568	965	48
	PD	6.28	6.09	8.66	9.46	6.5	2.44	0.121
	AREA_MN	3.88	3.11	1.57	1.31	2.57	0.73	48.62
	LSI	64.59	63.86	73.26	81.51	62.31	31.97	20.516
	ENN_MN	137.41	138.72	136.27	128.65	142.1	223.7	106.79
	IJI	69.44	80.1	77.24	65.18	74.66	68.72	66.835
	F	0.71	0.58	0.21	0.17	0.44	0.21	589
2010	TA	9836	7564	3333	3365	8002	996	2280
	NP	1523	1600	2620	1489	1811	338	56
	PD	3.95	4.15	6.8	3.86	4.7	0.88	0.54
	AREA_MN	8.76	4.72	1.27	2.26	4.41	1.35	5.11
	LSI	50.53	51.7	58.96	43.5	35.49	19.12	19.13
	ENN_MN	127.05	138.91	147.96	155.86	160.85	217.14	117.58
	IJI	73.75	69.78	74.27	57.78	69.53	74.61	62.09
	F	2.02	1.29	0.2	0.59	0.92	2.43	100
2019	TA	10951	11271	2176	2729	7188	1386	2236
	NP	1143	1216	1707	2064	1334	348	39
	PD	2.89	3.07	4.32	5.22	3.37	0.88	0.10
	AREA_MN	9.58	9.26	1.27	1.32	5.38	1.67	57.32
	LSI	39.08	37.17	46.39	50.75	38.06	20.18	14.65
	ENN_MN	149.84	139.19	178.83	162.20	167.69	269.81	108.49
	IJI	71.53	77.71	78.63	66.25	71.05	66.49	69.90
	F	3.48	3.40	0.26	0.25	1.50	2.63	841

Valuable assessments can be obtained from the yearly detailed analysis. The year 2010 deserves a special comment since the initial data used for the mapping of landscape units contained parts covered by snow (unit 6). Snow represents an important and common disturbance in vegetation and, in general, in mountain areas. It highlights changes in landscape heterogeneity in annual and sub-annual space. In this particular case, it causes notably high levels in the cohesion and proximity index, and a decrease in the interdispersion and juxtaposition index. In addition, the area occupied by bare soil (snow) increases significantly to the detriment of areas of scrub and grassland, typical of mountain areas where snow accumulates.

Finally, it should be noted that the evolution of urban infrastructures shows a stable, moderate growth and does not represent a problem for the landscape, with diffuse anthropic pressure being practically non-existent, apart from some isolated agricultural and livestock constructions.

4. Discussion

The identification of changes in land use over the last 50 years makes it possible to identify the environmental impact that has occurred in the territory. The decrease in the area of snow, the increase in crop and urban areas, demonstrate an impact generated by anthropogenic activity in the area, mainly by the cherry tree cultivation activity. In other landscape evolution simulation studies, remote sensing techniques were used [33], with classification metrics between 80 and 90% of the images studied for the classification of the extent of tropical forest fragmentation [3] and a supervised classification methodology and landscape indices for landscape fragmentation simulations [37].

The analysis of the areas of occupation of the different types of land use allows a preliminary evaluation of the evolution of the territory. As previously mentioned, the landscape is a dynamic entity in continuous spatial and temporal evolution [38]. The environmental mosaic present in the Jerte Valley is quite heterogeneous and, therefore, the agents that cause these changes are diverse. The most important are: geomorphological agents in the areas of rock and bare soil, where they are the main modelers of the landscape together with the climate; and, in the areas dominated by vegetation, (geobotanical landscape) it is the plants that modify the edaphoclimatic conditions of the area. In addition, we are in a territory with a growing human influence, which determines a much more accelerated time scale of changes due to the disturbances caused by their activities.

Curiously, as a consequence of the change in climatic conditions, the optimal areas for cherry cultivation are no longer the valley bottoms and slopes, but now occupy the higher areas (scrub and pasture). It is a phenomenon marked in other territories with other fruit crops and especially for grapevine [39]. Ultimately it is a consequence of climate change. Presumably, this trend will continue until the lithological and edaphological conditions mean that the land cannot be cultivated, even with the application of the current techniques of farming [40].

Another of the recurrent landscape disturbances in the Jerte Valley and in other areas with a Mediterranean climate are forest fires [41]. At an ecological level, they can be beneficial for the maintenance of the forest structure if they do not occur too frequently, in limited areas, or in areas with a subarid climate. In these cases, they cause simplification of the environmental mosaics [42]. In the study area, forest fires in recent years have mainly affected scrub and grassland areas (see the decrease in the area of occupancy for the year 2020), minimizing their ecological effects, compared to the foreseeable consequences of affecting complex tree formations.

The naturalization of pastures, together with the growth of cherry orchards, has caused the landscape mosaic to change and present more homogeneity and less diversity. It is an unstable and singular situation, since at the same time the natural and cultural landscapes are advancing, giving rise to a common result. This situation can lead to the generation of a cyclical dynamic based on the recurrent occurrence of fires, which is not typical of this territory and can lead to the loss of biological and environmental resources, increased erosion [43], ecological imbalances, etc.

The expansion of cherry trees towards the highlands could be a partial mitigation of the effects described above, enhancing the coexistence of valuable cultural and natural elements. This would favor the maintenance of anthropized areas with lower fire risk and greater economic and ecological potential. Another favorable measure would be the maintenance of extensive livestock farming, both in pasture areas and in mountain pastures, since it plays a fundamental role in the conservation of these habitats, as well as the geobotanical, genetic, and landscape diversity found in them [44].

Dehesa grassland systems have been found to present a high and changing fragmentation of the stippled or mottled type due to the type of management. They are very heterogeneous but harbor a greater diversity than other equivalent potential systems and are therefore a priority for conservation.

5. Conclusions

Previous work has been carried out combining landscape metrics and remote sensing, however, changes in land use have not been studied with this methodology. In the present study, the importance of land use changes in landscape dynamics is highlighted by systematically analyzing landscape evolution using spatial patterns and indices of diversity, dominance, form, and fragmentation. The scenarios analyzed reveal a trend towards homogenization of the territory with loss of diversity and changes in natural dynamics, largely caused by human influence (cherry tree cultivation) and major disturbances created by forest fires.

The research has shown that the analysis of landscape patterns in a multi-temporal study allows both the analysis of changes in past land uses and the promotion of actions for the future conservation of land uses that are positive for the landscape, allowing compliance with conservation strategies, as well as the evaluation of the causes and consequences of large-scale actions or disturbances in the landscape. In addition, the use of geographic information systems, remote sensing, and spatial analysis software allows the evaluation and integration of many geo-environmental parameters that form the landscape, enabling the development of a base cartography of superior detail and quality; and enables the spatio-temporal and statistical analysis of landscape units and their effect on the environment. This methodology can be implemented in different regions with similar characteristics, large areas with a large number of classes, remote regions, difficult to access, etc. For the future, we intend to apply this methodology for shorter time intervals and to see the gradual modification of the landscape for annual periods to observe whether the landscape modifications are gradual or abrupt. It is a fundamental tool for the proper management of land use, land use planning, and environmental conservation. Institutions and territorial managers will be able to adapt the policies and programs of each region to each present and future scenario, adapting this generic methodology to particular situations, thus achieving objective and comparable parameters, extremely useful for decision making.

Author Contributions: Conceptualization, Y.S.S. and J.L.R.R.; methodology, Y.S.S., J.L.R.R., and A.M.G.; software, Y.S.S. and J.L.R.R.; validation, Y.S.S. and F.S.-F.; formal analysis, A.M.G.; investigation, Y.S.S. and M.C.; resources, A.M.G.; data curation, Y.S.S. and A.M.G.; writing—original draft preparation, Y.S.S.; writing—review and editing, A.M.G.; visualization, M.C.; supervision, A.M.G. and F.S.-F.; project administration, A.M.G.; funding acquisition, A.M.G. All authors have read and agreed to the published version of the manuscript.

Funding: This research was funded by the project SA044G18 of the Regional Government of Castilla y Leon, and the GEAPAGE research group (Environmental Geomorphology) of the University of Sala-manca.

Acknowledgments: This research was funded by the project SA044G18 of the Regional Government of Castilla y Leon, and the GEAPAGE research group (Environmental Geomorphology) of the University of Salamanca.

Conflicts of Interest: The authors declare no conflict of interest.

References

1. Farina, A. *Ecología del Paisaje*; Universidad de Alicante: San Vicente de Raspeig, Spain, 2011; ISBN 978-84-9717-167-0.
2. Hunzkier, M.; Felber, P.; Gehring, K.; Buchecker, M.; Bauer, N.; Kienast, F. Evaluation of landscape change by different social groups: Results of two empirical studies in Switzerland. *Mt. Res. Dev.* **2008**, *28*, 140–147. [CrossRef]

3. Shi, X.; Feng, G.; Yi, Y.; Zou, Y.; Ge, H.; Su, P. Temporal and Spatial Evolution Characteristics and Ecological Risk Assessment of Land Use Landscape Patterns in Central Zhejiang Urban Agglomeration. *Nongye Jixie Xuebao/Trans. Chin. Soc. Agric. Mach.* **2020**, *51*, 242–251. [CrossRef]

4. Chhogyel, N.; Kumar, L.; Bajgai, Y. Spatio-temporal landscape changes and the impacts of climate change in mountainous Bhutan: A case of Punatsang Chhu Basin. *Remote Sens. Appl. Soc. Environ.* **2020**, *18*, 100307. [CrossRef]

5. Cornejo-Denman, L.; Romo-Leon, J.R.; Hartfield, K.; van Leeuwen, W.J.D.; Ponce-Campos, G.E.; Castellanos-Villegas, A. Landscape dynamics in an iconic watershed of Northwestern Mexico: Vegetation condition insights using landsat and planetscope data. *Remote Sens.* **2020**, *12*, 2519. [CrossRef]

6. Swan, M.; Christie, F.; Steel, E.; Sitters, H.; York, A.; Di Stefano, J. Ground-dwelling mammal diversity responds positively to productivity and habitat heterogeneity in a fire-prone region. *Ecosphere* **2020**, *11*, e03248. [CrossRef]

7. Li, H.; Reynolds, J.F. On definition and quantification of heterogeneity. *Oikos* **1995**, *73*, 280–284. [CrossRef]

8. Martínez-Graña, A.M.; Goy, J.L.; Zazo, C.; Silva, P.G.; Santos-Francés, F. Configuration and evolution of the landscape from the geomorphological map in the natural parks Batuecas-Quilamas (central system, SW Salamanca, Spain). *Sustainability* **2017**, *9*, 1458. [CrossRef]

9. Mariotti, G.; Spivak, A.C.; Luk, S.Y.; Ceccherini, G.; Tyrrell, M.; Gonneea, M.E. Modeling the spatial dynamics of marsh ponds in New England salt marshes. *Geomorphology* **2020**, *365*, 107262. [CrossRef]

10. Gardner, J. How water, wind, waves and ice shape landscapes and landforms: Historical contributions to geomorphic science. *Geomorphology* **2020**, *366*, 106687. [CrossRef]

11. Cantonati, M.; Segadelli, S.; Spitale, D.; Gabrieli, J.; Gerecke, R.; Angeli, N.; De Nardo, M.T.; Ogata, K.; Wehr, J.D. Geological and hydrochemical prerequisites of unexpectedly high biodiversity in spring ecosystems at the landscape level. *Sci. Total Environ.* **2020**, *740*, 140157. [CrossRef]

12. Clément, F.; Ruiz, J.; Rodríguez, M.A.; Blais, D.; Campeau, S. Landscape diversity and forest edge density regulate stream water quality in agricultural catchments. *Ecol. Indic.* **2017**, *72*, 627–639. [CrossRef]

13. Gao, P.; Kasimu, A.; Zhao, Y.; Lin, B.; Chai, J.; Ruzi, T.; Zhao, H. Evaluation of the temporal and spatial changes of ecological quality in the Hami Oasis based on RSEI. *Sustainability* **2020**, *12*, 7716. [CrossRef]

14. Rajakumari, S.; Sundari, S.; Meenambikai, M.; Divya, V. Impact analysis of land use dynamics on coastal features of Deshapran block, Purba East Medinipur, West Bengal. *J. Coast. Conserv.* **2020**, *24*, 1–10. [CrossRef]

15. Rodríguez-Medina, K.; Yañez-Arenas, C.; Peterson, A.T.; Ávila, J.E.; Herrera-Silveira, J. Evaluating the capacity of species distribution modeling to predict the geographic distribution of the mangrove community in Mexico. *PLoS ONE* **2020**, *15*, e0237701. [CrossRef]

16. Mantero, G.; Morresi, D.; Marzano, R.; Motta, R.; Mladenoff, D.J.; Garbarino, M. The influence of land abandonment on forest disturbance regimes: A global review. *Landsc. Ecol.* **2020**, *35*, 2723–2744. [CrossRef]

17. Navalgund, R.R.; Jayaraman, V.; Roy, P.S. Remote sensing applications: An overview. *Curr. Sci.* **2007**, *93*, 1747–1766.

18. Arora, A.; Pandey, M.; Mishra, V.N.; Kumar, R.; Rai, P.K.; Costache, R.; Punia, M.; Di, L. Comparative evaluation of geospatial scenario-based land change simulation models using landscape metrics. *Ecol. Indic.* **2021**, *128*, 107810. [CrossRef]

19. Du, P.; Xia, J.; Zhang, W.; Tan, K.; Liu, Y.; Liu, S. Multiple Classifier System for Remote Sensing Image Classification: A Review. *Sensors* **2012**, *12*, 4764–4792. [CrossRef]

20. Abdullah, H.M.; Mahboob, M.G.; Rahman, M.M.; Ahmed, T. Monitoring natural Sal forest cover in Modhupur, Bangladesh using temporal Landsat imagery during 1972–2015. *Int. J. Environ.* **2015**, *5*, 1–7.

21. Li, C.; Wang, J.; Wang, L.; Hu, L.; Gong, P. Comparison of Classification Algorithms and Training Sample Sizes in Urban Land Classification with Landsat Thematic Mapper Imagery. *Remote Sens.* **2014**, *6*, 964–983. [CrossRef]

22. Zhang, Y.; Qin, K.; Bi, Q.; Cui, W.; Li, G. Landscape Patterns and Building Functions for Urban Land-Use Classification from Remote Sensing Images at the Block Level: A Case Study of Wuchang District. *Remote. Sens.* **2020**, *12*, 1831. [CrossRef]

23. Peng, J.; Wang, Y.; Zhang, Y.; Wu, J.; Li, W.; Li, Y. Evaluating the effectiveness of landscape metrics in quantifying spatial patterns. *Ecol. Indic.* **2010**, *10*, 217–223. [CrossRef]

24. Zheng, X.; Wang, Y.; Gan, M.; Zhang, J.; Teng, L.; Wang, K.; Shen, Z.; Zhang, L.; Campbell, J.; Atzberger, C.; et al. Discrimination of Settlement and Industrial Area Using Landscape Metrics in Rural Region. *Remote. Sens.* **2016**, *8*, 845. [CrossRef]

25. Aguilera, F.; Valenzuela, L.M.; Botequilha-Leitão, A. Landscape metrics in the analysis of urban land use patterns: A case study in a Spanish metropolitan area. *Landsc. Urban Plan.* **2011**, *99*, 226–238. [CrossRef]

26. Voltersen, M.; Berger, C.; Hese, S.; Schmullius, C. Object-based land cover mapping and comprehensive feature calculation for an automated derivation of urban structure types at block level. *Remote Sens. Environ.* **2014**, *154*, 192–201. [CrossRef]

27. Zhang, Y.; Li, Q.; Huang, H.; Wu, W.; Du, X.; Wang, H. The Combined Use of Remote Sensing and Social Sensing Data in Fine-Grained Urban Land Use Mapping: A Case Study in Beijing, China. *Remote Sens.* **2017**, *9*, 865. [CrossRef]

28. Song, W.; Song, W.; Gu, H.; Li, F. Progress in the remote sensing monitoring of the ecological environment in mining areas. *Int. J. Environ. Res. Public Health* **2020**, *17*, 1846. [CrossRef] [PubMed]

29. Peng, Y.; Wang, Q. Spatial distribution and influencing factors of settlements in the farming–pastoral ecotone of Inner Mongolia, China. *Ecosyst. Health Sustain.* **2020**, *6*, 1771213. [CrossRef]

30. McGarigal, K.; Marks, B.J. *Fragstats: Spatial pattern analysis program for Quantifying Landscape Structure*; USDA Forest Service General Technical Report, PNW-GTR 351; USDA: Washington, DA, USA, 1995.

31. Carrasco Gonzalez, R.M. *Geomorfología del Valle del Jerte. Las Líneas Maestras del Paisaje*; Universidad de Extremadura: Caceres, Spain, 1999.
32. Martínez-Graña, A.M.; Silva, P.G.; Goy, J.L.; Elez, J.; Valdés, V.; Zazo, C. Geomorphology applied to landscape analysis for planning and management of natural spaces. Case study: Las Batuecas-S. de Francia and Quilamas natural parks. *Sci. Total Environ.* **2017**, *584–585*, 175–188. [CrossRef]
33. Anderson, J.R.; Hardy, E.E.; Roach, J.T.; Witmer, R.E. *A Land Use and Land Cover Classification System for Use with Remote Sensor Data*; US Government Printing Office: Washington, DC, USA, 1976. [CrossRef]
34. Olaya, F.V. *Sistemas de Información Geográfica*, 1st ed.; CreateSpace Independent Publishing Platform: North Charleston, SC, USA, 2020; ISBN 9781716777660.
35. Rivas-Martínez, S.; Armaiz, C. Bioclimatología y vegetación en la Península Ibérica. *Bull. Soc. Bot. Fr. Actual. Bot.* **1984**, *131*, 110–120. [CrossRef]
36. McGarigal, K.; Cushman, S.A.; Ene, E. FRAGSTATS v4: Spatial Pattern Analysis Program for Categorical and Continuous Maps. Computer Software Program Produced by the Authors at the University of Massachusetts, Amherst. 2012. Available online: www.umass.edu/landeco/research/fragstats/fragstats.html (accessed on 15 July 2021).
37. Pyngrope, O.R.; Kumar, M.; Pebam, R.; Singh, S.K.; Kundu, A.; Lal, D. Investigating forest fragmentation through earth observation datasets and metric analysis in the tropical rainforest area. *SN Appl. Sci.* **2021**, *3*, 705. [CrossRef]
38. Wu, J.-M.; Cai, C.-C.; Sun, H.; Yao, J.-H.; Chen, W.; Gu, J.-S.; Jiang, J.-B. Spatiotemporal evolution and driving force analysis of fractional vegetation coverage over the urban belt along the Yellow River in Ningxia. *Arid Zo. Res.* **2020**, *37*, 696–705. [CrossRef]
39. Sánchez, Y.; Martínez-Graña, A.M.; Santos-Francés, F.; Yenes, M. Index for the calculation of future wine areas according to climate change application to the protected designation of origin Sierra de Salamanca (Spain). *Ecol. Indic.* **2019**, *107*, 105646. [CrossRef]
40. Asins Velis, S.; Sánchez Díaz, J. Inclusión de las estrategias de control de la erosión de laderas aterrazadas en las políticas de paisaje. *Comun. Congresos.* **2011**, 565–568.
41. Sánchez, Y.; Martínez-Graña, A.; Santos Francés, F.; Mateos Picado, M. Mapping wildfire ignition probability using sentinel 2 and LiDAR (Jerte Valley, Cáceres, Spain). *Sensors* **2018**, *18*, 826. [CrossRef] [PubMed]
42. Moreno, J.M. Cambio global e incendios forestales: Una visión desde España. In Proceedings of the 4a Conferencia Internacional Sobre Incendios Forestales (CD) DGB/MMA, Sevilla, Spain, 13–17 May 2007.
43. Sholagberu, A.T.; Mustafa, M.R.; Yusof, K.W.; Hashim, A.M. Geo-statistical based susceptibility mapping of soil erosion and optimization of its causative factors: A conceptual framework. *J. Eng. Sci. Technol.* **2017**, *12*, 2880–2895.
44. Vicente Serrano, S.M.; Lasanta Martínez, T.; Cuadrat, J.M. Transformaciones en el paisaje del Pirineo como consecuencia del abandono de las actividades económicas tradicionales. *Cons. Super. Investig. Científicas* **2000**, *155*, 111–133. [CrossRef]

Article

Improving the Management of a Semi-Arid Agricultural Ecosystem through Digital Mapping of Soil Properties: The Case of Salamanca (Spain)

Marco Criado [1,*][iD], Antonio Martínez-Graña [2][iD], Fernando Santos-Francés [1] and Leticia Merchán [1]

1   Department of Soil Sciences, Faculty of Agricultural and Environmental Sciences, University of Salamanca, Avenue Filiberto Villalobos, 119, 37007 Salamanca, Spain; fsantos@usal.es (F.S.-F.); leticiamerchan@usal.es (L.M.)
2   Department of Geology, Faculty of Sciences, University of Salamanca, Plaza de la Merced s/n., 37008 Salamanca, Spain; amgranna@usal.es
*   Correspondence: marcocn@usal.es

Abstract: Soil protection and the increase and intensification of agricultural production require detailed knowledge of soil properties and their variability. On the other hand, the complexity associated with traditional soil mapping processes can lead to the implementation of inappropriate agricultural practices that degrade this resource. Therefore, it is necessary to use mapping techniques to provide more detailed information to farmers and managers. In this study, the geostatistical technique ordinary kriging was used to map the distribution of the most important edaphic properties (texture, nutrients content -N, P, K-, pH, organic carbon, water retention, COLE, carbonate content, and cation exchange capacity) from known sampled points, which allows inferring the value and distribution of the different edaphic parameters studied along the agricultural fields. The results obtained show after validation that the analysis of semivariograms is suitable for evaluating the distribution of the main soil parameters on a large scale, since it faithfully reflects their distribution and makes the ordinary kriging tool a suitable method for optimizing the resources available in soil mapping processes. In addition, the knowledge of these distributions made it possible to establish different recommendations for improving the management of the agricultural ecosystem, which will guarantee a higher agricultural yield as well as a better protection of the analyzed soils.

Keywords: agrarian ecosystem; GIS; geostatistics; kriging; soil mapping

Citation: Criado, M.; Martínez-Graña, A.; Santos-Francés, F.; Merchán, L. Improving the Management of a Semi-Arid Agricultural Ecosystem through Digital Mapping of Soil Properties: The Case of Salamanca (Spain). *Agronomy* **2021**, *11*, 1189. https://doi.org/10.3390/agronomy11061189

Academic Editor: Rose Brodrick

Received: 20 May 2021
Accepted: 9 June 2021
Published: 10 June 2021

Publisher's Note: MDPI stays neutral with regard to jurisdictional claims in published maps and institutional affiliations.

1. Introduction

Agriculture is currently facing multiple challenges with high global implications, especially the food supply of a growing population [1]. Although agricultural knowledge and techniques have experienced a high development in the last decades that has contributed to meet food demands [2], other problems have not been solved and put agricultural activity at risk in the current stage of productive intensification [3,4]. Poor agricultural practices and lack of agronomic information can lead to soil erosion [5], contribute to global warming [6], affect biodiversity [7], and cause contamination of agricultural land and water [8,9]. Specifically, the lack of information about soils is of concern, which is usually limited to isolated analytics that do not reflect soil variability and, therefore, the farmer does not know the different soil requirements in each location [10]. In-depth knowledge of soil characteristics is necessary to achieve sustainable agriculture that avoids malpractices [11].

Digital soil mapping (DSM) using statistical inference and GIS is a technique that allows researchers to accurately interpolate spatial patterns of soil properties [12]. The use of the geostatistical interpolation technique also reduces the costs of field sampling and laboratory analysis and allows the soil mapping process to be accelerated [13]. However, the reliability of spatial variability maps depends on adequate sampling data (such as the number of samples or the distance between them) and the accuracy of spatial interpolation [11]. There are several interpolation methods, including deterministic and stochastic ones, well compiled by Myers [14]. Among them, kriging has proven to be robust enough to estimate values at unsampled locations from the study of specific sampled locations. Moreover, kriging provides the best unbiased linear estimates and information on the estimation error distribution and shows strong statistical advantages over other geostatistical methods (Wang et al., 2012). Several authors already employed this technique in predictive soil mapping, studying different factors: various edaphic parameters [15,16]; nutrient distribution [17], erosion estimation [18]; organic carbon [19–23]; organic matter [24]; heavy metals [25]; chemical properties with importance in agriculture [10]; or soil quality indices [26].

Despite the good adaptation of these interpolation techniques to predictive soil mapping, there are no studies that represent the variability of the main edaphic properties in the study sector. In practice, this translates into the proliferation of agricultural practices inadequate to the needs of the soils. Therefore, the objectives of this work are: (1) to predict the distribution of soil properties that have a greater relevance in agricultural development from the sampled soils and the geostatistical technique of ordinary kriging interpolation; and (2) to improve agronomic knowledge of the soils in the study area and to promote increased crop yields through the use of sustainable agricultural management practices.

2. Materials and Methods

2.1. Study Area

The study area is located in the west of Spain and southwest of the region of Castilla y León, more specifically, in the northeast of the province of Salamanca (Figure 1). It covers a total area of 770 km^2 with its epicenter in the city of Salamanca. In the center and north is the region of La Armuña, famous for its agricultural products, and in the south the Campo Charro, characterized by extensive livestock farming. Geographically, the region studied is located within the Spanish Northern Plateau, with a flat to gently undulating topography, as evidenced by the small difference in altitude (182 m.) between the highest peaks (Los Montalvos mountain, 942 m.a.s.l., modeled on a synclinoric syncline) and the highest peaks (Los Montalvos, 942 m.a.s.l., modeled on a synclinorium formed by Armorican quartzite and slate of Ordovician-Silurian age) and minimum (corresponding to the fertile lowlands of the Tormes River, on alluvial deposits of the Pleistocene-Holocene with heights of 760 m.), predominating in the intermediate heights the plains carved on Palaeogene and Neogene horizontal sediments [27].

The climate is characterized by long, cold winters and hot, dry summers, with an average annual precipitation and temperature of 400 mm and 12 °C, respectively, and precipitation is concentrated mainly in the winter period (between October and March). The average temperature in the summer months (June, July, and August) is 21 °C and the average temperature in the winter months is 6 °C [27]. These temperatures define a climate of short, relatively cool summers and long, harsh winters, with the frost period lasting from mid-October to mid-May. The soils of this region have a xeric humidity regime and a mesic temperature regime [27].

The vegetation present in the studied area is scarce due to the high anthropization of the area, derived mainly from the intense agricultural activity that has led to the deforestation of the climactic Mediterranean forest for centuries. Currently, together with the predominant crops, small redoubts of holm oak groves and holm oak pastures, as well as areas with natural grasslands and riparian vegetation [28,29] constitute the general cover of the environment.

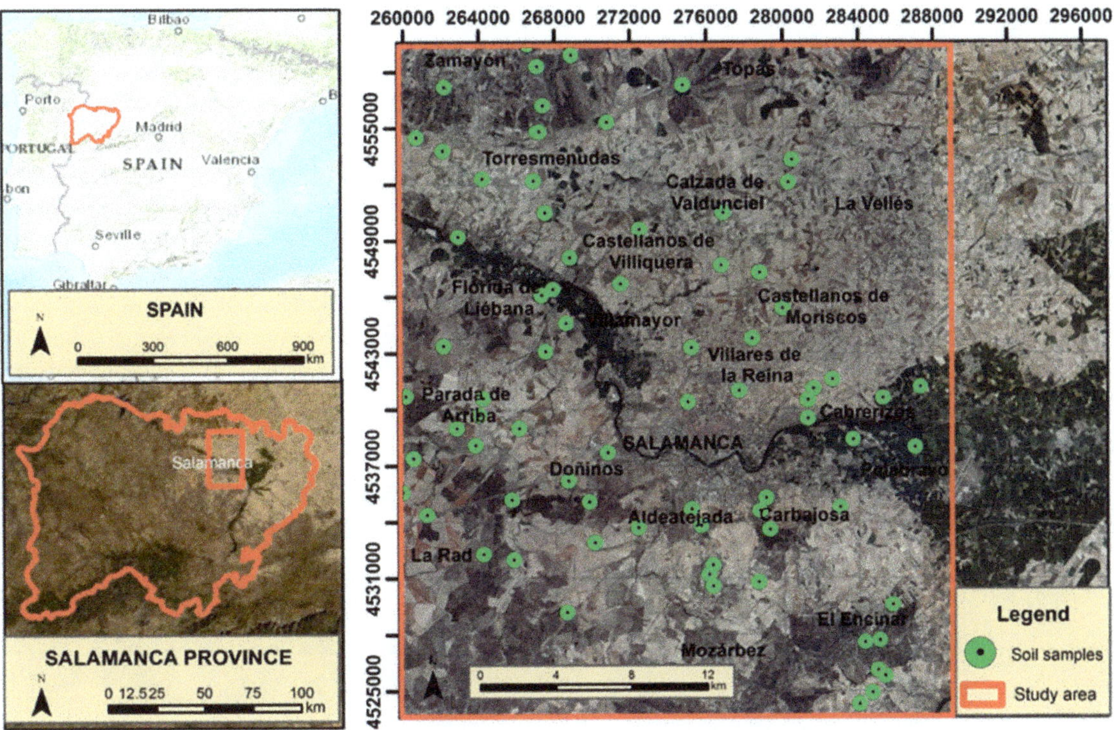

Figure 1. Location of the study area and soils sampled.

The soils of the study area [27], classified according to the soil taxonomy method World Reference Base for Soil Resources of the FAO [30], are mostly Luvisols and Cambisols and, to a lesser extent, Fluvisols, Arenosols, Regosols, and Leptosols. This region is dominated by plains tilled over Cenozoic sediments, with the presence of powerful and fertile soils, which have allowed the establishment of high-yield rain-fed agriculture since Roman times, which considered this region as "the breadbasket of Spain" [31]. The agricultural management carried out in this sector is linked to the traditional practices of the area, with deep ploughing with soil turning (cross ploughing with subsoilers or moldboards) and subsequent shallower ploughing with cultivators or disc harrows. These practices are also associated with the indiscriminate application of fertilizers and phytosanitary products. The main crops grown are rain-fed winter cereals, especially wheat, barley, and rye, some leguminous plants such as lentils and chickpeas, and different combinations of varieties suitable for fodder for livestock feed. Irrigated agriculture predominates in the soils of the Tormes valley, with maize and sugar beet as the main crops.

2.2. Sampling, Soil Analysis, and Selection of Soil Properties

For the general characterization of the soils of the region, 76 soils distributed by the different lithologies and physiographic positions were studied. Once the different sampling points were located, a sample was taken from the superficial horizon (0–25 cm), since it is the one that corresponds to the arable layer and, therefore, has the greatest importance in the development of crops. Subsequently, they were air-dried, crumbled, and sieved through a 2 mm sieve, before an analysis of the complete physicochemical properties was carried out.

From the total number of properties analyzed, available for the entire estimated area, a representative set was chosen for the study to include those parameters of greatest importance for determining the agricultural conditions of the soil. The choice of parameters, moreover, coincides with the opinions of several authors due to their influence on soil fertility, nutrient supply, root growth, and soil porosity [32–38]. The properties studied are 12: nitrogen (N), assimilable phosphorus (P), and assimilable potassium (K), due to their importance in soil fertility, and texture (sand and clay content), cation exchange capacity (CEC), organic carbon (OC), pH, carbonate content (CaCO$_3$), COLE and water retention at 33 and 1500 kPa (GWC) being the most relevant properties in soils as demonstrated in previous studies [26]. The methods used in the determination of the 12 selected properties are listed in (Table 1).

Table 1. Methods used for the analysis of the studied properties.

Parameter	Units	Method
Granulometric analysis (sand and clay)	%	Robinson pipette [39]
Organic carbon (OC)	%	Dichromate oxidation [40]
Water retention at 33 and 1500 kPa (GWC)	%	Pressure membrane [41]
Cation exchange capacity (CEC)	cmol kg^{-1}	Ammonium acetate at pH 7 [39]
Nitrogen (N)	%	Kjeldahl N [42]
Available Phosphorus (P)	mg 100 g^{-1}	Bray 1 [43]/Olsen [44]
Available Potassium (K)	mg 100 g^{-1}	Ammonium acetate at pH 7 [39]
pH	-	Potentiometric method (1:1 -soil-water-)
CaCO$_3$ equivalent (CaCO$_3$)	%	Bernard calcimeter [39]
COLE	mm cm^{-1}	Membrana de Richards [39])

2.3. Geostatistical Analysis

Initially, descriptive statistics were calculated for the properties studied (mean, median, standard deviation, maximum, minimum, kurtosis, and coefficient of variation) to broaden and complement the knowledge of the soils in the environment, for which the SPSS v.25 statistical software was used. After this, from the sample data, we proceeded to perform the interpolation, for which the ordinary kriging method was used.

Ordinary kriging is a linear geostatistical interpolation technique that performs the estimates based on the weighted sums of the sampled point values, and it is well detailed in other works [45–47]. Previously, the normality of the data for all parameters studied was studied using SPSS v.25 statistical software using the visual method (histograms, box-plots, and Q-Q plots). This was refuted by performing the Kolmogorov–Smirnov test at 0.01% probability. The kriging method uses semivariance to estimate the structure of the spatial distribution of soil properties [48]. Modeling and semivariogram estimation are essential for structural analysis and spatial interpolation [10], so the main geostatistical parameters were studied for each property: model type, rank, nugget, structural, threshold, and range. In addition, the spatial dependence (Sp. D), also known as RD (ratio of dependency), of the soil parameters was determined from the relationship between nugget and threshold variations [49]. To ensure spatial dependency, as a rule of thumb, the sampling interval (lag) should be less than half of the range of spatial variation. If the ratio between nugget and Sill (Sp. D) is less than 0.25, the variance has a strong spatial dependence, if the ratio ranges between 0.25 and 0.75, the variance has a moderate spatial dependence, and if it is greater than 0.75, it is considered as slight [46,50].

To perform the interpolation, the dbf file was inserted in ArcMap with the coordinates and sample data related to each property in the 76 profiles studied, the digital layers with the estimate of the distribution of each property were elaborated using the ordinary kriging method and the Geostatistical Analyst module of ArcMap v.10.5 (ESRI). For each parameter, 5 types of estimated distribution were established: very high, high, moderate, low, and very low. For this, the range of each property was divided by the number of desired intervals (5), and the result was used as the width of each interval. Adding this value to the lowest value of the corresponding property yielded the upper limit of the first interval, and so on, until the upper range of the property was reached. These subdivisions do not indicate a suitability or lack for a given property, but rather a lesser or greater separation from the regional mean.

Finally, the interpolation results were checked using cross-validation [51], for which the Geostatistical Analyst extension of ArcMap is also used. This method consists of eliminating a data location and then predicting the values at that point using the data from the other locations, so that the predicted value is compared with the observed value. Derived from this cross-validation, the mean standardized error (MSE), the root mean square error (RMSE), usually the most commonly used, and the root mean square standardized error (RMSSE) were estimated to verify the adequacy of the accuracy of the kriging tool. A value of MSE close to zero indicates that the interpolation method is unbiased. If the RMSE value is close to the standard deviation of the data, then the model has made an adequate prediction. The RMSSE should be close to one if the standard errors of prediction are valid and if it is greater than one, it is underestimating variability in its predictions, while if it is less than one, it is overestimating variability [52,53].

3. Results and Discussion

3.1. Descriptive Statistics for Soil Parameters

The compilation of the main descriptive statistics for each edaphic parameter is presented in Table 2. The coefficient of variation reflects the variability of the edaphic properties. It is observed that the CV data are high, which means that there is great edaphic diversity in the sampled area (linked to the lithological and geomorphological diversity of the sector), which is in line with what was observed during the field campaign and in other previous works [26,27]. In addition, the study of the normality of the data determined that the parameters sand, pH, and water retention at 33 kPa showed normal distributions. For the rest of the parameters, the logarithmic transformation was required in order to perform the interpolation. Table 3 shows the rating scales that have been made from the data obtained from the soil profiles studied in this report.

Table 2. Descriptive statistics of the edaphic studied parameters.

Parameters	Mean	Median	St. Desv.	Max.	Min	Kurtosis	CV
Sand	58.19	58.48	19.40	95.34	1.04	−0.88	33.34
Clay	16.76	13.29	13.59	76.11	0.67	−0.11	81.09
OC	1.47	1.47	1.32	7.32	0.25	24.77	89.80
N	0.10	0.07	0.10	0.61	0.01	15.76	100.0
P	4.23	3.12	3.55	15.2	0.50	1.61	83.93
K	10.06	8.95	7.42	58.4	3.00	17.87	73.76
GWC_{33kPa}	16.70	16.19	8.67	54.61	4.36	0.06	51.92
$GWC_{1500kPa}$	7.89	6.29	6.02	32.21	0.96	0.56	76.30
COLE	0.03	0.01	0.04	0.26	0.00	4.09	133.33
pH	5.91	5.85	0.86	7.9	4.30	−1.06	14.55
CEC	11.82	9.39	8.84	46.05	1.74	1.05	74.79
$CaCO_3$	1.37	0.17	3.13	24.17	0.00	14.07	228.46

3.2. Soil Properties Maps and Agricultural Management Recommendations

It is well known that plant development is closely related to a series of soil characteristics, such as texture, nutrient content, degree of saturation in bases, cation exchange capacity, organic matter content, salinity, etc. Therefore, these factors are the ones that have been used as a basis for evaluating and mapping soil fertility, except for salinity, because there are no saline soils in the region studied (the average conductivity of the saturation extract is very small −0.83 dSm^{-1}-). The results of the interpolations allow us to know the distribution of the main edaphic parameters in the soils of Salamanca and its surroundings. From this information, more efficient agricultural practices can be developed and recommendations can be made to local farmers [10,54,55].

Table 3. Established values for each of the property ranges.

Property	Classes				
	I(Very Low)	**II(Low)**	**III(Moderate)**	**IV(High)**	**V(Very High)**
Sand	1.04–19.90	19.91–38.76	38.77–57.62	57.63–76.48	76.49–95.34
Clay	0.67–15.76	15.77–30.85	30.86–45.93	45.94–61.02	61.03–76.11
OC	0.25–1.66	1.67–3.08	3.09–4.49	4.50–5.91	5.92–7.32
N	0.014–0.134	0.136–0.254	0.255–0.374	0.375–0.494	0.495–0.614
P	0.5–3.4	3.5–6.4	6.5–9.3	9.4–12.3	12.4–15.2
K	3.0–14.1	14.2–25.2	25.3–36.2	36.3–47.3	47.4–58.4
GWC (33 kPa)	4.36–14.41	14.42–24.46	24.47–34.51	34.52–44.56	44.57–54.61
GWC (1500 kPa)	0.96–7.21	7.22–13.46	13.47–19.71	19.72–25.96	25.97–32.21
COLE	0.000–0.052	0.053–0.104	0.105–0.156	0.157–0.208	0.209–0.260
pH	4.3–5.0	5.1–5.7	5.8–6.5	6.6–7.2	7.3–7.9
CEC	1.74–10.60	10.61–19.46	19.47–28.33	28.34–37.19	37.20–46.05
$CaCO_3$	0.00–4.83	4.84–9.67	9.68–14.50	14.51–19.34	19.35–24.17

3.2.1. Physical Properties

Texture is closely related to porosity, aeration, water retention, permeability, with the processes of waterlogging, hydromorphy, washing of salts and nutrients, etc., soil properties, and processes that have an enormous influence on the development of agricultural crops. Figure 2 shows the distribution of the main physical properties of the soil in the studied sector.

Figure 2a shows the distribution of total sand content. Moderate and high ranges predominate (sand between 40–80%), which translates into a predominance of loamy textures, characteristic of superficial horizons developed on sandstones. Loam soils have a "balanced" texture, i.e., the ideal texture for good crop development. On the other hand, it is worth noting the high content in the extreme northwest, which is related to the presence of sandy soils developed from weathered granites. Sandy soils have good aeration and are easy to till, but have low cation exchange capacity, are deficient in plant nutrients, have low water retention (they dry out very easily), and are very permeable (bases are easily washed out and transported out of the soil profile).

Figure 2b shows the clay distribution. In general, the predominant clay content is moderate. From the center to the northeast, there are soils with high clay content corresponding to the Luvisols typical of the region of La Armuña, while to the northwest, the clay content is minimal due to the presence of soils developed on granitic rocks. Regarding the south, the clay content in the analyzed profiles presents a low to moderate content, corresponding to soils formed on slate, in which textures with a high silt content predominate. In areas where the clay content is high, the storage of water and nutrients is high, which will have a positive effect on agricultural activities, which is enhanced by the presence in some areas of argillic subsurface horizons with clay illuviation (Bt). However, soils with an excess of clay (>40%) are impermeable, agricultural work becomes very difficult due to their strong plasticity in the wet state or excessive compaction in the dry state.

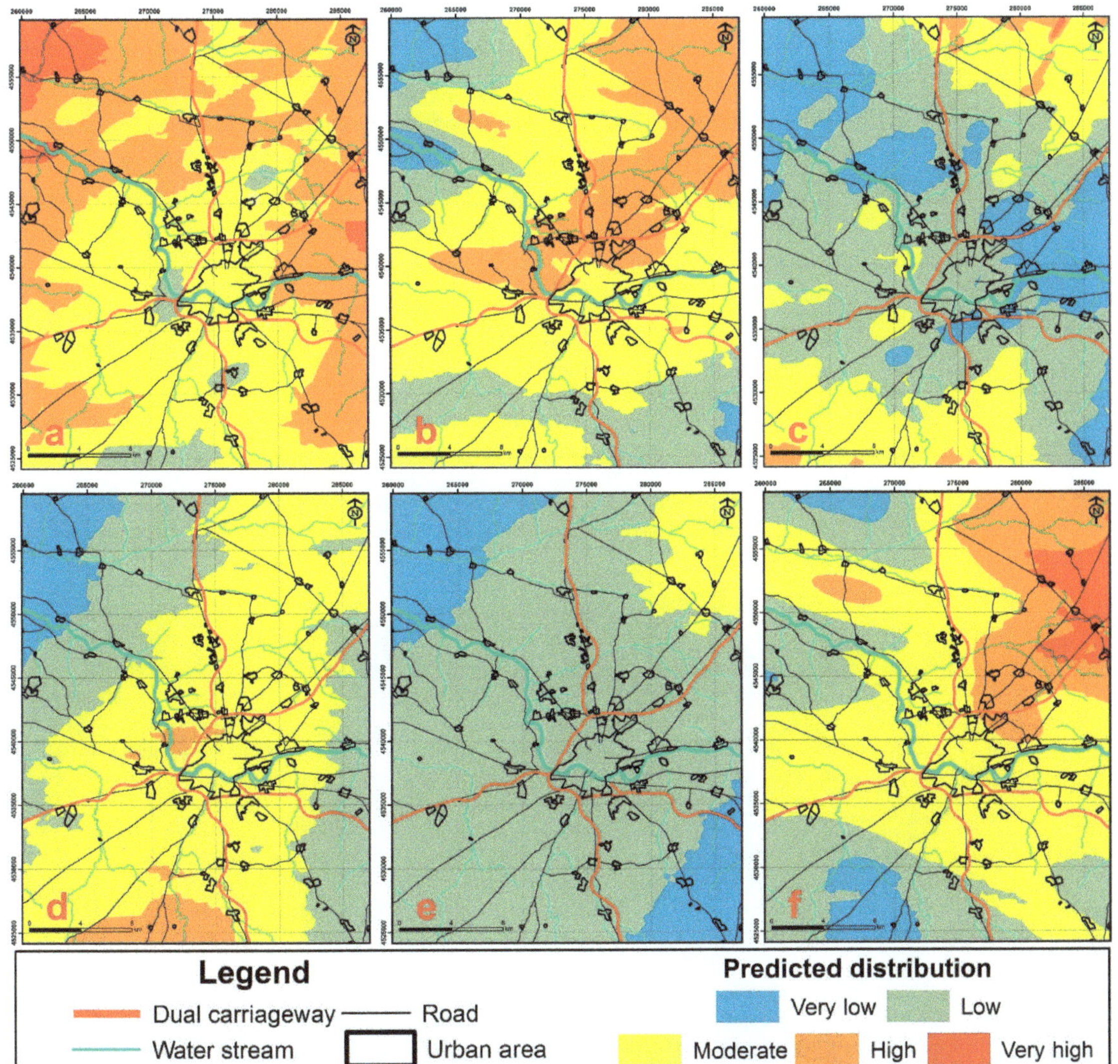

Figure 2. Distribution of the physical properties with the greatest impact on agricultural productivity in the soils studied:
(**a**) Sand; (**b**) Clay; (**c**) Organic carbon; (**d**) Water retention at 33 kPa; (**e**) Water retention at 1500 kPa; (**f**) COLE.

The organic carbon content is presented in Figure 2c. The organic carbon content in the study area is very low (<1%) for the most part, presenting a mean value of 1.5%. The distribution is very irregular and does not seem to follow clear patterns. This is because the main factor dominating the distribution of CO is the use to which the soil is subjected. The surface horizons of soils located in areas of natural vegetation (holm oak groves and pastures) have a double content (1.8%) in organic carbon than the Ap horizons of cultivated soils (0.8%), which shows the rapid degradation of organic matter in the surface horizons due to the effect of agricultural work. At present, in the face of climate change, arable soils generally have low organic carbon values, while values are higher under permanent vegetation cover. The conversion of natural land to cropland is one of the largest anthropogenic sources of carbon emissions and has led to the release of about 200 Pg C over the last 250 years worldwide [56,57]. It has recently been recognized worldwide that soil carbon sequestration can be of great importance as a climate change mitigation and adaptation measure. Faced with this new challenge, the agricultural soils of the region studied could act as potential carbon sinks. To this end, it is necessary to promote agricultural techniques that favor the conservation and increase of carbon, such as: conservation tillage, addition of exogenous organic matter (manure, compost, etc.) and cover crops and fallows with vegetation.

The physical properties of the soil inform us about the capacity of the soil to provide a suitable physical environment for crop root growth. For this purpose, the soil must present low compaction so that it does not oppose excessive mechanical resistance to root advance; as well as a porosity that facilitates drainage, in wet periods, and the storage or retention of water to cover the needs of the plant, in dry periods. The distribution of values for soil water retention at 33 and 1500 kPa (GWC_{33kPa} and $GWC_{1500kPa}$) are shown, respectively, in Figure 2d,e. The distribution of both properties is similar, however, the values are higher for the case of the GWC_{33kPa}, where moderate values predominate versus low values in the $GWC_{1500kPa}$. The minimum contents stand out in both cases in the extreme northwest, due to the sandy character of the soils in this granitic sector, since retention depends largely on the fine particle (clay) content of the soil. The soils with the highest water retention are the Vertisols, developed on loams and clays, and those with the lowest values are the Fluvisols, formed on sands. The use of crops with drought-resistant varieties is recommended in low GWC sectors.

Certain soils have the capacity to expand significantly when wet and to contract when dry, which is related to a relatively high content of montmorillonite type clays (smectites). This ability to expand and contract is quantified by using a coefficient called linear extensibility coefficient or COLE. Figure 2e shows the distribution of COLE values in the soils analyzed, ranging from 0.000 to 0.260. COLE is closely related to the clay content of the soils, since it assesses their expansibility or swelling. Moderate distribution predominates, being lower towards the south and northwest (presence of slate and granitic metasedimentary materials). The values are higher towards the northeast, coinciding with the clay-rich soils (Vertisols) of the region of La Armuña. Fluvisols and Leptosols are the soils with the lowest values. Soils with high COLE are very appropriate for the development of some leguminous plants (lentils) but can negatively affect other crops with low response capacity to conditions of excessive plasticity and dryness, during the winter and summer periods, respectively, and to the rupture of the root system of the plants, due to the processes of contraction and expansion of the clays.

3.2.2. Chemical Properties

The distribution of the main chemical properties of the soils of Salamanca and surrounding areas is presented in Figure 3. The cation exchange capacity (CEC) is a property by which anions or cations in the soil water can be exchanged with the anions or cations contained in the clay minerals (phyllosilicates) and organic matter, with which it is in contact. When an aqueous solution is brought into contact with certain soil substances, an exchange of ions takes place between the solid and the solution. The higher the CEC, the higher the fertility of the soil. The CEC ranges from 1.74 to 46.05 (cmol (+) kg^{-1}), the average value being 11.82 cmol (+) kg^{-1} (Figure 3a). Values in the sector are irregular, with moderate-high values in the more clayey soils (Vertisols) and low values in the more sandy textured soils (Leptosols and Fluvisols). An increase in the organic matter content of the soil would lead to higher CEC values and thus improve crop yields.

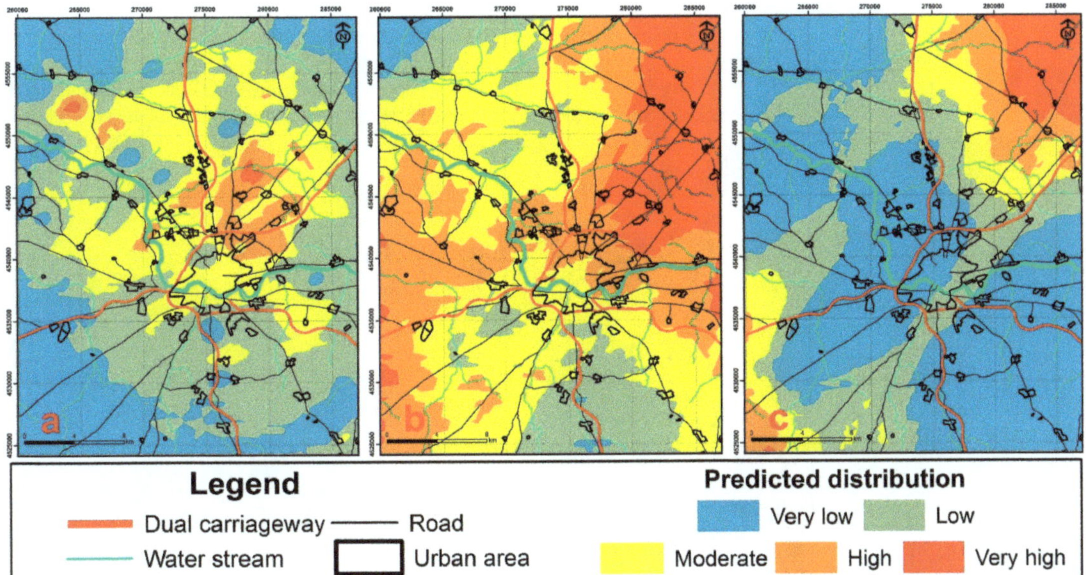

Figure 3. Distribution of the chemical properties with the greatest impact on agricultural productivity in the studied soils: (**a**) CEC; (**b**) pH; (**c**) CaCO$_3$.

The pH is a decisive factor in soil fertility since it can cause the immobilization (precipitation) of certain nutrients for plants; that is to say, at certain pH, the nutrients are not assimilated by plants (they are precipitated and are not soluble). On the other hand, at other pHs, these same nutrients are solubilized and assimilated. A low pH leads to a lack of nutrient availability, especially Ca and Mg [58,59]. It can be stated that a soil with intermediate pH (between 6.6 and 7.3) has the best conditions for most crops to assimilate soil nutrients. Figure 3b shows the pH distribution, with moderate and high pH values predominating. The highest values appear in the northeast and southwest quadrants, coinciding with the presence of highly developed soils with carbonate accumulations. The most acid soils are restricted to thin soils developed on geological materials of marked acid character: granites to the northwest, and slates to the south. The average pH value of the soils in the sector studied is slightly acidic (5.9), although there are actually two completely different populations of soils: acidic soils and calcareous or basic soils. The soil units with the highest pH correspond to the Calcisols and Vertisols. To produce an increase in crop yields, efforts must be made to increase pH, which can be addressed through the process of liming soils.

The carbonate content of soils is closely related to their pH (Figure 3c). The percentage of calcium carbonate equivalent of the samples studied ranges from 0.00 to 24.17%, with an average value of 1.37%. It is observed that low and very low values predominate in a large part of the study sector, and that the highest values correspond to the northeast and southwest. It should be noted that in the study area, the presence of carbonates is usually linked to washing and accumulation processes in the deep horizons of the soil, so that the presence in the superficial horizon is usually low. In the region studied, calcareous soils predominate over non-calcareous soils. Calcareous Regosols and Phaeozems and Calcisols are the soils with the highest percentages of carbonates. Soils developed on shales, quartzites, siliceous sandstones, etc. (Leptosols, Nitosols, and Acrisols) have negligible amounts of calcium carbonate equivalent.

3.2.3. Soil Nutrients

Properties related to soil chemical fertility refer to the soil's capacity to retain and supply the necessary nutrients for crop needs, such as nitrogen, phosphorus, and potassium (N, P, K). In a conventional agricultural system, nutrients come both from the mineralization of organic matter and weathering of minerals, as well as from inputs such as synthetic fertilizers. Regarding the fertility observed in soils, the distribution of the main nutrients is shown in Figure 4.

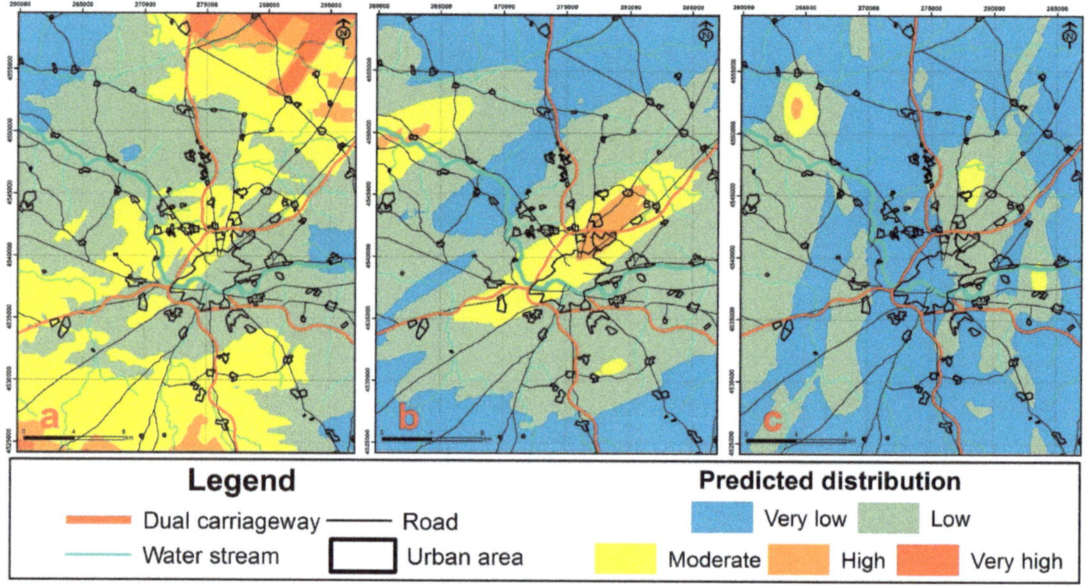

Figure 4. Distribution of the main soil nutrients in the study area: (**a**) N; (**b**) P; (**c**) K.

A productive soil must contain all the essential nutrients for plants in sufficient and proportional quantities. They must also be in an assimilable form so that plants can utilize them. The nutrients that have traditionally been studied in fertility programs are nitrogen, phosphorus, and potassium. They are used by plants in large quantities and are therefore referred to as "main or primary nutrients". The predominant nitrogen content (Figure 4a) is low at the regional level, which is clearly observed in large areas, reaching very low values in the poorer granitic soils. In the interior of the studied area, small areas with moderate N contents are also observed. Normally, N has a greater effect on crop growth, quality, and yield. However, N was deficient in most areas with values between 0.01 and 0.61% (low and very low). Acute N deficiency was due to low CO content, higher mineralization rate, and insufficient N fertilizer application to nutrient-depleting crops such as wheat. The rate of CO decomposition and soil N mineralization has complex interactions with the microbial population and other environmental factors, mainly soil moisture and temperature.

The assimilable phosphorus content ranges from 0.50 to 15.2 mg 100 g^{-1}, the average value being 4.23 mg 100 g^{-1}. It shows an irregular distribution characterized by two nuclei of moderate and high contents in the central and western sectors, with low and very low values predominating at the regional level. It is complex to establish distribution relationships for phosphorus. This usually comes from the parent rock, or it can be applied in phosphate fertilizers. In general terms, we can say that the soils studied have low phosphorus content, and that this can be a limiting factor for agricultural development if appropriate management measures are not taken. Phosphorus is more directly affected by soil pH than other major plant nutrients such as N and K. For example, at alkaline values, the pH of the soil is higher than the pH of the soil. For example, at alkaline values greater than pH 7.5, phosphate tends to react rapidly with calcium (Ca) and magnesium (Mg) to form less soluble compounds. At acidic pH values, phosphate ions react with aluminum (Al) and iron (Fe) to form less soluble compounds again. Soils with pH values between 6 and 7.5 are ideal for the availability of P. In addition to pH, the amount of CO and the supply of phosphate fertilizers also control the availability of phosphorus in the soil, while erosion and runoff are associated with its loss.

As in other parameters related to soil fertility, the study area presents distributions dominated by low and very low potassium (K) contents (Figure 4c). Only in isolated areas are the concentrations of this element higher. The percentage of assimilable potassium is between 3.00 and 58.4 mg 100 g^{-1}, the average value being 10.06 mg 100 g^{-1}. Agricultural practices may be the cause of this "stress" suffered by the nutrients in the soils of Salamanca and surrounding areas. Soil pH affects the availability of potassium in the soil. When the pH is greater than 7, the higher Ca concentration increases K availability through the displacement of exchangeable K by Ca. Conversely, when soil pH is lower than 5.5, the reduction in Ca concentration reduces K availability. In addition, low CO levels, low clay content, and possible nutrient losses through leaching and erosion also reduce K levels.

3.2.4. General Agricultural Recommendations

This region is dominated by plains carved out of Cenozoic sediments, with the presence of powerful and fertile soils, which have allowed the establishment of high-yield rain-fed agriculture. The best soils in this region are considered to be deep soils (Luvisols), with a loamy texture in the surface horizon and a low organic matter content due to degradation by cultivation over several centuries, with a subsurface horizon of argillic type (Bt), in many cases with clays of the smectite group, with high water retention (the soils have a xeric moisture regime, which means a deficit in the water balance in the soil between April and October), with the presence of a horizon of carbonate accumulation and a pH close to neutral (the optimum for most crops), with a high capacity for cation exchange and a high degree of base saturation. On the contrary, in the northern and southern limits of the studied area, a steeper landscape can be observed, developed on the Palaeozoic base, with thin soils (Leptosols and Regosols), with a lithic contact near the surface that prevents root development, with a high degree of erosion and low fertility, which have traditionally lent themselves to a land use dedicated to extensive livestock farming (pastures). It is interesting to note that in this region, the most fertile soils have been used, as is logical for agricultural use for several thousand years; on the contrary, shallow soils, with abundant rockiness or stoniness, with a very sandy texture, low CEC, etc., have been used solely and exclusively for pasture [26,27].

The bad practices, so widespread in conventional agricultural soil management, lead to the major environmental problems described above. An alternative to the traditional system to avoid these problems would be the application of conservation and ecological soil management techniques. These practices have in common a more efficient supply of organic matter (application of organic amendments such as manure, compost, biochar), lower tillage intensity (conservation tillage), use of higher yielding species/biomass, or limitation in the use of agrochemicals. Such practices increase soil quality, improve soil fertility, retain more water, and reduce susceptibility to compaction and erosion.

3.3. Geostatistics for Agricultural Land Management

The semivariogram model and the main geostatistical parameters of the soil properties studied are shown in Table 4. The semivariogram model for each property was chosen based on the one with the lowest RMSE [60]. For the parameters GWC$_{33kPa}$, GWC$_{1500kPa}$, pH, and COLE, the exponential model provided the best fit to the semivariogram, with the circular model being the best fit to the semivariogram for the parameters N, K, and CEC, while the Gaussian was the best fit for the rest. In relation to the Sp. D of the soil parameters, it ranged from 0.02 (in CEC) to 0.91 (in GWC$_{33kPa}$), being high (in CEC), moderate (in Sand,

Clay, OC, N, P, K, $GWC_{1500kPa}$, COLE, pH, and $CaCO_3$), or weak (in GWC_{33kPa}). The ranges of spatial dependencies were large and typically vary between 2947 m for N and 10,565 m for P, indicating that the optimal sampling interval varies greatly between different soil properties. An abnormally high value was obtained for the water retention parameters GWC with a value of 44,607 m. As for the nugget effect, the highest value observed corresponds to $CaCO_3$ (7.26), which indicates discontinuity between samples [61] and is in line with it being the parameter with the worst fit, while the lowest value resulted for pH and CEC (0.02 and 0.04, respectively).

Table 4. Semivariance analysis of spatial structure in soil parameters.

Parameters	Model	Range	Lag Size	Nugget	Partial	Sill	Nugget/Sill	Sp. D
Sand	G	4236	353.0	0.21	0.16	0.37	0.57	M
Clay	G	6510	542.5	0.33	0.22	0.55	0.60	M
OC	G	3001	250.0	0.16	0.29	0.45	0.36	M
N	C	2947	245.5	0.24	0.18	0.42	0.57	M
P	G	10,565	880.4	0.48	0.30	0.78	0.62	M
K	C	4639	386.6	0.12	0.25	0.37	0.32	M
GWC (33 kPa)	E	44,607	3717.3	0.20	0.02	0.22	0.91	L
GWC (1500 kPa)	E	44,607	3717.3	0.29	0.10	0.39	0.74	M
COLE	E	8918	743.2	0.88	0.81	1.69	0.52	M
pH	E	3730	310.9	0.01	0.00	0.02	0.74	M
CEC	C	2982	248.5	0.04	1.63	1.67	0.02	H
$CaCO_3$	G	8811	734.3	7.26	2.61	9.87	0.74	M

The mapping of specific soil parameters using geostatistics can represent a valid, fast, and inexpensive tool that provides useful and accurate information to evaluate the characteristics of agricultural land. Based on this information, agricultural practices and policies can be reformulated at local and regional levels with a consequent increase in agricultural yields as well as soil protection and improvement.

3.4. Validation of Results

The results of the cross-validation are summarized in Table 5. The MSE obtained can be considered, globally, as acceptable, due to their closeness to 0. With respect to RMSS, parameters with values close to 1 such as sand, COLE, and pH (0.9 < RMSS < 1.1) present very good precision. The parameters clay, phosphorus, water retention at 1/3 atm, and calcium were obtained with high precision, since they present values close to 1 (0.8 < RMSS < 0.9 and 1.1 < RMSS < 1.2). The properties organic carbon, nitrogen, potassium, water retention at 15 atm, magnesium, and exchange capacity showed acceptable accuracies (0.5 < RMSS < 0.8 and 1.2 < RMSS < 1.6). Finally, carbonates show a very low precision (RMSS close to 0), which is related to the two clearly differentiated populations that exist in the soils: soils on acidic ancient materials do not present carbonates, while in the case of tertiary materials, the presence is usual. In addition, the RMSS value shows that the model has underestimated the values for the parameters clay, organic carbon, nitrogen, potassium, water retention at 1/3 and 15 atm, COLE, pH, conductivity, and bulk density (RMSSE > 1), while for the remaining ones, it has overestimated them, although these errors are not high due to their closeness to 1 (except for carbonates). Finally, the RSME, the most commonly used meter, refutes these good accuracies obtained for most of the parameters, due to the closeness of its value with respect to the standard deviation of each parameter, which is reflected in Table 2.

Table 5. Errors obtained from cross-validation to estimate the interpolation accuracy.

Parameters	MSE	RMSE	RMSSE
Sand	0.0266	18.8581	0.9640
Clay	−0.1050	14.1105	1.1391
Organic Carbon	−0.1376	1.3265	1.2204
Nitrogen	−0.1236	0.1001	1.2902
Available phosphorus	−0.0699	3.4521	0.8164
Available potassium	−0.1036	8.3775	1.3638
Water retention (33 kPa)	−0.0753	8.6836	1.1145
Water retention (1500 kPa)	−0.1635	6.0435	1.5121
COLE	−0.1384	0.0446	1.0133
pH	−0.0031	0.8596	1.0221
Cation exchange capacity	−0.0095	12.5947	0.6881
CaCO3 equivalent	0.0019	5.3464	0.0062

Therefore, the results derived from the cross-validation show good and adequate accuracies for all parameters, except for carbonates, which is derived from the two very different populations existing in the studied area.

4. Conclusions

Knowledge of the distributions of the main physico-chemical properties of soils in an environment is essential for proper agricultural management. The application of geographic information systems and geostatistical methods, including descriptive statistics and semivariogram analysis, allows to improve the description of the spatial variability of the physicochemical properties of the arable layer of the soil. Descriptive statistics provide insight into data distributions and lay the foundation for geostatistical analysis. Geostatistical interpolation identified that the best fit of the semivariogram for each soil property can be achieved by different models (exponential, spherical, circular, or Gaussian) and, in general, showed moderate spatial dependence. In addition, the accuracy of the results was determined by cross-validation, with the interpolations generally showing low errors. The kriging maps of the properties studied were effective in explaining the distribution of soil properties in unsampled locations based on sampled data. These maps will help farmers to make efficient management decisions based on adequate knowledge of existing agricultural soil conditions. Therefore, these results, together with the agricultural measures that could be proposed based on them, show that geostatistical analysis using the kriging technique is an effective predictive tool to explore the spatial variability of soil properties.

Author Contributions: Conceptualization, M.C. and A.M.-G.; methodology, M.C.; software, M.C. and L.M.; validation, M.C. and F.S.-F.; formal analysis, A.M.-G. and F.S.-F.; investigation, M.C. and F.S.-F.; resources, A.M.-G.; data curation, A.M.-G.; writing—original draft preparation, M.C.; writing—review and editing, M.C. and A.M.-G.; visualization, L.M.; supervision, F.S.-F.; project administration, A.M.-G.; funding acquisition, A.M.-G. All authors have read and agreed to the published version of the manuscript.

Funding: This research received no external funding.

Acknowledgments: This research was supported by the project SA-044G18 of Regional Government of Castilla y Leon, and the GEAPAGE research group (Environmental Geomorphology and Geological Heritage) of the University of Salamanca.

Conflicts of Interest: The authors declare no conflict of interest.

References

1. Hunter, M.C.; Smith, R.G.; Schipanski, M.E.; Atwood, L.W.; Mortensen, D.A. Agriculture in 2050: Recalibrating Targets for Sustainable Intensification. *BioScience* **2017**, *67*, 386–391. [CrossRef]
2. Evenson, R.E.; Gollin, D. Assessing the impact of the Green Revolution, 1960 to 2000. *Science* **2003**, *300*, 758–762. [CrossRef] [PubMed]
3. Knickel, K.; Ashkenazy, A.; Chebach, T.C.; Parrot, N. Agricultural modernization and sustainable agriculture: Contradictions and complementarities. *Int. J. Agric. Sustain.* **2017**, *15*, 575–592. [CrossRef]
4. Rockström, J.; Williams, J.; Daily, G.; Noble, A.; Matthews, N.; Gordon, L.; Wetterstrand, H.; DeClerck, F.; Shah, M.; Steduto, P.; et al. Sustainable intensification of agriculture for human prosperity and global sustainability. *Ambio* **2017**, *46*, 4–17. [CrossRef]
5. Panagos, P.; Standardi, G.; Borrelli, P.; Lugato, E.; Montanarella, L.; Bosello, F. Cost of agricultural productivity loss due to soil erosion in the European Union: From direct cost evaluation approaches to the use of macroeconomic models. *Land Degrad. Dev.* **2018**, *29*, 471–484. [CrossRef]
6. Yadav, D.; Wang, J. Modelling carbon dioxide emissions from agricultural soils in Canada. *Environ. Pollut.* **2017**, *230*, 1040–1049. [CrossRef]
7. Benton, T.G.; Juliet, A.V.; Wilson, J.D. Farmland biodiversity: Is habitat heterogeneity the key? *Trends Ecol. Evol.* **2003**, *18*, 182–188. [CrossRef]
8. Matzeu, A.; Secci, R.; Uras, G. Methodological approach to assessment of groundwater contamination risk in an agricultural area. *Agric. Water Manag.* **2017**, *184*, 46–58. [CrossRef]
9. Silva, V.; Mol, H.G.; Zomer, P.; Tienstra, M.; Ritsema, C.J.; Geissen, V. Pesticide residues in European agricultural soils–A hidden reality unfolded. *Sci. Total Environ.* **2019**, *653*, 1532–1545. [CrossRef]
10. Panday, D.; Maharjan, B.; Chalise, D.; Shrestha, R.; Twanabasu, B. Digital soil mapping in the Bara district of Nepal using kriging tool in ArcGIS. *PLoS ONE* **2018**, *13*, e0206350. [CrossRef]
11. Kravchenko, A.N. Influence of Spatial Structure on Accuracy of Interpolation Methods. *Soil Sci. Soc. Am. J.* **2003**, *67*, 1564–1571. [CrossRef]
12. Webster, R. The Development of Pedometrics. *Geoderma* **1994**, *62*, 1–15. [CrossRef]
13. Panagopoulos, T.; Jesus, J.; Antunes, M.D.C.; Beltrao, J. Analysis of spatial interpolation for optimising management of a salinized field cultivated with lettuce. *Eur. J. Agron.* **2006**, *24*, 1–10. [CrossRef]
14. Myers, D.E. Spatial interpolation: An overview. *Geoderma* **1994**, *62*, 17–28. [CrossRef]
15. Lopez-Granados, F.; Jurado-Exposita, M.; Pena-Barragan, J.M.; Garcia-Torres, L. Using geostatistical and remote sensing approaches for mapping soil properties. *Eur. J. Agron.* **2005**, *23*, 279–289. [CrossRef]
16. Sun, B.; Zhou, S.; Zhao, Q. Evaluation of spatial and temporal changes of soil quality based on geostatistical analysis in the hill region of subtropical China. *Geoderma* **2003**, *115*, 85–99. [CrossRef]
17. Zhang, Q.; Yang, Z.; Li, Y.; Chen, D.; Zhang, J.; Chen, M. Spatial variability of soil nutrients and GIS-based nutrient management in Yongji County, China. *Int. J. Geogr. Inf. Sci.* **2010**, *24*, 965–981. [CrossRef]
18. Balkovič, J.; Rampašeková, Z.; Hutár, V.; Sobocká, J.; Skalský, R. Digital soil mapping from conventional field soil observations. *Soil Water Res.* **2013**, *8*, 13–25. [CrossRef]
19. Adhikari, K.; Hartemink, A.E.; Minasny, B.; Bou Kheir, R.; Greve, M.B.; Greve, M.H. Digital Mapping of Soil Organic Carbon Contents and Stocks in Denmark. *PLoS ONE* **2014**, *9*, e105519. [CrossRef]
20. Malone, B.; Styc, Q.; Minasny, B.; McBratney, A. Digital soil mapping of soil carbon at the farm scale: A spatial downscaling approach in consideration of measured and uncertain data. *Geoderma* **2017**, *290*, 91–99. [CrossRef]
21. Song, X.; Liu, F.; Zhang, G.; Li, D.; Zhao, Y.; Yang, J. Mapping soil organic carbon using local terrain attributes: A comparison of different polynomial models. *Pedosphere* **2017**, *27*, 681–693. [CrossRef]
22. Wang, B.; Waters, C.; Orgill, S.; Gray, J.; Cowie, A.; Clark, A.; Liu, D.L. High resolution mapping of soil organic carbon stocks using remote sensing variables in the semi-arid rangelands of eastern Australia. *Sci. Total Environ.* **2018**, *630*, 367–378. [CrossRef] [PubMed]
23. Lamichhane, S.; Kumar, L.; Wilson, B. Digital soil mapping algorithms and covariates for soil organic carbon mapping and their implications: A review. *Geoderma* **2019**, *352*, 395–413. [CrossRef]
24. Guo, P.T.; Li, M.F.; Luo, W.; Tang, Q.F.; Liu, Z.W.; Lin, Z.M. Digital mapping of soil organic matter for rubber plantation at regional scale: An application of random forest plus residuals kriging approach. *Geoderma* **2015**, *237*, 49–59. [CrossRef]
25. Santos-Francés, F.; Martínez-Graña, A.; Zarza, C.Á.; Sánchez, A.G.; Rojo, P.A. Spatial Distribution of Heavy Metals and the Environmental Quality of Soil in the Northern Plateau of Spain by Geostatistical Methods. *Int. J. Environ. Res. Public Health* **2017**, *14*, 568. [CrossRef]
26. Santos-Francés, F.; Martínez-Graña, A.; Ávila-Zarza, C.; Criado, M.; Sánchez, Y. Comparison of methods for evaluating soil quality of semiarid ecosystem and evaluation of the effects of physico-chemical properties and factor soil erodibility (Northern Plateau, Spain). *Geoderma* **2019**, *354*, 113872. [CrossRef]
27. Criado, M. Análisis Geoambiental Aplicado a la Evaluación Estratégica de la Ciudad de Salamanca y Alrededores. Cartografías Temáticas Mediante SIG. Ph.D. Thesis, University of Salamanca, Salamanca, Spain, 20 July 2020.

28. Criado, M.; Santos-Francés, F.; Martínez-Graña, A.; Sánchez, Y.; Merchán, L. Multitemporal Analysis of Soil Sealing and Land Use Changes Linked to Urban Expansion of Salamanca (Spain) Using Landsat Images and Soil Carbon Management as a Mitigating Tool for Climate Change. *Remote Sens.* **2020**, *12*, 1131. [CrossRef]
29. Criado, M.; Martínez-Graña, A.; Santos-Francés, F.; Merchán, L. Landscape Evaluation as a Complementary Tool in Environmental Assessment. Study Case in Urban Areas: Salamanca (Spain). *Sustainability* **2020**, *12*, 6395. [CrossRef]
30. World Reference Base for Soil Resources. *International Soil Classification System for Naming Soils and Creating Legends for Soil Maps, Update 2015*; World Soil Resources Report No. 106; FAO: Italy, Rome, 2015.
31. Peña Sánchez, M.; de Campos, T. *La Integración de un Espacio Rural en la Economía Capitalista*; Serie Geográfica no 5 Secretariado de Publicaciones: Universidad de Valladolid, Valladolid, Spain, 1987; p. 468.
32. Mukhopadhyay, S.; Maiti, S.K.; Masto, R.E. Development of mine soil quality index (MSQI) for evaluation of reclamation success: A chronosequence study. *Ecol. Eng.* **2014**, *71*, 10–20. [CrossRef]
33. Sánchez-Navarro, A.; Gil-Vazquez, J.M.; Delgado-Iniesta, M.J.; Marin-Sanleandro, P.; Blanco-Bernardeau, A.; Ortiz-Silla, R. Establishing an index and identification of limiting parameters for characterizing soil quality in Mediterranean ecosystems. *Catena* **2015**, *131*, 35–45. [CrossRef]
34. Cheng, J.; Ding, C.; Li, X.; Zhang, T.; Wang, X. Soil quality evaluation for navel orange production systems in central subtropical China. *Soil Tillage Res.* **2016**, *155*, 225–232. [CrossRef]
35. Das, B.; Chakraborty, D.; Singh, V.K.; Ahmed, M.; Singh, A.K.; Barman, A. Evaluating fertilization effects on soil physical properties using a soil quality index in an intensive Rice-wheat cropping system. *Pedosphere* **2016**, *26*, 887–894. [CrossRef]
36. Biswas, S.; Hazra, G.C.; Purakayastha, T.J.; Saha, N.; Mitran, T.; Roy, S.S.; Basak, N.; Mandal, B. Establishment of critical limits of indicators and indices of soil quality in rice-rice cropping systems under different soil orders. *Geoderma* **2017**, *292*, 34–48. [CrossRef]
37. Sione, S.M.J.; Wilson, M.G.; Lado, M.; Gonzalez, G.P. Evaluation of soil degradation produced by rice crop systems in a Vertisols, using a soil quality index. *Catena* **2017**, *150*, 79–86. [CrossRef]
38. Nabiollahi, K.; Golmohamadi, F.; Taghizadeh-Mehrjardi, R.; Kerry, R.; Davari, M. Assessing the effects of slope gradient and land use change on soil quality degradation through digital mapping of soil quality indices and soil loss rate. *Geoderma* **2018**, *318*, 16–28. [CrossRef]
39. United States Department of Agriculture (USDA). Soil survey laboratory methods manual. In *Soil Survey Investigations, Report No. 42*; U.S. Department of Agriculture, National Resources Conservation Services, National Soil Survey Centre: Washington DC, USA, 1996.
40. Walkley, A.; Black, I.A. An examination of the Degtjareff method for the determining soil organic matter and proposed modification of the chromic titration method. *Soil Sci.* **1934**, *37*, 29–38. [CrossRef]
41. Richards, L.A. Pressure membrane apparatus: Construction and use. *Agric. Eng.* **1947**, *28*, 451–455.
42. Bremmer, J.M.; Mulvaney, C.S. Nitrogen total. In *Methods of Soil Analysis, Part 2, Chemical and Microbiological Properties*; Page, A.L., Miller, R.H., Keeney, D.R., Eds.; American Society of Agronomy: Madison, WI, USA, 1982; pp. 595–624.
43. Bray, R.H.; Kurtz, L.T. Determination of total, organic and available forms of phosphorus in soils. *Soil Sci.* **1945**, *59*, 39–45. [CrossRef]
44. Olsen, S.R.; Cole, C.V.; Watanabe, F.S.; Dean, L.A. *Estimation of Available Phosphorus in Soils by Extraction with Sodium Bicarbonate*; U.S. Government Printing Office: Washington, DC, USA, 1954.
45. Wang, J.F.; Li, L.F.; Christakos, G. Sampling and kriging spatial means: Efficiency and conditions. *Sensors* **2009**, *9*, 5224–5240. [CrossRef]
46. Omran, E.E. Improving the prediction accuracy of soil mapping through geostatistics. *Int. J. Geosci.* **2012**, *3*, 574–590. [CrossRef]
47. Yao, X.; Yu, K.; Deng, Y.; Liu, J.; Lai, Z. Spatial variability of soil organic carbon and total nitrogen in the hilly red soil region of Southern China. *J. For. Res.* **2020**, *31*, 2385–2394. [CrossRef]
48. Zandi, S.; Ghobakhlou, A.; Sallis, P. Evaluation of spatial interpolation techniques for mapping soil pH. In *International Congress on Modeling and Simulation*; Modelling and Simulation Society of Australia and New Zealand: Perth, Australia, 2011; pp. 1153–1159.
49. Al-Omran, A.M.; Al-Wabel, M.I.; El-Maghraby, S.E.; Nadeem, M.E.; Al-Sharani, S. Spatial variability for some properties of the wastewater irrigated soils. *J. Saudi Soc. Agri. Sci.* **2013**, *12*, 167–175. [CrossRef]
50. Cambardella, C.A.; Moorman, T.B.; Parkin, T.B.; Karlem, D.L.; Novak, J.M.; Turco, R.F.; Konopka, A.E. Fieldscale variability of soil properties in central Iowa soils. *Soil Sci. Soc. Am. J.* **1994**, *58*, 1501–1511. [CrossRef]
51. Browne, M.W. Cross-validation methods. *J. Math. Psychol.* **2000**, *44*, 108–132. [CrossRef]
52. Smith, J.; Smith, P. *Introduction to Environmental Modelling*; Oxford University Press: New York, NY, USA, 2007.
53. Ohmer, M.; Liesch, T.; Goeppert, N.; Goldscheider, N. On the optimal selection of interpolation methods for groundwater contouring: An example of propagation of uncertainty regarding inter-aquifer exchange. *Adv. Water Resour.* **2017**, *109*, 121–132. [CrossRef]
54. Piccini, C.; Francaviglia, R.; Marchetti, A. Predicted Maps for Soil Organic Matter Evaluation: The Case of Abruzzo Region (Italy). *Land* **2020**, *9*, 349. [CrossRef]
55. Jurišić, M.; Radočaj, D.; Krčmar, S.; Plaščak, I.; Gašparović, M. Geostatistical Analysis of Soil C/N Deficiency and Its Effect on Agricultural Land Management of Major Crops in Eastern Croatia. *Agronomy* **2020**, *10*, 1996. [CrossRef]
56. Jarecki, M.; Lal, R. Crop management for soil carbon sequestration. *Crit. Rev. Plant Sci.* **2003**, *22*, 471–502. [CrossRef]

57. Fitzsimmons, M.J.; Pennock, D.J.; Thorpe, J. Effects of deforestation on ecosystem carbon densities in central Saskatchewan, Canada. *For. Ecol. Manag.* **2004**, *188*, 349–361. [CrossRef]
58. Tisdale, S.L.; Nelson, W.L.; Beaton, J.D. *Soil and Fertilizers*, 4th ed.; MacMillan Publications Inc: New York, NY, USA, 1985.
59. Dembele, D.; Traore, K.; Quansh, C.; Osei Jnr, E.M.; Bocar, D.S.; Ballo, M. Optimizing soil fertility management decision in Mali by remote sensing and GIS. *Donnis J. Agri. Res.* **2016**, *3*, 22–34.
60. Robertson, G.P. *GS+: Geostatistics for the Environmental Sciences*; Gamma Design Software: Plainwell, MI, USA, 2008.
61. Carvalho, J.R.; Vieira, S.R.; Marinho, P.R.; Dechen, S.C.F.; Maria, I.C.; Pott, C.A.; Dufranc, G. *Avaliação da Variabilidade Espacial de Parâmetros Físicos do Solo Sob Semeadura Direta em São Paulo, Brasil*; Embrapa: Campinas, Brazil, 2001; pp. 1–4.

 *agronomy*

Article

Remote Sensing Calculation of the Influence of Wildfire on Erosion in High Mountain Areas

Yolanda Sánchez Sánchez [1,*], Antonio Martínez Graña [1] and Fernando Santos- Francés [2]

1   Department of Geology, Faculty of Sciences, University of Salamanca, Plaza de la Merced s/n, 37008 Salamanca, Spain; amgranna@usal.es
2   Department of Soil Sciences, Faculty of Environmental Sciences, University of Salamanca, Avenue Filiberto Villalobos 119, 37007 Salamanca, Spain; fsantos@usal.es
*   Correspondence: yolanda.ss@usal.es

Abstract: Soil erosion is one of the most important environmental problems of the moment, especially in areas affected by wildfires. In this paper, we study pre-fire and post-fire erosion using remote sensing techniques with Sentinel-2 satellite images and LiDAR. The Normalized Burn Ratio is used to determine the areas affected by the fire that occurred on 18 August 2016 in the Natural Reserve of Garganta de los Infiernos (Cáceres). To calculate the erosion, the multi-criteria analysis is carried out from the RUSLE. Once all calculations were performed, there was a considerable increase in sediment production from 16 June 2016 (pre-fire) with an erosion of 31 T/ha·year to 16 June 2017 of 74 T/ha·year for areas of moderate fire severity, and an increase from 11 T/ha·year in 2016 to 70 T/ha·year for areas with a very high severity. From the NDVI, it was possible to verify that this also affected the recovery of post-fire vegetation, decreasing the NDVI index 0.36 in areas of moderate severity and 0.53 in areas of very high severity.

Keywords: vegetation dynamics; RUSLE; sentinel-2; soil erosion; wildfire

Citation: Sánchez Sánchez, Y.; Martínez Graña, A.; Santos- Francés, F. Remote Sensing Calculation of the Influence of Wildfire on Erosion in High Mountain Areas. *Agronomy* **2021**, *11*, 1459. https://doi.org/10.3390/agronomy11081459

Academic Editor: Massimo Fagnano

Received: 23 June 2021
Accepted: 20 July 2021
Published: 22 July 2021

Publisher's Note: MDPI stays neutral with regard to jurisdictional claims in published maps and institutional affiliations.

1. Introduction

Research on soil erosion has been carried out for decades [1–3] because it is perceived as one of the most important environmental problems in the world, especially in high mountain regions with a Mediterranean climate since they have a high rainfall intensity, frequent outcrops of soft and weatherable rocks and scarce vegetation cover. This combination constitutes a favorable framework for natural soil erosion, which is aggravated by strong human pressure from certain activities that imply a change in vegetation, land use and soil properties. Consequently, soil loss increases exponentially giving rise to so-called anthropogenic erosion [4]. Given this problem, a particularly pressing issue is the quantification of erosion; different methods have been studied for its calculation: MUSLE [5], PESERA [6,7], Eurosem [8], Artificial Neural Network [9] and USLE [10]. In recent years, these methods have been integrated into Geographic Information Systems (GIS) to increase the accuracy of the calculation [11].

One of the most important causes in the acceleration of erosion rates are forest fires [12] since they represent a very abrupt disappearance of the vegetation cover, leaving the soil bare for weeks or months, increasing the production of runoff and sediment [13]. For this purpose, different indexes have been studied to evaluate the damage caused by forest fires, among them the NBR (Normalized Burn Ratio) [14], applied to post-fire mapping [15], from low resolution satellites [16].

For the calculation of erosion the most used method is the USLE (Universal Soil Loss Equation) or the RUSLE (Revised Universal Soil Loss Equation) with the disadvantage of using the latter for forest fires being that the dynamic calculation of the C factor is very complicated and very costly because it is necessary to revisit the study area continuously, and taking into account the canopy and understory vegetation, together with the Covered

Cavity Fraction being very difficult to study with an accuracy of 10 m pixels. Therefore, different calculation methods have been used for this factor: vegetation cover [17], a radiometer [18] or from NDVI applied to other fields such as erosion in war zones [19].

Remote sensing has been used in other aspects related to wildfire: calculation of fuels models [20], post-fire recovery [21] and multitemporal study [22,23]. Satellite imagery has not only been used in wildfire but also in themes such as altimetry [24] bathymetry [25,26] and ice [27].

In this work, we related pre- and post-forest fire erosion production from remote sensing, observing the areas with a remarkable increase in erosion and those more susceptible to sediment production. For this purpose, two images obtained from the Sentinel 2 Satellite with a time difference of 1 year (June 2016–June 2017) were compared from the data of vegetation vigorousness and land uses; combining it with LiDAR data and the Digital Terrain Model (DTM) allowed us to calculate with a high accuracy the erosion from the multivariate model of the RUSLE. The study also intended to differentiate the different severity of the Garganta de los Infiernos forest fire in September 2016, from calculations obtained by remote sensing and to see how this evolved over a period of time with respect to vegetation, erosion and sediment production.

2. Materials and Methods

2.1. Study Area

The study area is the Garganta de los Infiernos nature reserve in the Jerte Valley in the north of the province of Cáceres (Spain), with an area of 73 km². It is predominantly steep slopes, starting from an altitude of 526 m to 2337 m. The main permanent watercourse is the stream of the "garganta de los infiernos". The average annual rainfall is 1300 mm, mainly between October and June. The lithology consists mainly of poorly developed soils due to the high slopes and the large influx of granite rocky outcrops.

The forest fire (Figure 1) studied took place on 18 August 2016 and remained at level 2 until 20 August; it was extinguished on 28 August having finally razed approximately 1000 ha. The fire occurred in a mountainous area with a large influx of granitic rock formations; the burned vegetation was scrubland and rocky vegetation.

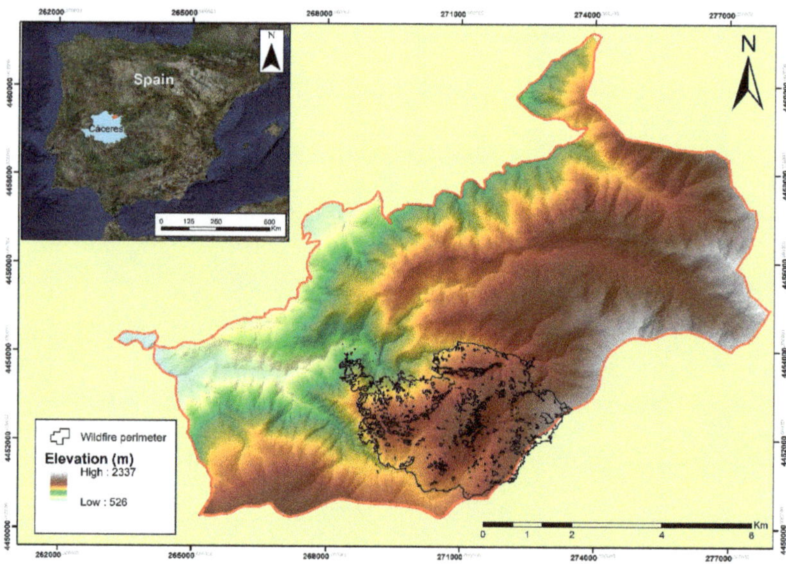

Figure 1. Situation map of Garganta de los Infiernos (Study area) of Digital Terrain Model (DTM) with the perimeter of forest fire.

2.2. Method

The methodology followed is summarized in Figure 2 and will be explained step by step in the following subsections.

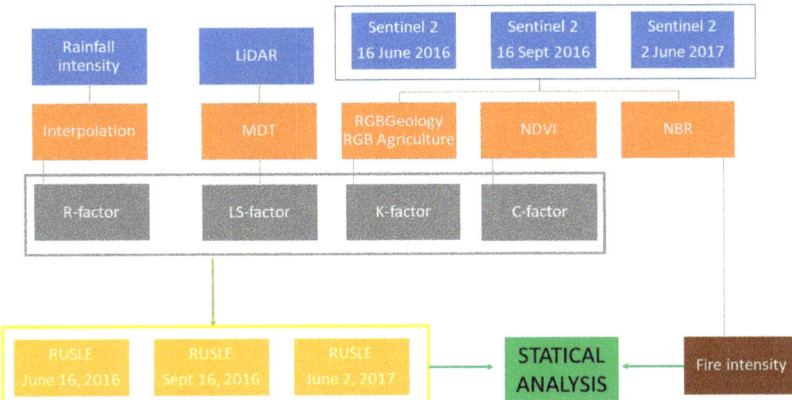

Figure 2. Workflow.

2.3. Remote Sensing

2.3.1. Remotely Sensed Image from Sentinel 2

Sentinel-2 is a mission of the European Space Agency composed of the Launch of two satellites: Sentinel 2A (launched in June 2015) and Sentinel 2B (Launched in March 2017). This satellite was chosen because it is the European satellite providing the highest spatial resolution (10 m) among those offering free services [28] and a temporal resolution of five days. The satellite has 13 spectral bands (Table 1) ranging from visible and near-infrared (VNIR) to shortwave infrared (SWIR) wavelengths over an orbital swath of 290 km [29].

Table 1. Radiometric and Spatial resolution of Sentinel-2.

Sentinel 2 Radiometric and Spatial Resolutions			
Band Number	**Name**	**Central Wavelength (nm)**	**Spatial Resolution (m)**
1	Aerosols	443	60
2	Blue	490	10
3	Green	560	10
4	Red	665	10
5	NIR	705	20
6	NIR	740	20
7	NIR	783	20
8	NIR	842	10
8a	NIR	865	20
9	Water vapor	945	60
10	Cirrus detection	1375	60
11	SWIR	1610	20
12	SWIR	2190	20

In this work, we have studied 3 Sentinel-2 satellite images of the dates: 16 June 2016 (Pre-fire), 4 September 2016 (Post-fire) and 2 June 2017 (spring post-fire). These dates were chosen because they were the images with the best conditions for obtaining the necessary indices. In addition, during the months of May and June, field campaigns were carried out for the mapping of fuel models [20], so the data were used to corroborate the calculated indices. The post-fire image chosen was in September since it was the first available image without smoke, without clouds and with high quality. So it is the closest available image to the date of interest.

From these 3 images, different band combinations were made between them: RGB Geology (band 12, 11 and 2) and RGB Agriculture (band 11, 8 and 2) which have been used for soil classification. Two indices were also calculated: NDVI (Normalized Difference Vegetation Index) [30] and NBR (Normalize Burn Ratio) [31].

2.3.2. LiDAR

LiDAR (Light Detection and Ranging) [32] information from Spanish National Geographic Institute of the PNOA (National Aerial Orthophotography Plan) with a resolution of 0.5 points/m was used, and from this information, and by interpolation, the Digital Elevation Model with a resolution of 5 m was obtained. The data of the study area were obtained in 2010 and the images had a calibration of the LiDAR sensor, a maximum of five returns per pulse, and a pre-classification of said returns. The flight height of the sensor was a maximum of 3000 m from the ground with a horizontal accuracy of 0.30 m and vertical accuracy of 0.20 m.

2.4. Normalized Burn Ratio (NBR)

This index allows us to delimit the perimeter of the wildfire [33] and also to delimit the areas of the fire according to its severity with great precision [34]. Fire severity is a critical factor because of its direct relationship with the biomass consumed, so it is linked to post-fire vegetation and hydrogeomorphological recovery [14]. This index integrates the two bands that best show the combustion B8 (NIR-842 nm) and B12 (SWIR-2190 nm) in the following Equation (1)

$$NBR = (NIR - SWIR)/(NIR + SWIR) \tag{1}$$

This gives the NBR for a given date. In order to appreciate the change derived from forest fire (dNBR), the post-fire NBR is subtracted from the pre-fire NBR (Equation (2))

$$dNBR = NBRprefire - NBRpostfire \tag{2}$$

Once this index is obtained, the study area is classified into different zones depending on the forest fire severity [35]. For this purpose, we will reclassify the output raster into 4 classes: high severity (dNBR > 680), moderate severity (dNBR > 275), low severity (dNBR > 90) and unburned (dNBR < 90) [16].

2.5. RUSLE (Revised Universal Soil Loss Equation)

In this work, the RUSLE (Revised Universal Soil Loss Equation) [36] was used to quantify the erosion rate of the Garganta de los Infiernos before the fire, just after the fire and one year after the fire. This equation was the most appropriate for this calculation since it is the one that has the widest application as it is a parametric and totally empirical model. It is a totally empirical formula (Equation (3)) that attempts to interpret the erosive mechanisms by their causes and effects.

$$A = R \times K \times LS \times C \times P \tag{3}$$

where A is the Average annual soil loss (T/ha·year), R is the rainfall erosivity (MJ·mm/ha·h·year), K is the soil erodibility (Tm/ha h/MJ·mm), LS is the hill slope length and steepness (dimensionless), C is the vegetation factor (dimensionless) and P is the support practice (dimensionless).

- Rainfall erosivity factor (R)

Rainfall erosivity factor is the factor that represents the effect of rainfall on soil erosion. It is calculated as the result of multiplying the kinetic energy of rainfall by the maximum intensity during 30 min of precipitation [37] (Equation (4))

$$R = E \cdot I_{30} \qquad (4)$$

where R is the rainfall erosivity (MJ·mm/ha·h·year), E is total storm energy (Mj/ha·year) and I_{30} the maximum 30 min rainfall intensity (mm/hour).

These data were obtained from the rainfall intensity recorded every 30 min from the Aldehuela del Jerte (CC04) and Valdeastillas (CC17) stations and an extrapolation was performed to calculate the R value at the peak with the highest altitude in the study area (Table 2) for the month of June 2016, the month of September 2016 and the month of June 2017.

Table 2. Location of weather stations and data of the R factor (Information elaborated using that obtained from the AEMET (Agencia Estatal de Meteorología).

Station	X	Y	Z	R Jun 2016	R Sep 2016	R Jun 2017
Valdeastillas (CC17)	255,607	4,447,376	495	208.05	33.33	101
Aldehuela del Jerte (CC04)	224,144.1781	4,434,510.37	262	76.076	46.11	23
Higher Elevation	277,054.3958	4,454,006.489	2300	842.8	0	669.52

- Soil erodibility factor (K)

This factor represents the response of the soil to a given erosive force or mechanism, or, in other words, the susceptibility of the soil to erosion [38]. It represents the quantified soil loss per unit of erosivity in a standard plot 22.6 m long with a slope of 9%. The data were obtained from the experimental plots of the national soil inventory of Cáceres [39] and an unsupervised classification of the RGBGeology and RGBAgriculture band composition was performed to separate the zone into rocky, bare soil and vegetation. The rocky zone was assigned the value 0.134 (mean measured R value for the plutonic rock lithofacies). For the other zones, the R value of the experimental plots was extrapolated.

- Topographic factor (LS)

The LS factor responds to the combined effect of slope length and slope angle. Its value is used to estimate the soil losses that occur on a sloping terrain compared to the losses per unit if the same rainfall were to fall on a standard plot with identical soil type, crop and management conditions.

This factor was calculated from the DTM-LiDAR with a resolution of 5 m; from the DTM, the slope and the flow direction were calculated to subsequently calculate the flow accumulation, and with the following formula the LS factor was calculated (Equation (5)) [40]

$$LS = (\text{flow accumulation cell size}/22.13)^{0.4} \; (\sin ([\text{slope}] \times 0.01745)/(0.0896)^{1.3} \qquad (5)$$

- Vegetation factor (C)

Factor C takes into account vegetation cover to protect the soil. Vegetation is a relevant factor in erosion, vegetation cover reduces the energy of rainfall by intercepting it and prevents it from falling directly to the ground in addition to promoting the infiltration of runoff water [41]. For this article, the C factor was calculated from the NDVI. This index is effective for quantifying green vegetation; it normalizes the scattering of green leaves in the near-infrared wavelength and the absorption of chlorophyll in the red wavelength [30]. The index has been used to see the fluctuations of vegetation at different times of the year and the difference of pre- and post-fire vegetation [18] by applying the formula (Equation (6)) [2]

$$C_{factor} = \exp\left[-\alpha \frac{NDVI}{\beta - NDVI}\right] \tag{6}$$

where α and β parameters determine the shape of the NDVI curve. Reasonable results are produced using values of $\alpha = 2$ and $\beta = 1$.

- Conservation practices (P)

Conservation practices are measures to reduce runoff, generally carried out in cultivated areas. In the study area, there are no cultivated areas, so conservation practices are discouraged.

2.6. Validation

Finally, 200 sampling points randomly distributed over the entire surface of the study area were used, from which the C-factor data obtained from the NDVI index by remote sensing were obtained and compared with the actual C-factor data and with the data obtained in the field. The mean square error (MSE) index was used to verify the validation of the cartography (Equation (7)).

$$MSE = \frac{1}{n} \sum_{i=1}^{n} (Xc - Xr)^2 \tag{7}$$

where n is the number of data points, Xc the value of the generated cartography and Xr the actual value for the data point.

2.7. Statistical Analysis

The correlation methods were used because they are the statistical tools that indicate the relationship between two variables. In this case, the variables that had a more direct affect in post-fire erosion were studied. The correlation method used was the Pearson coefficient that measures the degree of covariance between different linearly related variables. Therefore, a linear correlation procedure was performed to examine the relationship between the covariance of throat erosion post-fire with NDVI, factor C and factor R and to determine the relationship of each variable. The SPSS Statistics 25 software was used to carry out these analyzes.

3. Results

The process carried out to calculate the relationship between the pre-fire erosion of the Garganta de los Infiernos and the post-fire erosion was based on remote sensing data, so that different erosion values were obtained depending on the different fire severity zones.

In the first step of the process, calculation of the severity of the fire (Figure 3), four zones could be distinguished: where the fire had not affected (80%), zones near or with occurrence of other minor fires (7%) with a lower severity, zones of the perimeter where the severity of the fire has been moderate (10%) and the central focus of the fire with a high severity (3%).

Once the fire severity zones boundaries were marked, the NDVI index was calculated for the different dates (16 June 2016, 5 September 2016 and 2 June 2017) (Figure 4a–c) where the influence of the fire on the vegetation and subsequent regrowth of the vegetation was clearly verified.

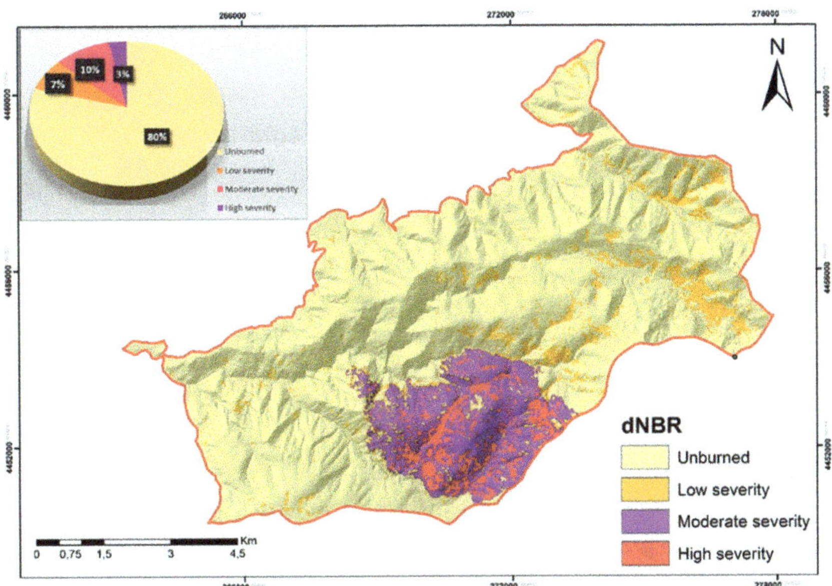

Figure 3. Map of areas of forest fires severity by the NBR and occupancy graph by severity.

Table 3. Table of the output of the SPSS with the results of the Pearson correlation between the factors that cause erosion and soil loss.

		Factor C	NDVI	Factor R	RUSLE
	Correlation of Pearson	−0.291 **	−0.291 **	0.226 **	1
RUSLE	Sig. (Bilateral)	0.000	0.000	0.001	
	N	200	200	200	200

** The correlation is significant at the level 0.01 (bilateral); The correlation is significant at the level 0.05 (bilateral).

In Figure 4d, it can be seen that in the area where the fire had no affect, the NDVI did not vary, however, in the area where the influence of the fire was low the NDVI varied from 0.52 pre-fire to 0.33 post-fire. Where the change is most noticeable is in the area where the fire severity had been very high: 0.55 pre-fire and 0.03 post-fire. Vegetation recovery also depended on the severity of the fire. In the medium severity zone, the difference in NDVI from 2016 to 2017 was 0.36, while in the high severity zone it was 0.52, which shows that vegetation recovery was slower in this zone.

Once the NDVI values were obtained, the C-factor mapping was created, this mapping was compared with the C-factor mapping produced from the values established by the Soil Conservation Service of the United States and with the values of the Second National Forest Inventory for the province of Cáceres [39]. The comparison of both C-factor mappings gave an MSE of 0.15, which indicates that the procedure for obtaining the C-factor mapping from the satellite images was highly accurate.

From the mapping of the C factor, R factor, LS factor and K factor, the RUSLE was calculated and which shows a clear difference between the month of June 2016 and the month of June 2017, in which the mapping of orography and lithology were the same; the mapping of climatology varied since the month of June 2017 was a rainier month than the month of June 2016. However, the most pronounced change was the vegetation, which, in the areas of high severity and moderate severity of the fire, almost disappeared. Therefore, a comparison was made between the erosion mapping before the fire and after the fire.

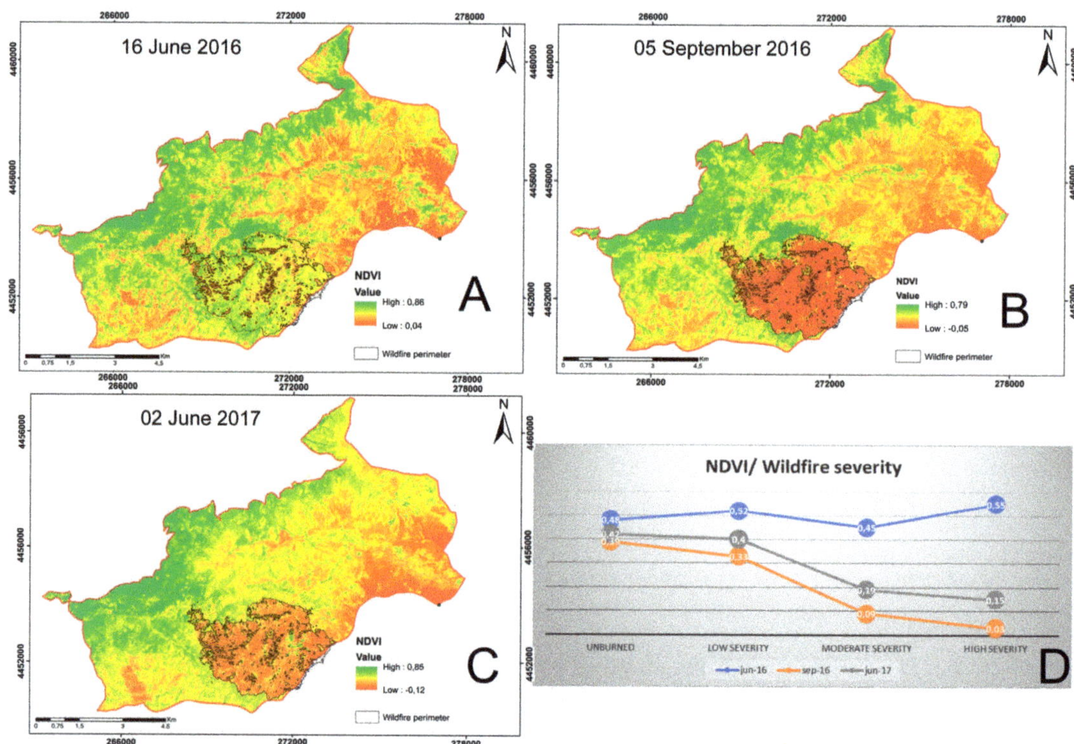

Figure 4. (**A**) NDVI 16 June 2016; (**B**) NDVI 5 September 2016; (**C**) NDVI 2 June 2017; (**D**) Graph of Table 3. the covariation of vegetation and climatology factors versus erosion data in 2017 was studied. The results of Pearson correlation between C factor, NDVI, R factor, erosion and fire severity are presented in Table 2.

The analysis shows that there was an inverse correlation between the C factor and erosion with a high significance, the same correlation value gives us the relationship between NDVI and erosion since the C factor was obtained from NDVI, which indicates a precision in the calculations. The R factor and RUSLE had a positive covariance, which indicates that the relationship between rainfall intensity and erosion was positive, but the significance, despite being high, was lower than in the case of vegetation; however, in areas where the soil was not affected by forest fire, rainfall was more important than vegetation.

In Figure 5 (Figure 5a,b), it can be seen that in the study area not affected by the fire, erosion was similar, while in the study area that was affected by the fire (Figure 5b,d) there are areas that went from low to high or very high erosion. The moderate severity zone in 2016 possessed an average erosion of 31 T/ha·year, while in 2017 it possessed an average erosion of 74 T/ha·year, 2.5 times higher. Even more remarkable was the change in the areas of high fire severity: in 2016 these areas had an erosion of 11 T/ha·year, however, in 2017, they reached an erosion of 70 T/ha·year, more than six times higher.

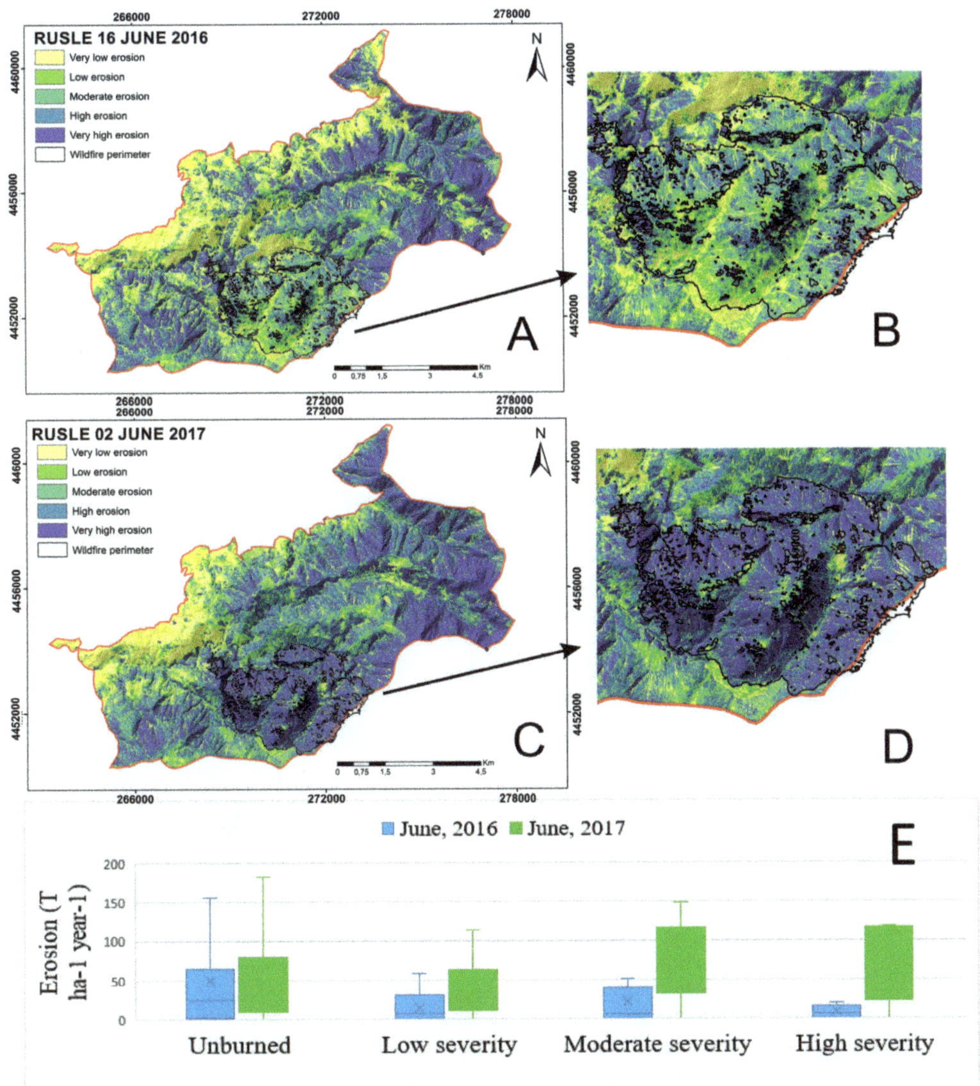

Figure 5. (**A**) Erosion map 16 June 2016; (**B**) Zoom of erosion map in forest fire area 16 June 2016; (**C**) Erosion map 02 June 2017; (**D**) Zoom of erosion map in forest fire area 02 June 2017; (**E**) Comparative graph of the erosion in the different months by area of forest fire severity.

4. Discussion

The NBR index has been used in several forest fire studies with lower resolution satellites [42], or with Sentinel-2 [43]. In all of them, the relationship between fire severity [44] and remotely sensed data from this index has been very good.

The resulting difference between the 2016 and 2017 NDVI is due to the fact that, after a wildfire, vegetation biomass and wattle layer are converted into ash, carbon and organic matter altered by the fire, which are partially carried away by runoff [45] and partially incorporated into the soil [46], retarding the growth of vegetation cover by modifying regrowth dynamics. The vegetation did not fully recover one year after the fire, thus limiting the input of waffles [47].

This analysis shows us that the increased erosion rate is due to the changes that wildfire ignition causes in soils and vegetation. Different authors have shown that the increase in temperatures up to 550 °C from a wildfire completely destroys soil hydrophobicity [48] and that soil water repellency.

Decreases enhancing splash erosion processes [49]. If to that is added the direct relationship with rainfall intensity, then this leads to an increase in runoff rates reducing the availability of nutrients such as organic carbon and water [50]. In other similar studies with similar vegetation and climate conditions, they concluded [7,43], as in this article, that the erosion rate increases greatly during the first two years after the fire and that the vegetation would largely recover after the fifth year [51]. Thus, we can highlight an exponential increase in erosion in mountain areas after the occurrence of a fire due to the loss of vegetation and the exposure of bare soil in possible high intensity rainfall events, and that, even after the following period of maximum vegetation reproduction, erosion is very high [52].

5. Conclusions

Once the results were analyzed, it wase concluded that soil loss for the same area, with the same lithology and orography, is much more abundant after a fire than after a burn, which implies soil and vegetation degradation. There is a clear relationship between the occurrence of a fire and the increase in soil loss, which can be up to 6.5 times higher in areas with a high fire affectation, as a consequence of the massive loss of vegetation. This could lead to contamination of downstream aquifers due to ash production, to saturation of reservoirs due to sedimentation and to a loss of soil productivity, which is why the subsequent treatment of a forest fire by the Spanish Ministry of the Environment is of great concern.

On the contrary, the NBR index is found to be effective in quantifying fire damage in the different affected areas and shows a very high spatial coincidence between the areas of highest fire severity and the areas of highest post-fire sediment production. It is also concluded that the NDVI index is very effective in calculating the RUSLE vegetation factor because it is able to calculate vegetation vigor dynamically, instantaneously and parametrically with high accuracy (MSE = 0.15). Using the NDVI index together with the NBR it can be seen that, in the areas most impacted by the fire, the vegetation takes longer to recover since the damage was greater, so the erosion rate increases linearly at the same time that the vegetation has decreased and the intensity of rainfall increases.

Finally, with remote sensing from Sentinel-2 images, it is possible to quantify and compare the erosion of a pre- and post-fire area in the different areas of severity of forest fires with high accuracy and at low cost. This methodology allows us to make a useful tool to map a burned area with a high spatial resolution (10×10 m) and to study the evolution of post-fire vegetation with a temporal resolution of only 10 days, which allows us to make better decisions when carrying out the soil study.

Author Contributions: Conceptualization, Y.S.S. and A.M.G.; methodology, Y.S.S.; software, Y.S.S.; validation, Y.S.S. and F.S.-F.; formal analysis, A.M.G. and F.S.-F.; investigation, Y.S.S. and F.S.-F.; resources, Y.S.S. and A.M.G.; writing—original draft preparation, Y.S.S.; writing—review and editing, A.M.G.; supervision, F.S.-F.; project administration, A.M.G.; funding acquisition, A.M.G. All authors have read and agreed to the published version of the manuscript.

Funding: This research was funded by the project SA044G18 of Regional Government of Castilla y Leon, and the GEAPAGE research group (Environmental Geomorphology) of the University of Salamanca.

Institutional Review Board Statement: Not applicable.

Informed Consent Statement: Not applicable.

Data Availability Statement: Not applicable.

Acknowledgments: This research was funded by the project SA044G18 of Regional Government of Castilla y Leon, and the GEAPAGE research group (Environmental Geomorphology) of the University of Salamanca.

Conflicts of Interest: The authors declare no conflict of interest.

References

1. Martinez-Graña, A.; Goy, J.; Zazo, C. Dominant soil map in 'Las Batuecas-Sierra De Francia' and 'Quilamas' nature parks (Central System, Salamanca, Spain). *J. Maps* **2014**, *11*, 371–379. [CrossRef]
2. Van der Knijff, J.M.; Jones, R.J.; Montanarella, L. *Soil Erosion Risk Assessment in Europe*; EUR 19044 EN; Office for Official Publications of the European Communities: Luxembourg, 2000; Volume 34, p. 38.
3. Panagos, P.; Ballabio, C.; Poesen, J.; Lugato, E.; Scarpa, S.; Montanarella, L.; Borrelli, P. A Soil Erosion Indicator for Supporting Agricultural, Environmental and Climate Policies in the European Union. *Remote. Sens.* **2020**, *12*, 1365. [CrossRef]
4. Garcia-Ruiz, J.M.; Lopez Bermudez, F. *La Erosión del Suelo en España*; Sociedad Española de Geomorfología (SEG), Ed.; Sdad. Coop. de Artes Gráficas: Zaragoza, Spain, 2009; ISBN 978-84-692-4599-6.
5. Flacke, W.; Auerswald, K.; Neufang, L. Combining a modified Universal Soil Loss Equation with a digital terrain model for computing high resolution maps of soil loss resulting from rain wash. *Catena* **1990**, *17*, 383–397. [CrossRef]
6. Meusburger, K.; Konz, N.; Schaub, M.; Alewell, C. Soil erosion modelled with USLE and PESERA using QuickBird derived vegetation parameters in an alpine catchment. *Int. J. Appl. Earth Obs. Geoinf.* **2010**, *12*, 208–215. [CrossRef]
7. Karamesouti, M.; Petropoulos, G.; Papanikolaou, I.; Kairis, O.; Kosmas, K. Erosion rate predictions from PESERA and RUSLE at a Mediterranean site before and after a wildfire: Comparison & implications. *Geoderma* **2016**, *261*, 44–58. [CrossRef]
8. Khaleghpanah, N.; Shorafa, M.; Asadi, H.; Gorji, M.; Davari, M. Corrigendum to "Modeling soil loss at plot scale with EUROSEM and RUSLE2 at stony soils of Khamesan watershed, Iran" [Catena (147C) (2016) 773–788]. *Catena* **2017**, *151*, 259. [CrossRef]
9. Gholami, V.; Booij, M.; Tehrani, E.N.; Hadian, M. Spatial soil erosion estimation using an artificial neural network (ANN) and field plot data. *Catena* **2018**, *163*, 210–218. [CrossRef]
10. Sanchez, Y.; Martínez-Graña, A.; Santos-Francés, F.; Yenes, M. Influence of the sediment delivery ratio index on the analysis of silting and break risk in the Plasencia reservoir (Central System, Spain). *Nat. Hazards* **2018**, *91*, 1407–1421. [CrossRef]
11. Terranova, O.; Antronico, L.; Coscarelli, R.; Iaquinta, P. Soil erosion risk scenarios in the Mediterranean environment using RUSLE and GIS: An application model for Calabria (southern Italy). *Geomorphology* **2009**, *112*, 228–245. [CrossRef]
12. AbdulKadir, T.S.; Muhammad, R.M.; Khamaruzaman, W.Y.; Ahmad, M.H. Geo-statistical based susceptibility mapping of soil erosion and optimization of its causative factors: A conceptual framework. *J. Eng. Sci. Technol.* **2017**, *12*, 2880–2895.
13. Esteves, T.; Kirkby, M.; Shakesby, R.; Ferreira, A.; Soares, J.; Irvine, B.; Ferreira, C.; Coelho, C.; Bento, C.; Carreiras, M. Mitigating land degradation caused by wildfire: Application of the PESERA model to fire-affected sites in central Portugal. *Geoderma* **2012**, *191*, 40–50. [CrossRef]
14. Llovería, R.M.; Cabello, F.P.; Martín, A.G.; Vlassova, L.; de la Riva Fernández, J.R. *La Severidad del Fuego: Revisión de Conceptos, Métodos y Efectos Ambientales. Geoecología, Cambio Ambient. y Paisaje: Homenaje al Profesor José María García Ruiz*; Instituto Pirenaico de Ecología: Huesca, Spain, 2014; pp. 427–440.
15. Miller, M.E.; Billmire, M.; Elliot, W.J.; Endsley, K.A.; Robichaud, P.R. Rapid response tools and datasets for post-fire modeling: Linking Earth Observations and process-based hydrological models to support post-fire remediation. *Int. Arch. Photogramm. Remote. Sens. Spat. Inf. Sci.* **2015**, *XL-7/W3*, 469–476. [CrossRef]
16. Epting, J.; Verbyla, D.; Sorbel, B. Evaluation of remotely sensed indices for assessing burn severity in interior Alaska using Landsat TM and ETM+. *Remote. Sens. Environ.* **2005**, *96*, 328–339. [CrossRef]
17. Iwuji, M.C.; Iroh, O.L.; Njoku, J.D.; Anyanwu, S.O.; Amangabara, G.T.; Ukaegbu, K.O.E. Vulnerability Assessment of Soil Erosion Based on Topography and Vegetation Cover in a Developing City of Orlu L.G.A, South East Nigeria. *Asian J. Environ. Ecol.* **2017**, *4*, 1–10. [CrossRef]
18. Verhegghen, A.; Eva, H.; Ceccherini, G.; Achard, F.; Gond, V.; Gourlet-Fleury, S.; Cerutti, P.O. The Potential of Sentinel Satellites for Burnt Area Mapping and Monitoring in the Congo Basin Forests. *Remote. Sens.* **2016**, *8*, 986. [CrossRef]
19. Abdo, H.G. Impacts of war in Syria on vegetation dynamics and erosion risks in Safita area, Tartous, Syria. *Reg. Environ. Chang.* **2018**, *18*, 1707–1719. [CrossRef]
20. Sánchez, Y.S.; Martínez-Graña, A.; Francés, F.S.; Picado, M.M. Mapping Wildfire Ignition Probability Using Sentinel 2 and LiDAR (Jerte Valley, Cáceres, Spain). *Sensors* **2018**, *18*, 826. [CrossRef]
21. Abney, R.B.; Sanderman, J.; Johnson, D.; Fogel, M.L.; Berhe, A.A. Post-wildfire Erosion in Mountainous Terrain Leads to Rapid and Major Redistribution of Soil Organic Carbon. *Front. Earth Sci.* **2017**, *5*, 99. [CrossRef]
22. Malowerschnig, B.; Sass, O. Long-term vegetation development on a wildfire slope in Innerzwain (Styria, Austria). *J. For. Res.* **2014**, *25*, 103–111. [CrossRef]
23. Mallinis, G.; Gitas, I.; Tasionas, G.; Maris, F. Multitemporal Monitoring of Land Degradation Risk Due to Soil Loss in a Fire-Prone Mediterranean Landscape Using Multi-decadal Landsat Imagery. *Water Resour. Manag.* **2016**, *30*, 1255–1269. [CrossRef]

24. Okeowo, M.A.; Lee, H.; Hossain, F.; Getirana, A. Automated Generation of Lakes and Reservoirs Water Elevation Changes From Satellite Radar Altimetry. *IEEE J. Sel. Top. Appl. Earth Obs. Remote. Sens.* **2017**, *10*, 3465–3481. [CrossRef]
25. Mateo-Pérez, V.; Corral-Bobadilla, M.; Ortega-Fernández, F.; Vergara-González, E. Port Bathymetry Mapping Using Support Vector Machine Technique and Sentinel-2 Satellite Imagery. *Remote. Sens.* **2020**, *12*, 2069. [CrossRef]
26. Mateo-Pérez, V.; Corral-Bobadilla, M.; Ortega-Fernández, F.; Rodríguez-Montequín, V. Determination of Water Depth in Ports Using Satellite Data Based on Machine Learning Algorithms. *Energies* **2021**, *14*, 2486. [CrossRef]
27. Gaur, M.K.; Goyal, R.K.; Raghuvanshi, M.S.; Bhatt, R.K.; Pandian, M.; Mishra, A.; Sheikh, S.I. Geospatially extracting snow and ice cover distribution in the cold arid zone of India. *Int. J. Syst. Assur. Eng. Manag.* **2019**, *11*, 84–99. [CrossRef]
28. Mandanici, E.; Bitelli, G. Preliminary Comparison of Sentinel-2 and Landsat 8 Imagery for a Combined Use. *Remote. Sens.* **2016**, *8*, 1014. [CrossRef]
29. ESA. Copernicus. (European Space Agency) Obtenido de Observing the Earth. Available online: http://www.esa.int/Our_Activities/Observing_the_Earth/Copernicus/Sentinel-2 (accessed on 14 July 2021).
30. Tucker, C.J. Red and photographic infrared linear combinations for monitoring vegetation. *Remote. Sens. Environ.* **1979**, *8*, 127–150. [CrossRef]
31. Lutes, D.C.; Keane, R.E.; Caratti, J.F.; Key, C.H.; Benson, N.C.; Sutherland, S.; Gangi, L.J. *FIREMON: Fire Effects Monitoring and Inventory System*; General Technical Report RMRS-GTR-164-CD; U.S. Department of Agriculture, Forest Service, Rocky Mountain Research Station: Fort Collins, CO, USA, 2006; Volume 1 CD, p. 164.
32. PNOA. CNIG (Centro Nacional de Información Geográfica). Available online: https://pnoa.ign.es/el-proyecto-pnoa-lidar (accessed on 15 July 2021).
33. Karmaker, A.C.; Hoffmann, A.; Hinrichsen, G. Influence of water uptake on the mechanical properties of jute fiber-reinforced polypropylene. *J. Appl. Polym. Sci.* **1994**, *54*, 1803–1807. [CrossRef]
34. Fernández-Manso, A.; Fernández-Manso, O.; Quintano, C. SENTINEL-2A red-edge spectral indices suitability for discriminating burn severity. *Int. J. Appl. Earth Obs. Geoinf.* **2016**, *50*, 170–175. [CrossRef]
35. Pérez-Cabello, F.; Fernandez, J.D.L.R.; Llovería, R.M.; Garcia-Martin, A. Mapping erosion-sensitive areas after wildfires using fieldwork, remote sensing, and geographic information systems techniques on a regional scale. *J. Geophys. Res. Space Phys.* **2006**, *111*. [CrossRef]
36. Renard, K.; Laflen, J.; Foster, G.; McCool, D. The Revised Universal Soil Loss Equation. *Soil Eros. Res. Methods* **2017**, 105–126. [CrossRef]
37. Wischmeier, W.H.; Smith, D.D. Rainfall energy and its relationship to soil loss. *Trans. Am. Geophys. Union* **1958**, *39*, 285–291. [CrossRef]
38. Martínez-Graña, A.; Goy, J.L.; Zazo, C.; Yenes, M. Engineering Geology Maps for Planning and Management of Natural Parks: "Las Batuecas-Sierra de Francia" and "Quilamas" (Central Spanish System, Salamanca, Spain). *Geosciences* **2013**, *3*, 46–62. [CrossRef]
39. Biodiversidad, D.G. *para la Inventario Nacional Erosion Suelos*; Ambiente, M.d.M., Ed.; Cáceres, Spain, 2005; ISBN 84-814-647-8. Available online: https://www.ine.es/dyngs/IOE/en/fichaInventario.htm?cid=1259930778148&inv=92005 (accessed on 20 June 2021).
40. Samanta, S.; Koloa, C.; Pal, D.K.; Palsamanta, B. Estimation of potential soil erosion rate using RUSLE and E30 model. *Model. Earth Syst. Environ.* **2016**, *2*, 149. [CrossRef]
41. Bellot, J.; Bonet, A.; Sanchez, J.; Chirino, E. Likely effects of land use changes on the runoff and aquifer recharge in a semiarid landscape using a hydrological model. *Landsc. Urban Plan.* **2001**, *55*, 41–53. [CrossRef]
42. Bastarrika, A.; Alvarado, M.; Artano, K.; Martinez, M.P.; Mesanza, A.; Torre, L.; Ramo, R.; Chuvieco, E. BAMS: A Tool for Supervised Burned Area Mapping Using Landsat Data. *Remote. Sens.* **2014**, *6*, 12360–12380. [CrossRef]
43. Navarro, G.; Caballero, I.; Silva, G.; Parra, P.-C.; Vázquez, Á.; Caldeira, R. Evaluation of forest fire on Madeira Island using Sentinel-2A MSI imagery. *Int. J. Appl. Earth Obs. Geoinf.* **2017**, *58*, 97–106. [CrossRef]
44. García, A.J.; Bakon, M.; Martínez, R.; Marchamalo, M. Evolution of urban monitoring with radar interferometry in Madrid City: Performance of ERS-1/ERS-2, ENVISAT, COSMO-SkyMed, and Sentinel-1 products. *Int. J. Remote. Sens.* **2018**, *39*, 2969–2990. [CrossRef]
45. Badía, D.; Martí, C.; Aguirre, A.J.; Aznar, J.M.; González-Pérez, J.; De la Rosa, J.; León, J.; Ibarra, P.; Echeverría, T. Wildfire effects on nutrients and organic carbon of a Rendzic Phaeozem in NE Spain: Changes at cm-scale topsoil. *Catena* **2014**, *113*, 267–275. [CrossRef]
46. Bodí, M.B.; Mataix-Solera, J.; Doerr, S.H.; Cerdà, A. The wettability of ash from burned vegetation and its relationship to Mediterranean plant species type, burn severity and total organic carbon content. *Geoderma* **2011**, *160*, 599–607. [CrossRef]
47. Hobley, E.U.; Brereton, A.J.L.G.; Wilson, B. Forest burning affects quality and quantity of soil organic matter. *Sci. Total. Environ.* **2017**, *575*, 41–49. [CrossRef]
48. Mataix-Solera, J.; Cerdà, A.; Arcenegui, V.; Jordán, A.; Zavala, L.M.M. Fire effects on soil aggregation: A review. *Earth Sci. Rev.* **2011**, *109*, 44–60. [CrossRef]
49. Jordán, A.; Zavala, L.M.M.; Mataix-Solera, J.; Nava, A.L.; Alanís, N. Effect of fire severity on water repellency and aggregate stability on Mexican volcanic soils. *Catena* **2011**, *84*, 136–147. [CrossRef]

50. Badía, D.; López-García, S.; Martí, C.; Perpiñá, J.O.O.; Girona-García, A.; Casanova-Gascón, J. Burn effects on soil properties associated to heat transfer under contrasting moisture content. *Sci. Total Environ.* **2017**, *601–602*, 1119–1128. [CrossRef] [PubMed]
51. Hueso-González, P.; Martínez-Murillo, J.F.; Ruiz-Sinoga, J.D. Prescribed fire impacts on soil properties, overland flow and sediment transport in a Mediterranean forest: A 5 year study. *Sci. Total Environ.* **2018**, *636*, 1480–1489. [CrossRef] [PubMed]
52. Samani, A.N.; Rad, F.T.; Azarakhshi, M.; Rahdari, M.R.; Rodrigo-Comino, J. Assessment of the Sustainability of the Territories Affected by Gully Head Advancements through Aerial Photography and Modeling Estimations: A Case Study on Samal Watershed, Iran. *Sustainability* **2018**, *10*, 2909. [CrossRef]

Article

Mapping the Risk of Water Soil Erosion in Larrodrigo (Salamanca, Spain) Using the RUSLE Model and A-DInSAR Technique

Antonio Martínez-Graña [1,*], Jerymy Carrillo [1], Lorena Lombana [1], Marco Criado [2] and Carlos Palacios [3]

1 Department of Geology, Faculty of Sciences, University of Salamanca, Plaza de la Merced s/n, 37008 Salamanca, Spain; idu20375@usal.es (J.C.); llombanag@usal.es (L.L.)
2 Department of Soil Sciences, Faculty of Environmental Sciences, University of Salamanca, Avenue Filiberto Villalobos 119, 37007 Salamanca, Spain; marcocn@usal.es
3 Department of Construction and Agronomy, Faculty of Agricultural Sciences, University of Salamanca, Avenue Filiberto Villalobos 119, 37007 Salamanca, Spain; carlospalacios@usal.es
* Correspondence: amgranna@usal.es

Abstract: The quantification of soil loss are studies driven by the importance of soil as a resource and are mainly due to risks of laminar and/or runoff water erosion. These problems directly affect the daily life of the population and serve as predictors of environmental effects. In this work, the quantification and calculation of the sheet water erosion caused mainly by rainfall has been carried out in a study area located in the municipality of Larrodrigo (Salamanca, Spain), based on the simultaneous application of the RUSLE model with GIS techniques. Thematic cartographies have been generated to determine soil loss in Tm/Ha/year and mm/year based on the use of parameters of the physical environment (lithology, rainfall, slopes …) where the erosive risk is quantified and its applicability to the study area by spatio-temporal extrapolation techniques. Simultaneously, the use of the A-DInSAR technique was implemented to calculate average ground deformation velocities in mm/year associated with water erosion. Two sectors with greater vulnerability to water erosion have been detected within the area of interest: one of them called main, which corresponds to the slopes near the Larrodrigo stream, with soil losses showing values of 0.3- > 12 mm/year, and a secondary sector belonging to the tributaries or channels derived from the mainstream with values of 0.3- > 12 mm/year. This type of study makes it possible to manage and organise human support practicesin order to subsequently establish measures that can prevent, mitigate and/or correct those areas with the greatest damage.

Keywords: erosion risk; RUSLE; GIS; runoff; A-DinSAR

Citation: Martínez-Graña, A.; Carrillo, J.; Lombana, L.; Criado, M.; Palacios, C. Mapping the Risk of Water Soil Erosion in Larrodrigo (Salamanca, Spain) Using the RUSLE Model and A-DInSAR Technique. *Agronomy* **2021**, *11*, 2120. https://doi.org/10.3390/agronomy11112120

Academic Editor: Camilla Dibari

Received: 9 September 2021
Accepted: 20 October 2021
Published: 22 October 2021

Publisher's Note: MDPI stays neutral with regard to jurisdictional claims in published maps and institutional affiliations.

1. Introduction

The risk of soil erosion is an issue that has become very important over the years, because soil degradation can cause important environmental impacts and high economic costs, through its effects on the agricultural production, infrastructures and water quality that, in turn, affect the well-being of citizens, potentially threatening their security and representing a serious problem for the sustainable development of the population. On the other hand, organic carbon emissions can be generated from the soil into the atmosphere in the form of CO_2, thus accentuating the effect of global warming [1]. These erosive processes are characterized by being relatively slow and intermittent, although periodic over the years. However, as a consequence of inadequate soil management, erosion from agricultural land can be very intense. Furthermore, soil degradation by erosive agents (water and air) is considered progressive and irreversible since, on the one hand, the volume of soil lost is usually irrecoverable and, on the other, the time required for it to form again, the floor is extremely long [2–4].This work will focus on the risks of sheet water erosion or surface rill due to precipitation, which consists of the loss of a more or less uniform layer of soil on a sloping terrain, which mainly affects the released particles by splash. This occurs mainly in situations where the intensity of precipitation exceeds infiltration or when the soil becomes saturated with water, resulting in excess water on the surface. Surface runoff transports the finest particles and causes a decrease in soil productivity (loss of clay, organic matter and nutrients). When the irregularities of the terrain and the greater dolawnstream flow cause the laminar flow to become concentrated, we speak of erosion by furrows, gullies and ravines that represent the three degrees of development of this process. The grooves are centimetric in size and can disappear when carving, the gullies are decametric to metric and generally cannot be removed with ordinary tillage and finally, the gullies are incisions of several meters, even dozens. The result of erosion by gullies and ravines is the dissection of the affected land [2,5]. The area affected by water erosion, as well as its total intensity and magnitude, can be increased by deforestation, forest fires, traditional tillage systems, etc. The absence of plant cover means that the impact of the raindrop not only causes disaggregation, but generally results in the formation of a superficial crust that affects the first millimeters or centimeters of the tillage horizon. This phenomenon is a consequence of the displacement of clay and silt particles that occupy the pores of the soil, causing their occlusion. With this, the infiltration intensity decreases, which favors the initiation of surface runoff [6]. In turn, the same soil, exposed to the action of the same rains, undergoes different erosion intensities depending on whether it is in the upper, middle or lower part of a hilside, and depending on slope (relief effect). Another thing to keep in mind is that the resulting erosion also varies depending on the type of vegetation that protects the soil, the cultivation practices or the use of said vegetation, its disposition with respect to the slope of the slope, and so on. These last two factors, relief and vegetation cover, are what make the erosive action of rain vary on the erodibility of each soil, resulting in different erosion rates in each case, which can be evaluated through the estimation of the effect of each of these factors mentioned.

On the other hand, the application of techniques through Geographic Information Systems (GIS) in environmental studies, allows analysis of the different geospatial databases of the territory to be studied. It is commonly used in areas such as environmental planning and protection, natural resources, natural risk analysis (geomorphological and hydrological modeling, etc.), or the generation of georeferenced thematic cartographies, establishing different analysis methods and models of process simulation [7,8]. The quantification of erosion in the last decades has been carried out by various methods such as USLE, SWAT, MUSLE, PESERA, EUROSEM, etc. which analyze the erosive process considering a static natural environment. For this reason, GIS and remote sensing techniques have made it possible to develop and quantify the parameters that intervene in the erosive process dynamically and with more exact techniques and platforms that have currently been developed (lidar data, satellite images resolution, cartography landuse CORINE-SIOSE, etc). In this article, to detect, monitor and model the erosion risks found, GIS techniques have been used together with the Revised Universal Soil Loss Equation (RUSLE) model, allowing the generation of different thematic cartographies of the territory, for the subsequent analysis of the possible emerging erosion risks (erosive, hydrological, ground movements, geotechnical) and environmental analysis of anthropic actions in the natural environment, at different scales. For the analysis of mass movements or sediment flow there are numerous techniques. The detection process can be carried out: (i) on a local scale, using monitoring systems with sensors such as deformometers, biaxial inclinometers, load cells, flow meters, piezometers, etc. (SIAP MICROS, 2003), which allow alerting in real time, each small landslide or, (ii) at a regional scale through the combined use of the techniques: Geographic Information Systems (GIS) and Advanced Differential Interferometry SAR (A-DInSAR). The Differential Synthetic Aperture Radar Interferometry (DInSAR) technique is based on the calculation of displacement maps of individual events calculating the phase difference (interferogram) of two SAR images obtained over the same acquisition area and orbit, separated in time, allowing the study of the deformation produced by the displacements of the terrain of any area of the planet in a given period of time. Since the beginning of this century, different advanced techniques (such as A-DInSAR) have been developed that allow analysis and studies of slope instabilities and subsidence phenomena through the acquisition and processing of a large number of SAR images, and therefore, of a large number of interferograms [9–12]. One of the most widely used techniques is the Parallel Small Baseline Subset (P-SBAS), developed and incorporated into the free-to-use platform Grid-Processing On Demand (G-POD) [13]. The SAR images used by this platform come from the ERS-1 and 2 (1991–2011) and Envisat ASAR (2002–2012) satellites. In our case it has been applied to be able to detect, monitor and model differential flows combining A-DinSAR and SIG [14,15].

Due to the importance of soil as a fundamental resource for life and the development of anthropic activities, the objective of this work is to quantify the risk of water erosion in an area such as Larrodrigo, with a high capacity for agricultural activity, simultaneously applying the RUSLE method and GIS techniques (ArcGis 10.9, Esri, Madrid, Spain), in order to know the high potential for water erosion risks in the study area. On the other hand, it is intended to detect trends of sediment flows within and in the surroundings of the study area, from the automation of thematic cartographies by means of algorithms obtained from DTMs in order to detect ground movement by applying DinSAR techniques. Finally, for the quantification and calculation of the risk of water erosion, different thematic cartographies generated and mechanized with GIS will be obtained.

2. Materials and Methods

2.1. Study Area

The study area is located in the southeast of the province of Salamanca in Spain and represents an agricultural and livestock exploitation area of almost 2300 hectares in which extensive cattle ranching and organic cereal agriculture coexist. The study area constitutes an area susceptible to water erosion risks, located in an area belonging to the southwestern Atlantic slope of the Iberian Peninsula (Figure 1), with a series of risks of medium and serious magnitude of erosion, especially in agricultural lands, and with tendency to accentuate towards the south. The effects of the physical losses of the soil in the study area are mainly of environmental and economic importance. As mentioned above, this occurs mainly due to surface water erosion of the sheet or gutter type due to its influence on the degradation of natural systems, the loss of land productivity and the alteration of hydrological processes, especially when the anthropically accelerated erosion is considered, which is what causes the great losses of soil. This is mainly caused by the clearing of sloping land, the indiscriminate application of inappropriate agricultural practices, deforestation or large public works. The presence of the Larrodrigo stream and heavy rainfall at certain times of the year, make it the main agents of this type of problem [16].

From a geological point of view, the study area is located on the Central System, close to the northern sub-plateau of the Iberian Peninsula. Formations of tertiary age appear fundamentally which are part of the set of continental facies that fill the Duero Basin. A homogeneity can be appreciated both in the petrography and in the textural context of the differentiated facies, which makes the separation between them very difficult at the outcrop scale, being made from broader criteria based on sedimentological considerations and abundance of structures. There is also the presence of Quaternary age formations superimposed on the Miocene materials. In general, within the study area they can be differentiated from oldest to most modern [17,18] (Figure 2). In the first place, there are deposits corresponding to arches of medium-coarse grain, which include edges of various natures and sizes, which are located scattered in the sands or well grouped, giving more or less continuous levels. Tertiary-age material that is formed mainly by medium-thick quartz grains and, to a lesser extent, by feldspars. The clayey fraction has a reddish tonality and its mica content is abundant. They are located on the southwestern edge of the study area and present hydromorphic processes such as vertically discolorations. The abundance of these discolorations indicates the subaerial exposure of wide areas and the development of vegetation, which requires some time and stability.

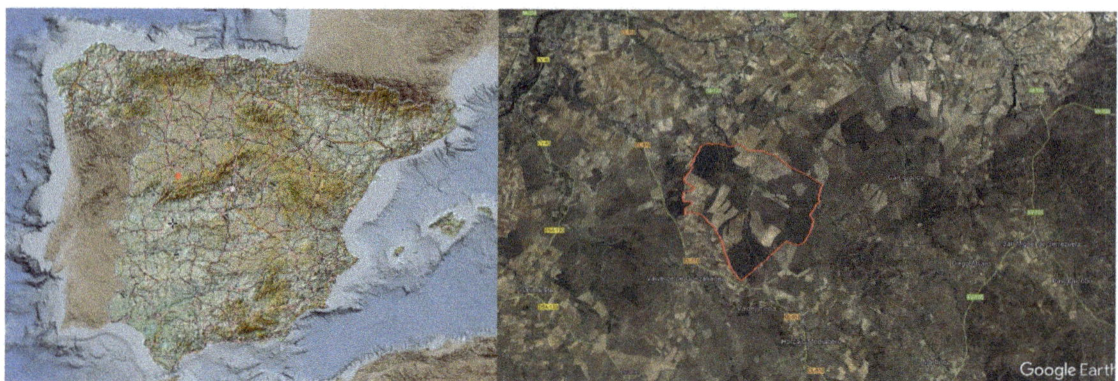

Figure 1. Situation map of Study area (**left**). The satellite image (**right**) shows the great importance of the conservation of the arboreal roots (dark green areas) scattered throughout a region in which bare soils are abundant (light brown color).

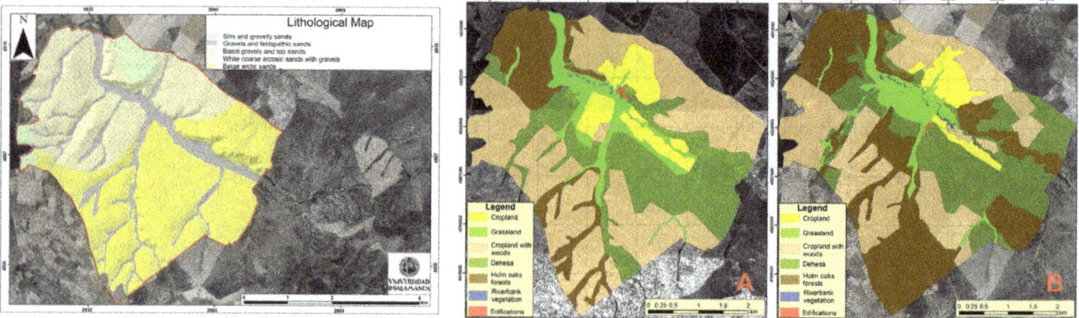

Figure 2. Left. Geological Map. **Right**: Changes in land use observed in the study area: 1956 (**A**) and 2020 (**B**) based on SIOSE (https://www.siose.es/) (accessed 20 October 2021).

On the other hand, arcosic beige sands of Miocene-Pliocene age can be found, which may be affected by a fault and/or be discordant towards the direction SW. They occur throughout the southern half of the study area, and may contain levels of songs that are par-conglomerate with a very abundant matrix, although there may be cases with only one layer of songs or isolated songs. There is very low-angle planar cross-bedding and the- ensemble slopes 5–10° to the direction NNW. In addition, there are large outcrops that correspond to thick white arkotic sands with few clayey intercalations positioned in the northern half of the area of interest. From a lithological point of view, the composition is very similar to that of the San Mamés facies, with a predominance of white-greenish or white microconglomerate sands of low potency and without lateral continuity. They can present some metamorphic rock in the form of an accessory such as slates and more or less altered schists. They contain a monotonous structure, with parallel lamination by alternating sheets of different granulometry, although, locally, very low-angle cross lamination can be observed. They are also from the Tertiary period.

Finally, the quaternary deposits are highlighted. The main generator of this type of formation is the Tormes River and its tributaries such as the Larrodrigo stream, which runs through the study area, due to its large flow. These deposits refer to important terraces located from +18 to +10 m above sea level. These are gravels at the base and predominant sands with edges towards the roof (terraces). There are also fluvial deposits that correspond to the bottoms of valleys and river plains of Holocean age. It extends from north to south and from east to west through the study area following the bed of the Larrodrigo stream and its small tributaries, corresponding mainly to gravels and feldspathic sands. In general, the main load of the riverbed is made up of sand-sized elements, with quartz and feldspar as the majority components, caused by the gentle slope of the riverbeds, making it difficult to remove the thicker edges. Finally, located at the southwestern edge of the study area and superimposed on the tertiary sediments, are the current or sub-current glacis. Its composition is silty-sandy with some sub-rounded edges of variable lithology. The full thickness is not visible, although it sometimes forms a thin layer on the underlying deposits.

The study area constitutes a pilot area in which sustainable agriculture and livestock are developed, supported by the quality and quantity of the existing soil in a fragile ecosystem of Dehesa. Figure 1 shows the changes in land use that have occurred in the sector from 1956 to the present. It is observed how the area occupied by holm oaks has increased considerably, especially in the southwestern sector, while the dehesa areas (term used nationally to describe pastures with holm oaks) have increased slightly. On the contrary, there is a decline in the pasturelands that were cultivated in 1956 and currently host pastures, which is why they are now considered as pastures. Regarding cereal fields and meadows, they have hardly changed.

2.2. Methods

To determine and detect erosion risks that cause soil loss, which are developed in the study area, there are numerous methodology and techniques. There is a first division based on those models that focus on physical processes that take place in erosion (the use of mathematical equations that consider the laws of conservation of mass and energy) and those empirical models that use a series of mathematical algorithms and correlations of erosive factors (although a direct cause-effect relationship with soil losses cannot be established). Due to the complexity of the natural phenomena involved in erosion, there are certain models that have a more probabilistic approach, associating a probability to each erosive event. -The Universal Soil Loss Equation (USLE). Model with greater diffusion and that is used in this project. Proposed by Wischmeier and Smith [19,20] very useful when making decisions about the use and conservation of the soil. It considers six factors and estimates the mean annual soil losses at the level of the agricultural plot or moderately sloping hillside. There is an update and expansion of the model called Revised Universal Soil Loss Equation (RUSLE).

The RUSLE model has been considered the best tool when estimating annual averages of soil losses, in order to inventory and map water erosion in the study area, and is focused on specific restoration plans. environmental and soil conservation. The technique used to develop the RUSLE model is based on the large amount of data collected. In addition, it is a model recognized throughout the world and its application is widespread within the scientific community and in the area of conservation of natural resources. The equation of the model focuses on considering rainfall as the main active agent of this surface erosion, establishing that annual soil losses are directly proportional to the erosivity index of rainfall, related to the kinetic energy of each downpour and its maximum intensity. The same conditions of erosion of the rains can produce different erosions depending on the characteristics and properties of the soil in which they act. With this method it is possible to be able to recognize a series of characteristics of the soil itself that determine its erodibility or vulnerability to erosion, related to its texture, structure, organic matter content and permeability [21].

To calculate the water erosion of the study area, a compilation of cartographic information was previously carried out to later calculate each factor of the RUSLE method. The basic equation of the RUSLE model for estimating average soil losses as a consequence of sheet and trickle water erosion, is $A = R \times K \times L \times S \times C \times P$ [19,20,22]; where: A are the soil losses per unit area for the period of time considered in Tm/Ha/year. It is obtained by the product of the following factors: R: rain factor (rain erosion index). It is the number of units of the erosion index (E I_{30}) in the period considered, where E is the kinetic energy of a given precipitation and I_{30} is the maximum intensity in 30 min of it. The erosion index is a measure of the erosive force of a given precipitation given in MJ·mm/Ha·year. K: soil erodibility factor. It is the value of soil losses per units of the pluvial erosion index in $t·m^2·h/Ha·J·cm$, for a determined soil in continuous fallow, with a slope of 9% and a slope length of 22.1 m. L: slope length factor, or the relationship between the loss of soil for a given slope length and the loss for a length of 22.1 m of the same type of soil and vegetation or use. S: slope factor, defined as the relationship between the losses for a given slope and the losses for a slope of 9% of the same type of soil and vegetation or use. C: cover factor and handling. It is the relationship between the losses of soil in a land cultivated under specific conditions or with certain natural vegetation and the corresponding losses of a soil in continuous fallow. P: Factor for soil conservation practices. It is the relationship between the soil losses with cultivation at level, in strips, on terraces, on terraces or with subsurface drainage, and the soil losses corresponding to work on the line of maximum slope [23–25]. The equivalence of Tm/Ha/year to mm/year is obtained considering the apparent den-sity, whose value is expressed in gr/cm^3. To calculate the thickness 1 Tm/Ha of each soil in mm, the relationship between soil loss (gr/Ha) and apparent density (gr/Ha) is considered. The methodology followed is summarized in Figure 3 and will be explained step by step in the following subsections.

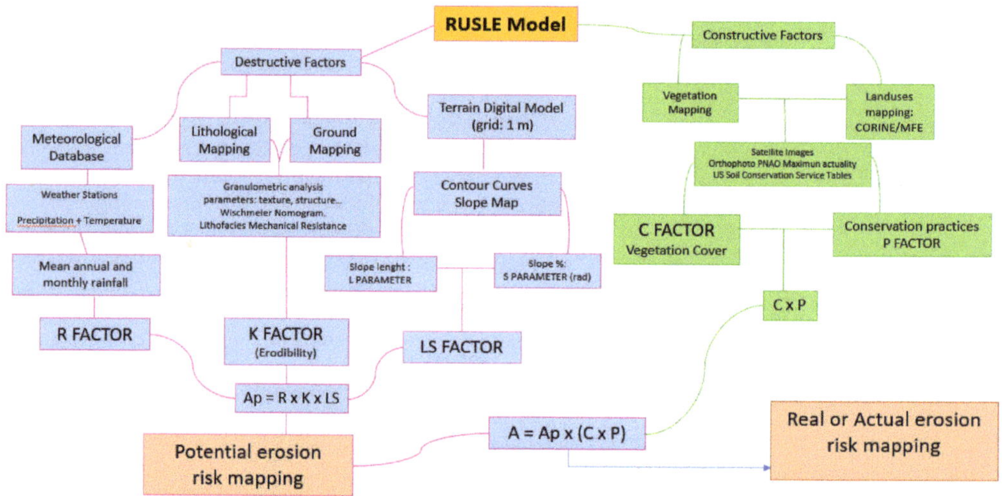

Figure 3. Workflow.

In the Figure 3 with the RUSLE model, a digital terrain model of the area of interest was created from LIDAR data and the application of A-DinSAR techniques was carried out to be able to observe some movement of the terrain.

2.2.1. Compilation of Cartographic Information of the Study Area

Relevant information was collected from the study area that refers to topography, lithology and relief: Orthophotography (Download Center of the National Geographic Institute -IGN-), a set of four raster files that correspond to the most current PNOA orthophotos from part of the province of Salamanca covering part of the geological sheets 1:50,000 of Salamanca and Ávila. LIDAR images (Spatial Data Infrastructure Download Center of Castilla y León, Idecyl), LAZ file package. corresponding to the LIDAR data of sheet 212 of 2010 of Castilla y León covering the study area. Raster maps at a scale of 1:50,000 and in greater detail at 1/25,000 IGN-, which present toponymic information of the study area.

Information on pluvio-metric and thermopluviometric meteorological stations (Geo Portal, Ministry of Ecological Transition), with a total of 58 files that show information about the geographical position, altitude, average monthly rainfall and maximum rainfall in 24 h of the stations close to the study area. Reference geographic information (BCN200) such as the regional and national road network, elevations, urban centers, drainage network, etc.

2.2.2. Field work

A field campaign was carried out in order to visualize the erosive state of the areas most affected by water erosion, especially those that are very close to the main channel. Several field campaigns have been carried out and a series of images was captured of the most significant footprints that verify the passage of water erosion in the area of interest. This type of campaign helps to have a preliminary visualization to carry out the analysis and geospatial modeling of the risk of water erosion in order to contrast what is observed in the field with the results obtained from the complete study (Figure 4).

Figure 4. Left. (A): Overview of spill cones associated with gullies. **(B)**: Detail of front of fans. **(C)**: Erosive tracks due to sediment flow. **(D)**: Contrast of vegetation present in the study area. **(E)**: Visualization of the material carried by the water flow. **(F)**: Detail photo corresponding to **(E)**. **Right**. Location in Google Earth of erosive marks such as gullies and their associated spill lobes.

2.2.3. LIDAR Data Application for the Creation of the Digital Terrain Model -DTM-

LIDAR data collects altitude data allowing to define the terrain surface and generate digital elevation models. They are massive point cloud datasets that can be managed, visualized, analyzed, and shared using GIS. In order to work with this information, it is necessary to download LIDAR data clouds, for this there are several sources of information worldwide such as the online LIDAR project that seeks to group this type of data and offer it free of charge. In the case of Spain, there is a spatial data infrastructure (www.idee.es) where data, metadata, services and geographic information produced at the state level are integrated through the internet regional and local. In addition, there are projects such as PNOA-LIDAR which can be purchased from the download center of the National Geographic Institute (IGN) or from the Autonomous Communities' portals. In this case, when the study area is located in Castilla y León, the LIDAR data will be obtained from the Idecyl. The file package or batch LIDAR_CyL_2015_h10_0529.laz belonging to the study area was chosen. A conversion is needed in the LAZ file format (LIDAR data) in LAS format so that ArcGIS can visualize it, for this, the LASTools tool package is used. The DTM is generated from the LAS files. It is necessary to filter the LIDAR data in such a way that only the terrain height data is kept and not the surface data (which includes, for example, the vegetation height). Such a filter can be done using other LASTools software or directly from the GIS ArcMap as in this case. A DTM is created with the highest possible resolution due to the size of the study area, so the assignment of the values to the cells and the cell size of the raster will be 1 (to create cells with 1mx1m resolution). Finally, the DTM of the study area is obtained, which will be the basis for obtaining different factors from the RUSLE model (Figure 5).

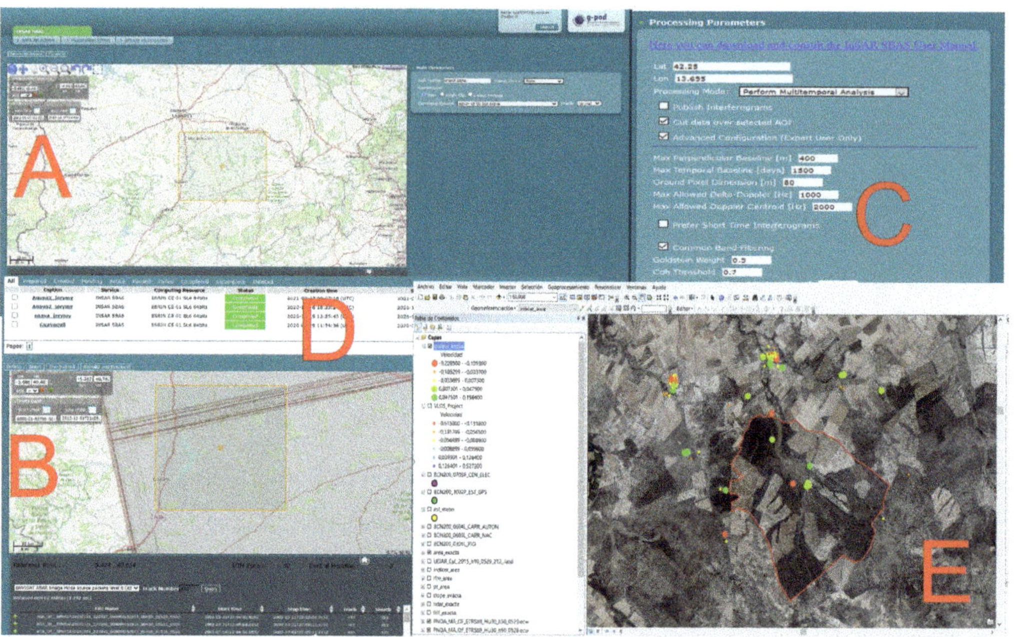

Figure 5. (**A**). Selection of the study area for the subsequent search and processing of SAR images. (**B**). Choice of the track number and display of the images to be processed. (**C**) Parameters chosen for image processing. (**D**). Process control and processing status. (**E**) Displacement mapping obtained in the study area.

2.2.4. Use of the Free use Platform G-POD for the Application of the A-DInSAR Technique

Access to the free-to-use platform G-POD is done through the web https://gpod. eo.esa.int/ (accessed 20 October 2021), where a user account is created by registering in the system. Within the platform, the A-DInSAR, SBAS-InSAR technique will be used, which is the one of interest in question for the study. Once within the service, the user selects the study area and checks the availability of the SAR images for the area and the dates to be analyzed. (Figure 5A). The objective of this methodology is to be able to develop a SBAS-InSAR technique linked to the G-POD platform, based on the formation of displacement maps of individual events calculating the phase difference (interferogram) of two SAR images obtained on the same acquisition area and orbit, but temporarily separated by acquiring a series of satellite images for subsequent processing in order to obtain information about the average deformation speeds of different points on the ground in the study area.

This technique allows directly the acquisition of satellite images that focuses on obtaining displacements generated in a determined interval of time to visualize the temporal evolution of the displacements of the terrain of the study area, calculating the difference between each displacement in time. The selected images that will later follow the SBAS-InSAR processing follow this procedure: first the choice of the dates of the desired images. The search carried out was ten years from 1 January 2002 to 31 December 2010. But the images that were processed were between March 2003–January 2010. Second, determination of the coordinates of the study area (longitude and latitude) indicating the type of virtual file 4 (VA4) or G-POD. In our case, a VA4 file will be obtained with a RAM label. The track number is a parameter that limits the number of images of interest in the delimited area. When searching for images, the system offers several image packages, using track number 459. The data used will be of the ASAR type because it automatically suppresses possible duplicate images and correctly correlates SAR data that are part of different image packages. the same sequence. This number is chosen based on the availability of the number of images each track contains, the orbital of the satellite and the conditions of the study area. To determine the appropriate track, the platform provides information about all available track numbers of the desired delimited area (http://esar-ds.eo.esa.int/socat/ASA_IM__0P_Scenes) (accessed 20 October 2021).

The images are processed from the SBAS-InSAR database, and for this study 62 SAR images were acquired from the Envisat ASAR satellite, with dates from April 2003 to January 2010. The main characteristics of the images are as follows: Level 0 ASA_IM_OP, track 459 and ascending orbits images (Figure 5B). Subsequently, with the desired image package, the reference point of the displacement measurements generated by the SBAS-InSAR algorithm is established by manually inserting the geographic coordinates of the location or by moving the halo found in the viewer of the platform. To facilitate the process, the largest number of urbanized areas and towns are included to reference the points. Subsequently, the Perform multitemporal analysis that generates time series of displacements through SBAS-InSAR. By selecting this option, the Publish Inter-ferograms check box is available, which allows the user to download the interferogromas generated during the SBAS-InSAR processing chain. Once the type of processing and the Cat Data Over Selected AOI check box have been chosen, the system delimits a square of the area of interest that will be the data that the identified AOI automatically processes (Figure 5B). There is an advanced configuration box in which you can modify the different parameters it contains, to specify the type of images package with greater precision. Normally this type of box is used by specialized people and dedicated to this type of platform G-POD and ABAS-InSAR and not beginner users (Figure 5C). Once the processing has started, the user can view the evolution of the different stages that take place during the A-DInSAR processing: co-registration of the SAR images, formation of the rolled differential interferograms, unrolling of the phase of the interferograms, estimation of the linear velocity, geocoding of velocity points, etc. The processing time of the images is totally dependent on the platform system, so it may vary depending on the availability of the G-POD. For our area, the processing time has been 48 h. A total of 64 interferograms were processed (Figure 5D). The final result is a set of points that shows the displacement speed of a given place within the area of interest in mm/year and the time series of deformation (accumulated deformation over time) of the study area in part ticular. After finishing the process, you can see the evolution of the processing and the time interval that it takes in the workspace, in which you can also see the summary of the process (task ID, Service used, task status, progress, time of creation, presentation time and completion time). The folder with the final result is the most important of the entire process since it is the result of the entire DinSAR methodology showing the average speed of land displacement in the study area mm/year.

The points obtained show a coherence scale of 0–1, with 1 being the maximum possible coherence and 0 the minimum, referring to this concept to the validity of the point as data, that is, to the degree of quality of co-registration between the images. HE. In this way, the algorithm implemented in P-SBAS directly eliminates the points that contain a coherence below 0.7, considering them null values of low validity, keeping all the points that are above that threshold. The points considered valid, contain a very high coherence, so when studying them they are very reliable. To facilitate the processing of the images and to obtain a good coherence of the data, it is necessary to verify that the coordinates entered are always on land and within the selected area. To make the process easier, urbanized areas are usually the most useful for locating reference points, since in them anthropic structures with the presence of planes, corners and straight forms, in general, are easily recognizable. The A-DInSAR information acquired is provided in a kmz file (for viewing in Google Earth) and a document in txt format where the characteristics of the acquired points are presented: point IDs, geographical coordinates in WGS 84, topographic elevation, mean velocity (mm/year), coherence value and deformation time series. This document is exported to a GIS (Figure 5E). Before finalizing the A-DInSAR methodology, all the information about the images processed in the G-POD platform will be collected in a database provided by the European Space Agency (http://esar-ds.eo.esa.int/ socat / ASA_IM__0P_Scenes) (accessed 20 October 2021).

2.2.5. Calculation R Factor or Climatic Aggressiveness/Pluvial Erosivity

Based on the monthly mean rainfall and the maximum rainfall in 24 h, the rainfall erosivity is calculated using the R index. Information was obtained from a total of 28 meteorological stations indicating the name, code or code of the station, its UTM 30N coordinates, its monthly average rainfall values. With all this data, an Excel table is created that is imported into ArcGIS. From the geodatabase, a point layer is created where some projected coordinates ETRS_1989_UTM_Zone_30N are entered. From this spreadsheet a point layer is obtained in vector format that reflects the geographical position of each one of the meteorological stations as its average monthly rainfall and maximum rainfall in 24 h. (Figure 6). With the vector file of the stations, the precipitation erosivity factor (R) is calculated according to the modified Fournier index (MFI). To obtain the aggressiveness of the rain, you must have the precipitations in raster format for each month. To do this, an interpolation is carried out, using the inverse of distance-IDW method. Through this method, the value of each cell of the raster is calculated as the weighted average of the values of the environment as a function of the inverse of the distance, so it is interpreted that the closest points will contain greater relevance. Interpolation is performed for each month (from January to December) with a cell size of 1 × 1 pixel to have the highest possible resolution. With the raster cartography of each month, the rain erosivity index is calculated from the Equation (1) and for the calculation of the IFM we use the Equation (2), where Pi is the precipitation of each month in mm and Pt the mean annual precipitation in mm. Each of the previously generated rasters correspond to each of the 12 values of Pi, so that from them the final raster that represents the value of the mean annual precipitation (Pt) can be obtained, being the sum of all the monthly values of precipitation. Applying map algebra, we do the summation of the monthly layers obtaining the map called Pt. Finally, with all the calculated values of the Equation (1) is applied with the raster calculator and the final result is a raster that shows the different values of the R index in the study area and its surroundings. As a final action, it is necessary to cut the relevant information that only refers to the study area (Figure 6A):

$$R = 2.56 \times IFM^{1.065} \tag{1}$$

$$IFM = \sum_{i=1}^{12} \frac{Pi^{(2)}}{Pt} \tag{2}$$

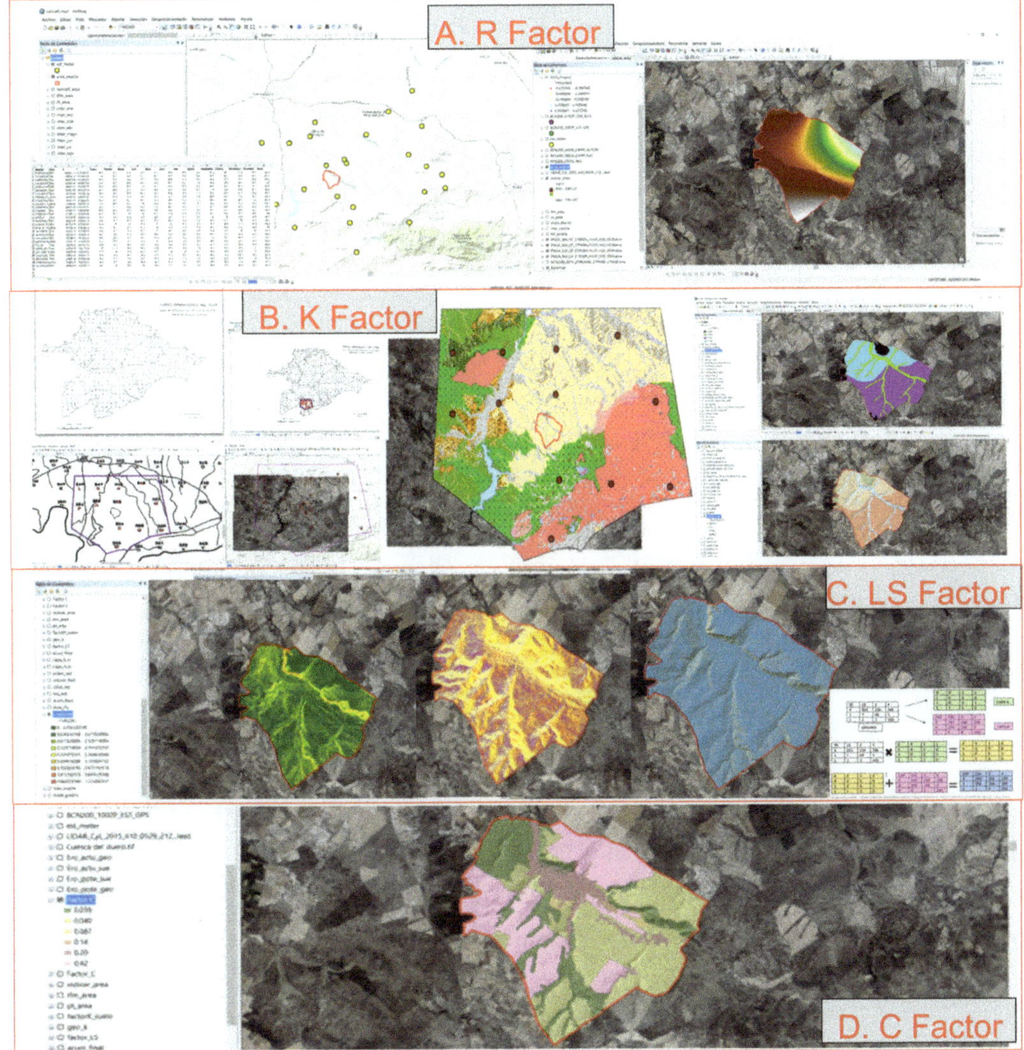

Figure 6. Calculation procedures for R factor (File **A**), K factor (File **B**), LS factor (File **C**) and C factor (File **D**).

2.2.6. Calculation of the K Factor or Resistance to Soil Erosivity

The K factor is a direct indicator of the erodibility or erodibility of the soil, specifically it reflects the lithological and edaphic vulnerability to erosion, calculated from the physical properties of the soil and the rocky substrate [26–29]. Linking the lithology of the study area with the K factor is key, since each type of existing lithology will have a specific value of the K factor, varying the vulnerability to erosion. In the calculation of the K factor, the lithological analysis is carried out according to the lithological resistance in those areas where the rock outcrops and the edaphological analysis by means of experimental plot stations and with analytical samplings.

Twelve stations of 20 m^2 that overlap with the geological cartography have been considered to determine their lithology. The analysis and improvement of the value of the K factor is carried out next with the edaphic analytical study located on each rock substrate [30]. For this we use the nomogram of Wischmeier and Smith and the Wischmeier regression equation (Equation (3)) [22] from the experimental data obtained with simulated rains, where: M = texture (100% clay) × (% silt + very fine sand), a = % organic matter, b = type of structure, c = permeability class. Finally, the K factor is obtained according to each soil, which reflects the difference in resistance to erodibility and ero River hydrographic basi the edaphic vulnerability to erosion of the desired area as a function of the soil. (Figure 6B). A field sampling has been carried out analyzing the upper 20 cm of the soil to analyze the percentage of organized matter and thickness of the A horizon:

$$100 \ldots K = [10 - 4 \ldots 2.71 \ldots M^{1.14} \ldots (12 - a)] + 4.2 \ldots (b - 2) + 3.2 \ldots (c - 3)] \quad (3)$$

2.2.7. Calculation of the LS Factor

This type of factor encompasses two topographic parameters that refer to the length of the slope (L) and its slope (S). To calculate its value, the Equation (4) [31]:

$$LS = [\![\text{Flow accumulation} \times (\text{cell size}/22.13)]\!]^{0.4} \times [\![(\text{SineSlope}/0.0896)]\!]^{1.3} \quad (4)$$

Where the flow accumulation is the number of cells that contribute to the flow in each given cell, the cell size is the length of the cell size used throughout the methodology and the SineSlope is the sine of the slopes in radians. First, the slopes (S) are calculated in degrees from the digital terrain model and the values are transformed into radians by multiplying each pixel of the raster map by $\pi/180$ applying map algebra. Subsequently, we continue with the calculation of the slope length (L) from obtaining the flow accumulation raster and the flow direction. The accumulation of flow is then multiplied by the cell size, which represents the length of the water path. A maximum slope length of 250 m is established in a standardized way, which is equivalent to 250 cells or pixels and the special resolution of the raster is 1 m × 1 m. Taking this into account, we elaborate a flow accumulation map whose maximum value is 250, this means that the cells in the tracker or accumulation map whose value is less than 250 must retain their original value, but instead, those that present higher cell values, they must all take the value of 250 (maximum value set). Two new layers of accumulation of the flow will be created from the original calculated, using reclassification algorithms, where: Layer A: If the accumulation flow is ≤ 250, it takes value 1; if the accumulation flow > 250 takes value 0; y Layer B: If the accumulation flow is ≤ 250, it takes value 0; if the accumulation flow > 250 takes value 250. Next, the accumulation of the original flow will be multiplied by Layer A, obtaining a file that preserves the original values except for those cells where the original value was greater than 250, which now It will be 0. Once this is done, Layer B is added to the new calculated layer so that those cells that have a value of 0 will take the value 250, which is the maximum that we wanted to set. The final mapping of the LS factor is obtained by applying the Equation (4) using raster calculator (Figure 6C).

2.2.8. Calculation of C Factor or Vegetation Cover

Factor C or Vegetation Cover considers the management of plant and crop masses; Therefore, based on the land uses obtained from the most current PNOA orthophoto and the forest map of the area, we can identify the type of vegetation and its density (Figure 6D). This factor analyzes the influence of the type of land use according to its plant species cover, the alternation of crops influencing the degree of erosive susceptibility of the land, which will influence its productivity. To obtain Factor C we use the reference tableand the values established by the United States Soil Conservation Service [3,21,32–36], for arboreal, shrub and mixed wooded formations, analyzing the percentage of tree and shrub cover; type of herbaceous cover and thickness of plant offal and extension. Regarding the herbaceous formations, the Wischmeier Table has been considered.

2.2.9. Obtaining the Potential and Current Erosion Risk Cartographies

Once we have all the calculated factors involved in the RUSLE model, the risk of potential erosion can be established, conditioned by the combined effect of the factors rain, runoff, soil and slopes. This cartography is obtained from the product of the factors R, K and LS generating a potential erosion map. To determine the degree of existing soil loss, the "present moment" is considered by analyzing the generating and protective factors of the soil; as well as their spatial distribution (types of cultivation and autochthonous plant masses, conservation practices). Basically, it is obtained from the product between the potential erosion calculated previously with the C factor of the RUSLE model, since the vegetation plays a fundamental role when it comes to the development of the erosion that may be generated [37–39]. Conservation practices have not been considered (RUSLE P factor = 1) given their lack of consistency and also the interest in calculating natural erosion without human support practices (both positive and negative) to establish the best land uses favoring edaphic recovery. In the erosion risk cartographies can be seen that soil losses per unit area increase with the increasing length of the slope [40]. The soil loss rate is affected by slope steepness more than slope length.

3. Results and Discussion

A series of results have been obtained from all the calculations carried out for each of the relevant factors of the study area by applying the RUSLE model, as well as the estimated values of average velocities (mm/year) of sediment movements and coherence to from the A-DInSAR processing, related to geological parameters of the study area.

3.1. Analysis of Ground Deformation Velocities (mm/year)

A series of points has been obtained from the application of the A-DInSAR technique of average ground deformation speeds in mm/year (Figure 7). Due to the limitations of the technique, it has been determined that the results obtained from are not valid and/or insufficient because the scale of the study area is too small for the application of tools of this style. In addition, the study area is located in areas with few essential urban reflectors for taking satellite data, which further limits its use. Finally, the presence of dense vegetation accentuates the difficulty of penetrating the satellite laser when collecting data, so this technique definitely contains numerous limitations for this type of study area. In any case, points with average ground velocities of −24.09 mm/year have been found, although very few points have been detected within the perimeter of interest. The positive data found in blue color can indicate the satellite's approach to the ground, bulging or swelling of the ground or simply atmospheric noise, making this uncertainty invalidate this type of data obtained.

3.2. R Factor or Pluvial Erosivity

It can be seen that the pluvial erosivity values of the study area oscillate between 1951.47 MJ·mm/Ha year–2061.41 MJ·mm/Ha year. There is an increasing trend of the aggressiveness of the rain from the NE zone towards the W-SW, marked by a maximum located in the SW corner where the greatest pluvial erosivity occurs, registered with values in 2061.41 MJ·mm/Ha/year. This maximum is due to the fact that it is positioned on flat areas that contain a gentle slope, causing it to favor the flow of sediments dragged by the water due to precipitation, which the fall impact due to the splash effect easily disintegrates and unstable the sediment of the water ground. On the other hand, the plots with less rainfall erosivity, located to the NE, are located on the flat terraces of the main river, which due to their morphology receive less volume of lateral water and therefore suffer less erosion, since they transport less sediment, some of which can be decanted in times of lower energy, with values between 1.951–1.987 MJ·mm/Ha/year. These types of areas eventually receive rainfall varying its intensity. Ultimately, the key to the R factor is that its increase is due to the presence of areas with steep decline, which favors the flow of water and the dragging of sediment to act as the main erosive factor (Figure 8).

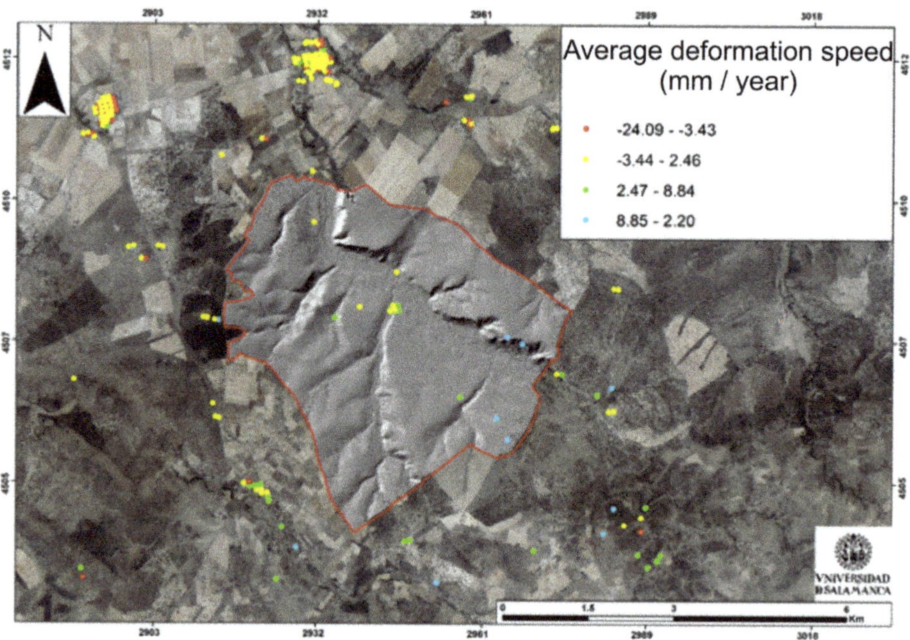

Figure 7. Cartography of mean deformations of the terrain indicating with colors the different speeds obtained from the satellite difference analyzed between images.

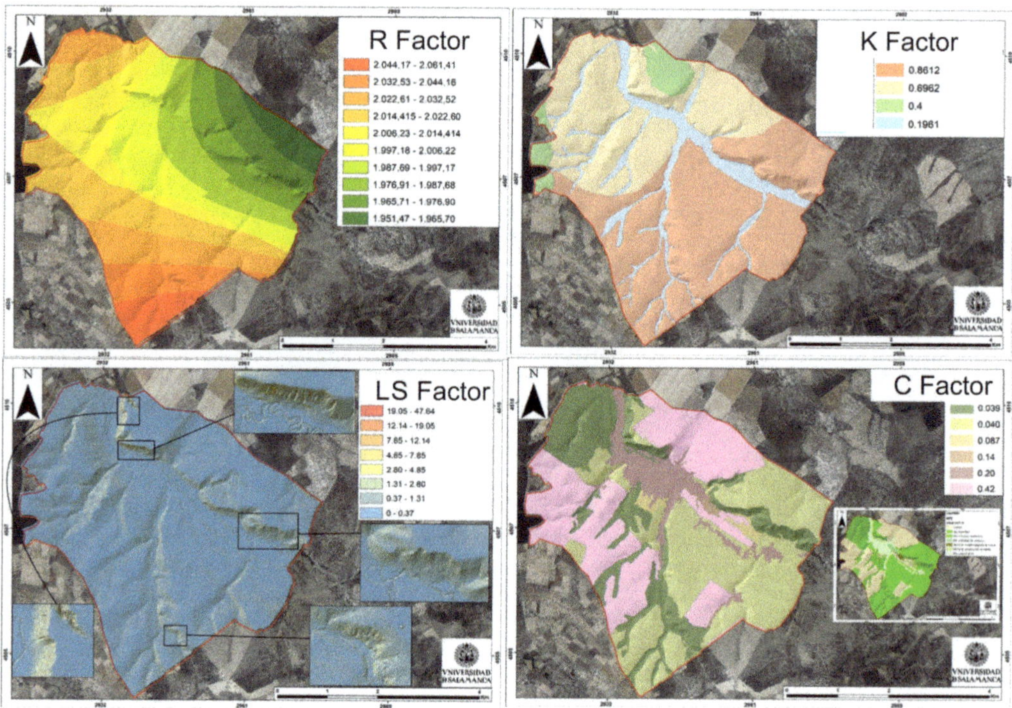

Figure 8. Cartographies obtained from the different factors of the RUSLE in the study area (red border line).

3.3. K Factor or Resistance to Erosion

This factor shows the erodibility or erodibility of the soil from physical properties, considering the lithology (Table 1) and validating our results with the cartography of erosive landscapes of the Duero River hydrographic basin.

Table 1. Values of the K factor depending on the lithologies validated for the study area.

Station	Plot	K Factor	Litology
(1) 248A	45.29-SA	0.213	Arcosic sands beige/Arcosic with gravel
(2) 249	45.30-SA	0.284	Gravels at the base and predominant sands with gravels towards top (terraces)
(3) 252A	45.38-SA	0.499	Gravels and feldspathic sands (valley bottoms and floodplains)
(4) 253	45.39-SA	0.133	White coarse arkotic sands with sparse clayey intercalations

W (soil surface mostly with broad-leaved herbaceous plants without plant offal) and G (surface of turf soil with decomposing plant offal at least 2 inches deep).

This factor is directly related to the geological map of the studied plot, so the resistance to erosion varies depending on the existing lithology. For this specific case, it must be considered that the value of the K factor varies between 0–1, where 0 are conditions more susceptible to erosion and 1 condition are less susceptible to erosion. Values ranging from 0.133 to 0.499 $Tm \cdot Ha \cdot h \cdot MJ^{-1} \cdot Ha^{-1} \cdot mm^{-1}$ are found, where the zone with the greatest resistance to erosivity is found on the alluvial deposits and valley bottoms since they contain granulometries of clay size which are compacted and adhere, that is, they coalesce, forming a film that is difficult to erode. On the other hand, the most vulnerable areas are those areas with sand-sized granulometries (almost the entire study area, except for the terraces and valley bottoms) which are easily transported by sheets of water causing significant erosion in their wake. The northern area of the study area can be highlighted as the one that suffers the highest soil erosion depending on the rocky substrate with a value of 0.133 $Tm \cdot Ha \cdot h \cdot MJ^{-1} \cdot Ha^{-1} \cdot mm^{-1}$ as it contains granulometries of coarse sand with ridges, causing greater erosion due to its size. If we consider the existing soils on the lithological substrate and their analytical data of texture and structure from the samplings carried out in the study area. K values vary between 0 (condition least susceptible to erosion) and 1 (condition most susceptible to erosion) contrary to lithology. The resistance to soil erosion contains values ranging from 0.1961 to 0.8612 $t \cdot Ha \cdot h \cdot MJ^{-1} \cdot Ha^{-1} \cdot mm^{-1}$, fluvisols-type soils being the most resistant to soil erosivity due to the fact that they are cohesive soils, with a greater amount of silts and clays, difficult to erode while passing of a sheet of water, showing values of 0.1961 $t \cdot Ha \cdot h \cdot MJ^{-1} \cdot Ha^{-1} \cdot mm^{-1}$. On the other hand, the soils located on arches: fundamentally cambisols, are of less resistance to erosion, since they are detrital soils easily eroded by the flow of water, because the sand grains are not connected and are easily transported, being visualized values of 0.6962–0.8612 $t \cdot Ha \cdot h \cdot MJ^{-1} \cdot Ha^{-1} \cdot mm^{-1}$ (Figure 8).

3.4. LS Factor

Both the length of the slope and its slope influence the erosion rates of the soils in our study area considerably, being the geomorphological aspects related to the forms of the relief one of the main factors that determine the emission of sediments catchment areas. Regarding the length of the slope, it is known the existence of a zone practically without erosion in the highest parts of the slope, the appearance of erosive phenomena of greater intensity in the middle part and sedimentation as the dominant process in the lower part of the slope, where in general its slope decreases. This distribution of erosion occurs in a generalized way on the slopes close to the channels through which the water flows from the study area. Observing the results obtained, it can be seen that soil losses per unit area increase with increasing length of the slope, being greater in its lower part, due to the fact that the runoff sheet accumulates downstream, increasing its drag force as it descends the slope by the action of gravity. On the other hand, on longer slopes the appearance of rills is more frequent, with which erosion rates are considerably increased, as the waters concentrate in these small channels, increasing their speed and transport capacity of the particles eroded soil. The L and S factors are dimensionless and the results obtained vary between 0–47.64, the areas with the greatest erosivity being those with the greatest slope and length of the slope located mainly in the north and northeast of the study area, the existing slopes being that there are between the terraces and the valley bottoms of the main stream. There are also secondary zones with significant erosion potential located to the south and west, caused by secondary channels generated from the main stream (Figure 8).

3.5. C Factor

In our study area, the analysis of the vegetation shows the indicated values (Table 2). In this factor, stonyness must be considered, which is the proportional decrease in erosion due to the percentage of soil covered by fragments of rock or gravel. Vegetation cover is a determining natural element when studying soil protection against the erosive force of precipitation, since it not only controls the energy with which raindrops reach the soil surface, but also the speed of surface runoff.

Table 2. Determination of factor C for arboreal, shrub, mixed wooded formations and herbaceous plant covers.

Arboreal, Shrubby and Mixed Wooded Formations					
Cover Type	Type of roof and average drop height of water drip	% Tree and shrub cover	Herbaceous cover and plant offal	% Herbaceous cover and plant offal	C FACTOR
Crops	Trees without low shrub understory (4 m height of fall)	25	W	0	0.42
Wooded mount	Trees without low shrub understory (4 m height of fall)	75	G	60	0.039
Wooded forest pasture	Trees without low shrub understory (4 m height of fall)	50	G	60	0.040
Mount with scattered trees of pasture	Trees without low shrub understory (4 m height of fall)	25	W	40	0.14
Sparse wooded mount of dehesa	Trees without low shrub understory (4 m height of fall)	25	W	60	0.087
Herbaceous Vegetation Covers					
Cover Type	% coverage	Establishment or consolidation			C FACTOR
Plantation wooded mount	20	Very poorly			0.20
Treeless mount	20	Very poorly			0.20

The study area is covered mainly by permanent vegetation, causing that the calculated C values are only related to the coverage percentages of the fraction of the room covered by the vegetation, and the herbaceous vegetation, of lower height, varies the protection of the soil throughout the year as a consequence of management (tillage, pruning, etc.) or the development of the plant itself (leaf fall, flowering, etc.). It must be considered that the values of the C factor obtained express the relationship that exists between the annual average soil losses of a plot with a certain vegetation and the losses that that same plot would have under conditions of continuous fallow and tillage according to the maximum slope.

In our sector there is great variability in the values of factor C. The areas where there is a high density of vegetation cover and in contact with the ground, are the areas most resistant to water erosion caused by rainfall, since this cover acts of main protective shield, referring to sectors with values of the C factor of 0.42, 0.20 or 0.14 corresponding to crops, bare forest (formed by grasslands and/or grasslands) and scattered wooded forest of pasture. Being in contact with the ground, the height of fall of the raindrop is lower and therefore contains less erosive energy, which causes greater protection, on the other hand, this type of vegetation obstructs the flow of surface runoff drastically reducing its speed causing it to significantly reduce the erosive power of the water flow, causing the edaphic surface to suffer less damage. In addition, this type of vegetation may have plant remains provided by the existing cover in the study area, which cover the soil and therefore make it more resistant to water erosion, reducing the effect of aggressive rain. On the other hand, the sectors that contain values of the C factor of 0.087, 0.040 and 0.039 corresponding to wooded mountains, wooded forests pastures and wooded forests of thin dehesa formed by hardwood forests and a combination of vegetation mainly, contain less protection against water erosivity due to the fact that they are large vege-tations with crown heights and raindrops falling several meters, making the incidence of rain aggressiveness higher, since the raindrop falls with greater energy accentuating its erosivity. In addition, the speed of the flow of surface runoff through this type of vegetation is higher than in the previous case, since the arboreal permeability and without almost any plant barrier through the soil, cause them to have greater vulnerability to erosion generated by the flow of water.

3.6. Potencial Erosión Risk

Mapping potential erosion (Figure 9) shows the susceptibility of this area to erosion, considering the existing conditioning factors. To generate the potential risk of soil loss, these factors of the physical environment that condition the erosion processes are multiplied, that is, the factor R, K and LS. Taking these factors into account, it is classified and mapped based on the erodability indices (lithology-soil science and slopes) and erosivity indices (rain aggressiveness). The results in our area show that areas near the main channel (Larrodrigo River) suffer greater processes of water erosion. The main erosive zones are located in the N-NE zone with values of 9,534.735–347,322.90 Tm/Ha/year, leaving certain significant erosive traces such as gullies and even ravines with their associated spill lobes. These areas with high potential for erosion are due to the fact that they contain the steepest slopes of the entire perimeter of interest with the presence of sand-type granulometries, which act as erosive agents when dragged by surface runoff. On the other hand, other areas with significant vulnerability to water erosion are located in secondary tributaries that originate from the main one that are positioned in the south-southwest part of the study area, focusing the erosion on the slopes formed on both sides of These secondary channels, although it contains less intensity due to the presence of a lower slope, exists. In the case of the areas with greater resistance to erosivity, they are distributed throughout almost the entire study area with values ranging between 0–1326.05 Tm/Ha/year, in horizontal flat areas where erosion hardly develops due to to the morphology of the terrain and the little influence that soil erosive agents have. In the field analysis, several videos have been made with drones that show the high degrees of incision of the erosive processes (see video gullies in Supplementary Material) (Supplementary Materials: Video S1: Badlands-gullies).

3.7. Actual Erosion Risk

In this section, the current erosion risk is analyzed and mapped (Figure 9), starting from the potential erosion risk and subtracting the protection offered by the vegetative cover based on its characteristics (height, density, stratification and spacing). territorial). The classification of the degrees of water erosion obtained with these thematic cartographies is regrouped in the intervals established by the Food and Agriculture Organization of the United Nations (FAO) expressed in Tm/Ha/year and mm/year.

The results of the current erosion show that the areas that suffer the greatest processes of water erosion fall mainly in areas close to the main stream that runs from top to bottom in the study area (Larrodrigo river), with current water erosion ranging from important-moderate to very severe or irreversible with values ranging between $5 \rightarrow 200$ Tm/Ha/year with soil losses ranging between $0.3 \rightarrow 12$ mm/year, mainly differentiating the N-NE and S-SW sectors of the study zone. These are sectors that contain high slopes where surface runoff flows at high speed, dragging sand and gravel, causing intense erosion and incision in its path. In addition, they contain vegetation that has little protective power, since they are wooded mountains with multi-meter crowns with hardly any vegetation on bare soils, further accentuating their erosive vulnerability. The rest of the study area corresponds to areas with great resistance to erosivity, since the morphology of its terrain with absence of slope, finer granulometries such as clays and its vegetation in contact with the ground makes it well protected to water erosion caused by rainfall, with values of <0.5–5 Tm/Ha/year with losses of >0.03–0.3 mm/year mainly, being a weak and/or slight soil loss.

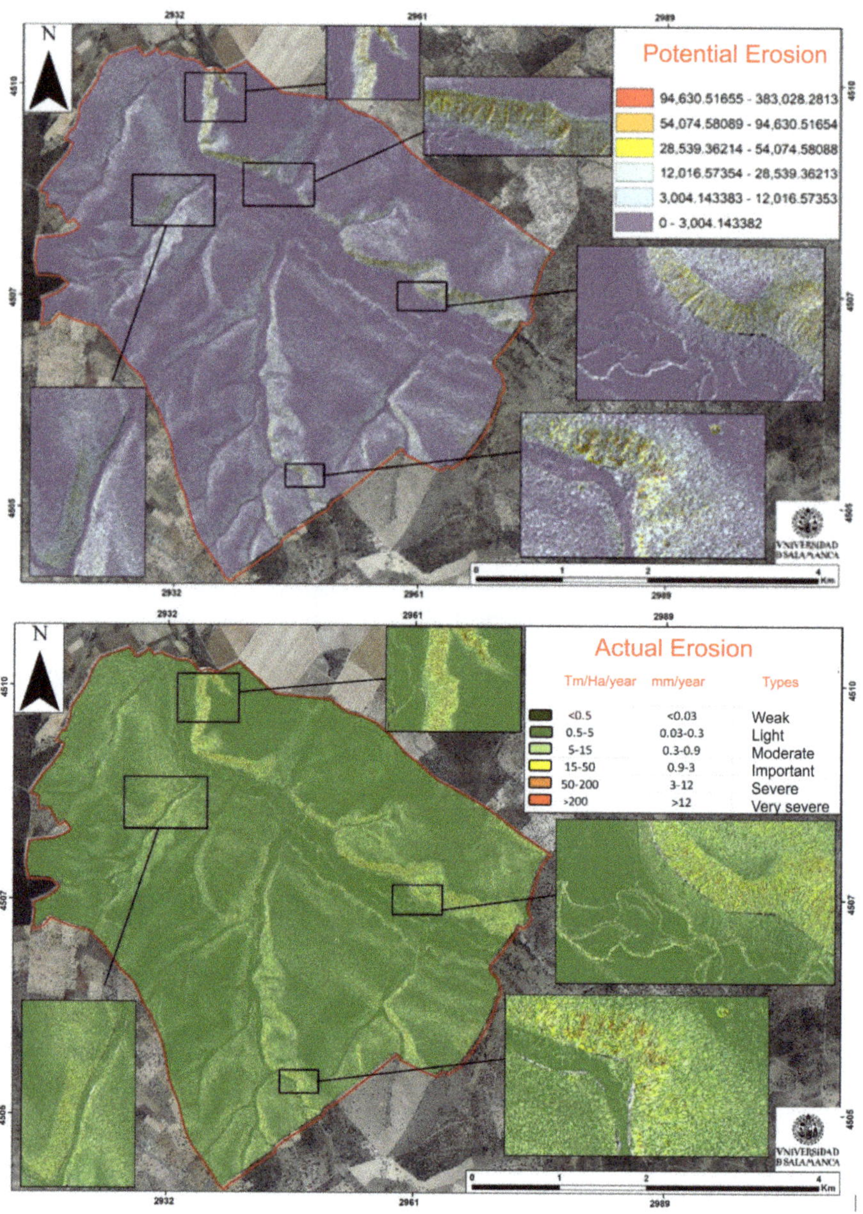

Figure 9. Cartographies with the distribution of the potential erosion risk expressed in Tm/Ha/year (**Top**) and the actual erosion risk (**Bottom**).

4. Conclusions

The modeling and automation of the RUSLE model with GIS allows a detailed and exhaustive multiparametric analysis that facilitates the analysis of the risk of water erosion in any territory where active processes are produced by sediment movements. The generation of a geodatabase, based on the determining components of the territory, has allowed obtaining the potential and actual erosion maps, being able to establish the volume of the problem in the study area, in such a way that soil losses are calculated and quantified by water erosion. Simultaneously, the combined use of A-DInSAR and GIS techniques was used to identify the possible movements of the sediment from differences between satellite images and observing the deformation speedb of the terrain associated with mass movements due to instability of slope and subsidence phenomena within the study area. It has clearly been shown that the use of this type of DInSAR techniques linked to the size or scale of the study area are not compatible when it comes to obtaining valid and consistent results, being more accurate in larger areas and where there is a greater number of receivers or urban helmets essential for taking satellite data. Two different sectors are defined by their erosive degree. The first presents levels that go from moderate to very serious or irreversible with values that oscillate between 5 → 200 Tm/Ha/ year with losses of the land of 0.3 → 12 mm/year. This sector corresponds to areas near the Larrodrigo river, encompassing the slopes that connect with the terraces positioned to the N and NE of the study area, with a high degree of water erosivity due to the presence of very high slopes and where the vegetation is scarce, generating erosive forms of great incision in the land (gullies) that put at risk the sustainable livestock and agricultural activity of this sector. The other sector corresponds to secondary channels perpendicular to the main one, located in the W and SW part.

According to the erosion rates observed among the different land uses, it should be noted that the main conversions in land use occurred between croplands with trees and open areas of pasture, which became holm oaks with dense areas shrubs, mainly the result of the lack of forest management, being observable in the southern and western areas of the studied sector. As a result, the current rate of erosion on the farm is lower due to the protection that the holm oaks and permanent pasture provide to the soil. The open areas dedicated to cultivation have higher rates than the holm oaks. On the other hand, the prairie areas present low rates of erosion, similar to holm oaks, the result of the dense herbaceous cover that covers them.

However, particularizing within each type of land cover, important differences were observed. The case for the use of the land of holm oaks stands out, in which erosion rates are usually minimal due to the protection provided by the vegetation and the smooth orography in which they are usually found. However, the wooded areas on the right bank of the Larrodrigo stream (also observable on a smaller scale in other smaller streams), with predominant land use of holm oaks, present the highest erosion rates, so the use of the soil Holm oak brings together the upper and lower extreme values of the erosion rate on the farm. These high rates of erosion in the escarpment areas are mainly due to their steep slope, which is further enhanced by a significant absence of herbaceous vegetation due to the absence of fertile soil, the loss of which has been accelerated by the trampling of the cattle found here over time.

The cartography of the potential and current erosion risk of water erosion allows to establish in a simple way, as applied in the methodology presented in this article, the classes of degrees of erosion obtained according to the FAO, expressed in Tm/Ha/year and mm/year. This thematic mapping constitutes in itself a low-cost non-structural measure that helps to identify the areas or sectors where the implementation of a soil management and conservation plan is necessary and urgent, such as management in the change of land use. land and reforestation, as well as detecting restoration measures and adaptation of land uses. In addition, it is possible to identify areas with high rates of sediment production that constitute areas susceptible to edaphic loss and to determine the location of sediment retention structures and other measures determining the true effect on anthropic activities of first need since the Soil is a resource that is difficult to remove at the speed with which it is lost, as shown in the ratios of the cartographies carried out.

Supplementary Materials: The following are available online at https://www.mdpi.com/article/10.3390/agronomy11112120/s1, Video S1: Badlands-gullies.

Author Contributions: Conceptualization, A.M.-G. and J.C.; methodology, A.M.-G.; software J.C..; validation A.M.-G. and J.C.; formal analysis, A.M.-G. and J.C.; investigation, A.M.-G. and J.C.; resources L.L. and M.C.; writing—original draft preparation, A.M.-G. and J.C; writing—review and editing, A.M.-G. and J.C.; supervision, A.M.-G.; project administration, A.M.-G. and C.P.; funding acquisition, C.P. All authors have read and agreed to the published version of the manuscript.

Funding: This research received no external funding.

Acknowledgments: This research was funded by the project This research was funded by Diputación de Salamanca, grant number 2018/00349/001 and the GEAPAGE research group (Environmental Geomorphology) of the University of Salamanca.

Conflicts of Interest: The authors declare no conflict of interest.

References

1. Nájera González, O.; Bojórquez Serrano, J.I.; Flores Vilchez, F.; Murray Núñez, R.M.; Areli González García-Sancho, A. Riesgo de erosión hídrica y estimación de pérdida de suelo en paisajes geomorfológicos volcánicos. En México. Ministerio de Educación Superior. Cuba Instituto Nacional de Ciencias Agrícolas. *Cultiv. Trop.* **2016**, *37*, 45–55.
2. Morgan, R.P.C. *Erosión y Conservación del Suelo*; Ediciones; Mundi-Prensa: Madrid, Spain, 1997; 354p.
3. Morgan, R.P.C. A simple approach to soil loss prediction: A revised Morgan–Morgan–Finney model. *Catena* **2001**, *44*, 305–322. [CrossRef]
4. Martínez-Graña, A.M.; Goy, J.L.; Zazo, C. Cartographic procedure for the analysis of eolian erosion hazard in Natural Parks (Central System, Spain). *Land Degrad. Dev.* **2015**, *26*, 110–117. [CrossRef]
5. Lal, R.; Sobecki, T.M.; Iivari, T.; Kimble, J.M. Soil Degradation in the United States. Extent, Severity and Trends. *Lewis Publ.* **2003**, *204*, 5–9.
6. FAO. Erosion de suelos en américa Latina. In Proceedings of the Taller Sobre la Utilización de un Sistema de Información Geográfica (SIG) en la Evaluación de la Erosión Actual y de Suelos y la Predicción del Riesgo de Erosión Potencial. Santiago, Chile, 27 June–1 August 1992. 1993 92–3001-5.
7. Gustavsson, M.; Kolstrup, E.; Seijmonsbergen, A.C. A new symbol-and-GIS based detailed geomorphological mapping system: Renewal of a scientific discipline for understanding landscape development. *Geomorphology* **2006**, *77*, 90–111. [CrossRef]
8. Martinez-Grana, A. *Estudio Geológico Ambiental Para la Ordenación de LOS espacios Naturales de "las Batuecas-Sierra de Francia". Aplicaciones Geomorfológicas al Paisaje, Riesgos e Impactos*; Universidad de Salamanca: Salamanca, Spain, 2010.
9. Ferretti, A.; Prati, C.; Rocca, F. Permanent scatterers in SAR interferometry. *IEEE Trans. Geosci. Remote Sens.* **2001**, *39*, 8–20. [CrossRef]
10. Berardino, P.; Fornaro, G.; Lanari, R.; Sansosti, E. A new algorithm for surface deformation monitoring based on small baseline differential sar interferogram. *IEEE Trans. Geosci. Remote Sens.* **2002**, *40*, 2375–2383. [CrossRef]
11. Mora, O.; Mallorquí, J.J.; Broquetas, A. Linear and nonlinear terrain deformation maps from a reduced set of interferometric SAR images. *IEEE Trans. Geosci. Remote Sens.* **2003**, *41*, 2243–2253. [CrossRef]
12. Sánchez-Sánchez, Y.; Martínez-Graña, A.; Santos-Francés, F. Remote Sensing Calculation of the Influence of Wildfire on Erosion in High Mountain Areas. *Agronomy* **2021**, *11*, 1459. [CrossRef]
13. Casu, F. SBAS-DInSAR Parallel Processing for Deformation Time-Series Computation. *IEEE J. Sel. Top. Appl. Earth Obs. Remote Sens.* **2014**, *7*, 3285–3296. [CrossRef]
14. Herrera, G.; Tomás, R.; López-Sánchez, J.M.; Monserrat, O.; Cooksley, G.; Mulas, J. Sistemas radar aplicados a la investigación de subsidencia y movimientos de ladera. *Enseñanza de las Ciencias de la Tierra* **2009**, *17*, 316–324.
15. Barra, A.; Solari, L.; Béjar-Pizarro, M.; Monserrat, O.; Bianchini, S.; Herrera, G.; Crosetto, M.; Sarro, R.; González-Alonso, E.; Mateos, R.M.; et al. A methodology to detect and update active deformation areas based on sentinel-1 SAR images. *Remote Sens.* **2017**, *9*, 19. [CrossRef]
16. Martinez-Graña, A.; Goy, J.; Zazo, C. Dominant soil map in 'Las Batuecas-Sierra De Francia' and 'Quilamas' nature parks (Central System, Salamanca, Spain). *J. Maps* **2014**, *11*, 371–379. [CrossRef]
17. Instituto Geológico Minero de España (IGME). *Mapa Geológico de España 1:50.000. Alba de Tormes*; Segunda Serie-Primera Edición; Ministerio de Industria y Energía: Madrid, Spain, 1982.
18. Instituto Geológico Minero de España (IGME). *Mapa Geológico de España 1:50.000. Santa María del Berrocal*; Ministerio de Industria y Energía: Madrid, Spain, 2008; ISBN 978-84-7840-747-7.
19. Wischmeier, W.H.; Smith, D.D. *Predicting Rainfall Erosion Losses: A Guide to Conservation Planning*; No. 537; Department of Agriculture, Science and Education: College Park, MD, USA, 1978.

20. Borrelli, P.; Alewell, C.; Alvarez, P.; Anache, J.A.A.; Baartman, J.; Ballabio, C.; Bezak, N.; Biddoccu, M.; Cerdà, A.; Chalise, D.; et al. Soil erosion modelling: A global review and statistical analysis. *Sci. Total. Environ.* **2021**, *780*, 146494. [CrossRef] [PubMed]
21. González del Tánago, M. *La Ecuación Universal de Perdidas de Suelo. Pasado, Presente y Futuro. Ecología, n. 5*; ICONA: Madrid, Spain, 1991; pp. 13–50.
22. Wischmeier, W.H. Use and Misuse of The Universal Soil Loss Equation. *J. Soil Water Conserv.* **1976**, *31*, 5–9.
23. Van der Knijff, J.M.; Jones, R.J.; Montanarella, L. *Soil Erosion Risk Assessment in Europe*; European Commission: Florence, Italy, 2000; p. 52. Available online: https://esdac.jrc.ec.europa.eu/content/soil-erosion-risk-assessment-italy (accessed on 20 October 2021).
24. Panagos, P.; Ballabio, C.; Poesen, J.; Lugato, E.; Scarpa, S.; Montanarella, L.; Borrelli, P. A Soil Erosion Indicator for Supporting Agricultural, Environmental and Climate Policies in the European Union. *Remote Sens.* **2020**, *12*, 1365. [CrossRef]
25. Garcia-Ruiz, J.M.; Lopez Bermudez, F. *La Erosión del Suelo en España*; Sociedad Española de Geomorfología (SEG), Ed.; Sdad. Coop. de Artes Gráficas: Zaragoza, Spain, 2009; ISBN 978-84-692-4599-6.
26. Flacke, W.; Auerswald, K.; Neufang, L. Combining a modified Universal Soil Loss Equation with a digital terrain model for computing high resolution maps of soil loss resulting from rain wash. *Catena* **1990**, *17*, 383–397. [CrossRef]
27. Meusburger, K.; Konz, N.; Schaub, M.; Alewell, C. Soil erosion modelled with USLE and PESERA using QuickBird derived vegetation parameters in an alpine catchment. *Int. J. Appl. Earth Obs. Geoinf.* **2010**, *12*, 208–215. [CrossRef]
28. Khaleghpanah, N.; Shorafa, M.; Asadi, H.; Gorji, M.; Davari, M. Corrigendum to "Modeling soil loss at plot scale with EUROSEM and RUSLE2 at stony soils of Khamesan watershed, Iran" [Catena (147C) (2016) 773–788]. *Catena* **2017**, *151*, 259. [CrossRef]
29. Gholami, V.; Booij, M.; Tehrani, E.N.; Hadian, M. Spatial soil erosion estimation using an artificial neural network (ANN) and field plot data. *Catena* **2018**, *163*, 210–218. [CrossRef]
30. Renard, K.G.; Foster, G.R.; Weesies, G.A.; McCool, D.K.; Yoder, D.C. *Predicting Soil Erosion by Water: A GUIDE TO COnservation Planning with the Revised Universal Soil Loss Equation (RUSLE)*; US Government Printing Office: Washington, DC, USA, 1997; Volume 703.
31. Moore, I.D.; Burch, G.J. Physical Basis of the Length-slope Factor in the Universal Soil Loss Equation. *Soil Sci. Soc. Am. J.* **1986**, *50*, 1294–1298. [CrossRef]
32. Sanchez, Y.; Martínez-Graña, A.; Santos-Francés, F.; Yenes, M. Influence of the sediment delivery ratio index on the analysis of silting and break risk in the Plasencia reservoir (Central System, Spain). *Nat. Hazards* **2018**, *91*, 1407–1421. [CrossRef]
33. Terranova, O.; Antronico, L.; Coscarelli, R.; Iaquinta, P. Soil erosion risk scenarios in the Mediterranean environment using RUSLE and GIS: An application model for Calabria (southern Italy). *Geomorphology* **2009**, *112*, 228–245. [CrossRef]
34. AbdulKadir, T.S.; Muhammad, R.M.; Khamaruzaman, W.Y.; Ahmad, M.H. Geo-statistical based susceptibility mapping of soil erosion and optimization of its causative factors: A conceptual framework. *J. Eng. Sci. Technol.* **2017**, *12*, 2880–2895.
35. Esteves, T.; Kirkby, M.; Shakesby, R.; Ferreira, A.; Soares, J.; Irvine, B.; Ferreira, C.; Coelho, C.; Bento, C.; Carreiras, M. Mitigating land degradation caused by wildfire: Application of the PESERA model to fire-affected sites in central Portugal. *Geoderma* **2012**, *191*, 40–50. [CrossRef]
36. Martínez-Graña, A.M.; Goy, J.L.; Zazo, C. Water and wind erosion risk in natural parks. A case study in "Las Batuecas-Sierra de Francia" and "Quilamas" protected parks (Central System, Spain). *Int. J. Environ. Res. IJER* **2014**, *8*, 61–68.
37. Alcañiz, J.M. *Erosión: Evaluación del Riesgos Erosivo y Practicas de Protección del Suelo*; Universidad de Girona: Girona, Spain, 2008; ISBN 978-8496742-37-6.
38. Lal, R. *Soil Erosion and the Global Carbon Budget*; School of Natural Resources, The Ohio State University: Columbus, OH, USA, 2002.
39. Martinez-Graña, A.; Goy, J.L.; Cruz, R.; Forteza, J.; Zazo, C.; Barrera, I. Cartografía Del Riesgo De Erosión Hídrica Mediante SIG En Los Espacios Naturales De Candelario–Gredos (Salamanca, Avila). *Edafología* **2006**, *13*, 11–20.
40. Ganasri, B.; Ramesh, H. Assessment of soil erosion by RUSLE model using remote sensing and GIS—A case study of Nethravathi Basin. *Geosci. Front.* **2016**, *7*, 953–961. [CrossRef]

MDPI

St. Alban-Anlage 66

4052 Basel

Switzerland

Tel. +41 61 683 77 34

Fax +41 61 302 89 18

www.mdpi.com

Agronomy Editorial Office

E-mail: agronomy@mdpi.com

www.mdpi.com/journal/agronomy